Secrets
of the
Universe

PAUL MURDIN

Secrets *of the* Universe

HOW WE DISCOVERED THE COSMOS

The University of Chicago Press

Chicago

On the case: Schiaparelli hemisphere of Mars,
USGS Astrogeology Research Program, 1980.

Paul Murdin is a senior fellow at the Institute of Astronomy at the
University of Cambridge and editor in chief of the *Encyclopedia of
Astronomy and Astrophysics*. Formerly, he was head of astronomy at the
Particle Physics and Astronomy Research Council and director of
science at the British National Space Centre. He is the author of
*Full Meridian of Glory: Perilous Adventures in the Competition to Measure
the Earth* and coauthor of *The Firefly Encyclopedia of Astronomy*.

The University of Chicago Press, Chicago 60637
The University of Chicago Press, Ltd., London
Published by arrangement with Thames & Hudson Ltd., London
© 2009 Thames & Hudson Ltd., London

18 17 16 15 14 13 12 11 10 09   1 2 3 4 5

ISBN-13: 978-0-226-55143-2 (cloth)
ISBN-10: 0-226-55143-1 (cloth)

Library of Congress Cataloging-in-Publication Data

Murdin, Paul.
  Secrets of the universe : how we discovered the cosmos / Paul Murdin.
      p. cm.
  Includes bibliographical references and index.
  ISBN-13: 978-0-226-55143-2 (alk. paper)
  ISBN-10: 0-226-55143-1 (alk. paper)
  1. Cosmology—History. 2. Astronomy—History. I. Title.

  QB982.M87 2010
  520—dc22
                          2009016382

Printed and bound in China by Toppan Printing

Great Discoveries

1543
The Sun is the centre of the Solar System
Nicolaus Copernicus

1572
The stars are not eternal, but change
Tycho Brahe

1610
Our world is a planet, like others
Galileo Galilei

1687
Everything is subject to the pull of universal gravitation
Isaac Newton

1868
Helium, a new element found in the Sun
Norman Lockyer & Jules Janssen

1929
The Universe has been expanding since the Big Bang
Edwin Hubble

1827
The greenhouse effect on Venus and on the Earth
Joseph Fourier

1953
The chemistry of emergent life in a flask
Stanley Miller

1933
Dark matter: the grip of the unseen and unknown
Fritz Zwicky

1959
The evolving Universe – it had a beginning
Martin Ryle

1965
The afterglow of the Big Bang: the Cosmic Microwave Background
Arno Penzias & Robert Wilson

1998
Dark energy: the push of the unseen unknown
The Supernova Cosmology Project & the High-Z Supernova Search Team

1957
Where the elements come from
Margaret Burbidge, Geoffrey Burbidge, William Fowler & Fred Hoyle

In questions of science the authority of a thousand is not worth the humble reasoning of a single individual.
Galileo Galilei

While I venture out beyond this tiny globe
Into reaches past the bounds of starry night,
I leave behind what others strain to see afar.
Giordano Bruno

It was the moment every astronomer, every planetary scientist lives for. I had the sense that I was seeing something that no one else had seen before.
Linda Morabito

I have a deep-seated faith in the boundless resourcefulness of nature, which so often leaves the most daring imagination of man far behind.
Bruno Rossi

1887
Chaos: the limits
of calculation
Henri Poincaré

1905–7
The nature of space and
time: the theories of Special
and General Relativity
Albert Einstein

1927
Nuclear energy, the
energy of the Sun and
stars: our saviour
or our nemesis?
*Fritz Houtermans
& Robert Atkinson*

1968
Cosmic neutrinos
from the Sun:
a new radiation
from the Universe
*Ray Davis &
Masatoshi Koshiba*

1919
Gravitational lenses
bend the path
of light
Arthur Stanley Eddington

Cosmic discoveries

Unseen forces

Atomic nuclei and the stars

Revolutionary calculations

The birth and death of the Universe

Our planet and our Solar System

New windows on the Universe

Mapping the Universe

1908
How to measure the
distances of galaxies
Henrietta Leavitt

1963
Quasars: beams from
across the universe
Maarten Schmidt

1969
Supermassive black
holes in the centre
of every galaxy
Donald Lynden-Bell

1995
Planets outside
the Solar System
*Michel Mayor &
Didier Queloz*

1781
Uranus, the first new
planet discovered
since antiquity
William Herschel

1783
The concept of
black holes: stars
that must be dark
*John Michell &
Pierre-Simon Laplace*

1846
The planet Neptune,
discovered by
calculation
Urbain Le Verrier

1969
First human landing
on the Moon
Apollo 11 astronauts

2004–8
Mars has water and an
active atmosphere
*Mars Global Surveyor,
Mars Express*

1910–25
White dwarfs, stars
made of a new state
of matter
*Henry Norris Russell
& Walter Adams*

1971
Mars had a past that
was wet and warm
Mariner 9

1962
X-ray stars: a window
on a violent universe
*Riccardo Givacconi
& Herb Gursky*

1932
Radio waves, the first
new window into space
Karl Jansky

Contents

Introduction

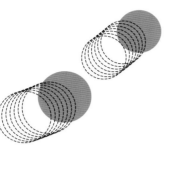

Discoveries in astronomy challenge our fundamental assumptions about the Universe. They alter our perception of matter, time and distance; they transform how we view our history and future as a species. Where the astronomers of antiquity spoke of fixed stars, we speak of whirling galaxies and the death and birth of stars in supernovae. Where we once considered the Earth to be the centre of the Universe, we now see it as a small planet among millions of similar systems, a few of which might also hold life. These dramatic shifts in perspective hinge on thousands of individual moments of discovery, moments when it became clear to an observer that a component of the Universe – from a tiny subatomic particle to a supermassive black hole – was not as it once seemed. Each is a revelation that unlocks yet another of the infinite secrets of the Universe.

The Anglo-Australian Observatory is located in the Warrumbungle mountain range, on the edge of a national park in rural New South Wales. The plains of the outback seem to stretch endlessly to the western horizon; in the foreground are the volcanic hills, dykes and plugs of the range, with fanciful names – the Breadknife, Belougery Spire, Crater Bluff. The hills are covered in eucalyptus trees and are home to kangaroos, koalas and brightly coloured birds. The whole area lies within a light pollution protection zone and the skies above the telescopes are brilliant with stars, especially in the southern winter when the centre of the Milky Way arches across the zenith.

In 1975 I was lucky enough to join the first group of six scientific staff of the Observatory. Its 4-m telescope was at that time the largest in the southern hemisphere, built to a high specification, and equipped with sensitive instrumentation, including computer-controlled electronic detectors, then a cutting-edge technology. Wherever the telescope pointed, it revealed a new discovery.

Over the next three years I was assigned to the telescope for about 150 nights. Colleagues from all over the world, including co-workers I had left behind in the UK, offered suggestions for what I should do with the telescope. During that period I co-authored about 150 scientific papers, each representing a new scientific discovery (editors of scientific journals use this as their selection criterion).

I remember one discovery in particular. I had been working through the night on the identification of an X-ray source; that is, I was trying to track down the optical star that was responsible for a beam of X-rays that had been detected by a satellite. After twelve hours of non-stop searching, I found the star – and something else about it. The star was the result of a relatively recent supernova explosion that had taken place perhaps 3,000 years ago. I was also able to estimate that the star was about 2,000 light years away.

As the day dawned, I helped the telescope operator shut down the equipment and began walking back to the lodge for a welcome sleep. In the golden light of the rising sun, kangaroos and wallabies were finishing their night's grazing, kookaburras were greeting the dawn with mad laughter and large black birds called currawongs were waking up in the gum trees with a chorus of melodious warbling. I was tired but it was a lovely morning and I was the only person in the world who knew what I knew about the star.

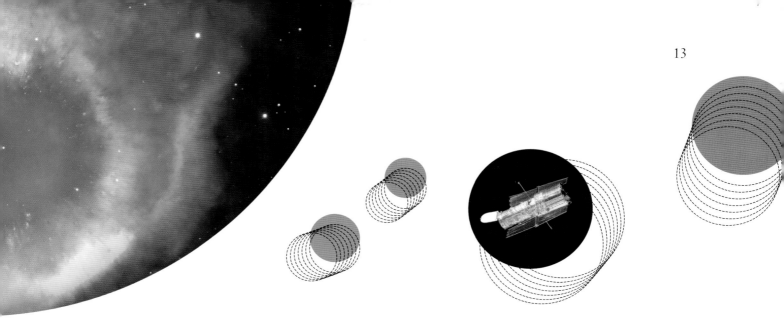

As I walked along the path an even more thrilling thought struck me. Light from the explosion of the supernova was only visible to observers within a sphere in space with a 5,000 light-year radius. Outside this sphere the supernova had, for all practical purposes, not yet happened. Now, 5,000 light years sounds like a large distance, and it is, but such a sphere is not large compared to the size of our Galaxy and it contains only a minute fraction of its stars. If there was only one habitable planet with an astronomy-curious civilization – ours – within that sphere, I was not just the only person in the world who knew what I knew about the star, but the only person in the Galaxy, or perhaps even in the Universe. I floated happily to bed and slept soundly, treasuring the secret that I had uncovered.

I was glad to learn during the research for this book that other scientists felt the same sense of exhilaration when they cracked a problem and made a cosmic discovery. Henry Norris Russell similarly recalled the thrill of being one of the privileged few in on a great secret: 'At that moment, Pickering, Mrs Fleming and I were the only people in the world who knew of white dwarfs' (**34**). Einstein's doubts about his theory of General Relativity were assuaged when he discovered the reason for a twist in the orbit of Mercury and, for a few days, he was beside himself, 'with joyous excitement' (**30**). Watching the transit of Venus in 1639 (**6**), William Crabtree 'stood for some time motionless, scarcely trusting his own senses, through excess of joy'.

My own discoveries were at the scale of the quanta of astronomy – over 50,000 pages describing astronomical discoveries of a similar scale are published every year. But this book is about the big ones: discoveries that unlocked major secrets of the Universe. I have selected them not only because they are important, but also because the people involved in them were interesting, and their stories illustrate how the science of astronomy works.

Science is a cyclic process that oscillates between the real world ('observation' or 'experiment') and the picture of it in the scientist's head ('theory'). A scientific discovery can take the form of new evidence in the physical world (such as a laboratory experiment or the discovery of a new star), or it can be a compelling new picture in one person's head that becomes accepted by most people as an illustration of something in the real world. Galileo saw mountains on the Moon. Copernicus pictured the Sun at the centre of the Solar System. Both were discoveries: one of them an observation, one of them a theory. For the man in the street, the word 'theoretical' sometimes carries a sense of derision – you can't trust something that is 'just a theory'. For scientists, the word can mean something that is as solid as the chair on which I am sitting and can definitely be counted as a discovery. Such a theory brings together a number of previously unrelated phenomena, expressed with such clarity that everybody is convinced. Or it predicts some phenomenon that has not yet been seen but turns out, when looked for, to be there.

Sometimes astronomical discoveries are serendipitous: the right person is in the right place at the right time. Tycho Brahe was returning home from an evening dinner at the time that the supernova of 1572 appeared in the sky (**41**); 400 years later, Ian Shelton happened to be pointing his telescope in the right direction when the supernova 1987A exploded (**42**). The crucial factor was that both discoverers knew about astronomy and understood what they were seeing. Other cosmic discoveries were unexpected by-products of investigations set up for entirely different purposes. Herb Gursky and Riccardo Giacconi made a chance discovery when an X-ray detector on a rocket they had launched to look at the Moon saw a bright source behind it (**32**). Jocelyn Bell discovered pulsars as a source of 'noise' during the observation of quasars (**35**). In both cases the scientists were remarkably persistent in systematically tracking down the origin of the inconsistency.

The word 'discovery' implies an element of surprise, but, like the discoveries that I made with the Anglo-Australian telescope, many 'surprise' discoveries have actually been made possible by an improvement in technology that was the result of meticulous engineering. Galileo learned how to make a telescope and pointed it at the sky. What he discovered – the satellites of Jupiter (**08**), the phases of Venus (**09**) and star clusters (**40**) – confirmed that the Sun, not the Earth, was at the centre of the Solar System; that we on Earth are not apart *from* the Universe, but are a part *of* the Universe. William Herschel built bigger telescopes that opened the window wider and wider on the Universe (**10, 38**); the Hubble Space Telescope blasted open the doors. The development of radio astronomy (**31**), X-ray astronomy (**32**), and infrared and millimetre wave astronomy (**51, 53**) in the 20th century allowed us to see objects in the Universe that are invisible to the naked eye. Martin Ryle's invention of the technique of aperture synthesis interferometry in radio astronomy made it possible to investigate radio galaxies, which showed that the Universe had a discrete beginning (**59**). When they reach the right level of sensitivity, gravitational wave detectors will open up a completely new view of the Universe (**64**). Since 1957, the possibility for spacecraft to carry equipment to the distant reaches of the Solar System has offered new perspectives on the planets.

Discoveries with new equipment can be unexpected but in a sense they are planned, because the equipment has to be made and deployed. That means having the right idea, gathering together the resources, and carrying out a plan to use the equipment for a specific purpose. William Herschel built a new telescope and used it to search the sky systematically; he found the planet Uranus (**10**) and his sister, Caroline, found her comet by applying the same technique (**07**). In modern times the new equipment has to be bought, and that takes great deal of money, so the scientist must make a detailed application for funding, predicting what will be discovered using this expensive new telescope or satellite. If you do no more than tell the simple truth: that the Universe is full of exciting things, and you can find something interesting with every instrumental advance – well, you won't get funding. At the very least you have to scope the range of potential discoveries in order to be taken seriously.

Certainly in some cases scientists did set out to find something specific. Urbain Le Verrier perceived the planet Neptune 'at the end of his pen; he determined it by the mere force of calculation' (**11**), while Daniel Barringer became obsessed with the idea that the Coon Butte crater in Arizona was meteoritic, spurred on by the thought of finding a profitable mass of iron and nickel (**20**). Subrahmanyan Chandrasekhar calculated the structure of white dwarf stars as a student exercise that he set himself to pass the time on an ocean voyage (**34**) and uncovered the reason for black holes. Raymond Davis spent over ten years searching for neutrinos from inside the Sun (**47**);

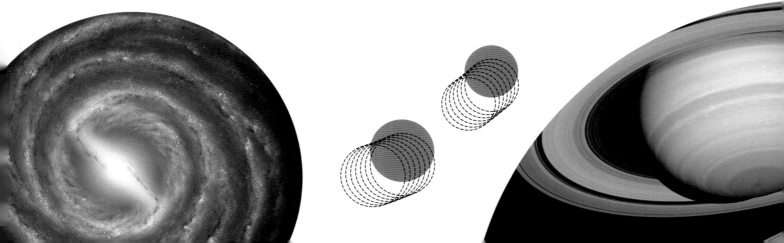

his discovery led to the development of a new kind of physics, properly deserving of the award of the Nobel Prize.

Computer modelling has shed new light on known phenomena, making surprising new astronomical discoveries possible. The phrase 'Garbage in, garbage out' is well known; you might think that its corollary is 'Put in the truth that you know, get out the truth that you know.' But when there is a lot of data or the calculations are complex, computers can reveal unexpected or previously unnoticed features about the real Universe. Computer simulations of the interactions of asteroids and comets led to our present understanding of the Oort Cloud and the Kuiper Belt (**14**). Satellites have great difficulty in probing the magnetosphere of the Earth (**17**) because it is so large that they can only investigate particular parts of it, like the people in the fable who grasp the tail, foot, tusks and trunk of an elephant, but fail to envisage the whole creature; computers are able to assemble these fragments into a complete picture. The Universe is hard to study because you cannot compare and contrast it with other real universes, but the Millennium Simulation models universes that are different from ours, which helps us to estimate how much dark matter and dark energy are present in the real Universe (**60**, **63**).

In a lecture in Lille in 1854 the French scientist Louis Pasteur perceptively noted that 'in the field of observation, chance favours only the prepared mind.' Usually in astronomy this means the prepared multi-disciplinary mind. Astronomy as a subject encompasses the study of everything in space. Physics, mathematics, chemistry, computing, instrument making, statistics – all these sciences, and others, are deployed by astronomers to understand what they see and to unlock cosmic secrets.

Some of the most important discoveries are collective, the product of investigations made by many different people over several generations, although there is usually one last genius who ties it all together. The laws of the motion of the planets engaged the minds of a stellar collection of giant talents before the proverbial apple dropped and Isaac Newton discovered the theory of gravity (**29**). 'If I have seen further, it is by standing on the shoulders of giants,' he wrote. The discovery of the greenhouse effect in the atmospheres of Venus and the Earth (**23**) took 150 years of investigation by dozens of scientists; there was really no one person who made the discovery, which is now recognized as so momentous that nothing less than the survival of life on Earth may depend upon our understanding of it. By contrast, Special Relativity and General Relativity (**30**) were the ideas of a single individual, Albert Einstein, working over only a few years.

Science-fiction writer Isaac Asimov described the most important feature of a scientific discovery as the open mind and curiosity of the person who makes it: 'The most exciting phrase to hear in science, the one that heralds new discoveries, is not "Eureka!", but "That's funny ..."'. In this book I try to explain what lies behind some of the great discoveries in astronomy, the train of events and thoughts that brought the scientist to exclaim 'Eureka' or 'That's funny…' as he or she unlocked one of the major secrets of the Universe. This book is therefore mostly scientific history – my top 65 cosmic discoveries. However, I also identify four major prospects for future discoveries: to uncover the secrets of dark matter (**62**) and dark energy (**63**), to detect gravitational radiation (**64**) and discover life elsewhere in the Universe (**65**) – although we may not find what we expect. The challenge for the next generation of astronomers will be to put themselves in the position to uncover these momentous secrets; together with them, I hope soon to learn what they are.

Paul Murdin

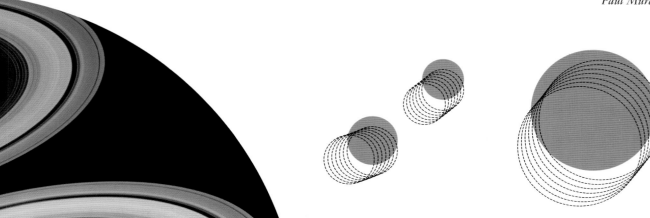

Discoveries Before the Telescope

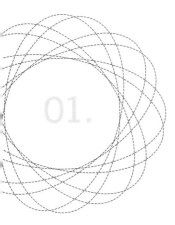

The Seven Planets

Wandering stars

The human fascination with the movement of the planets is almost as old as humanity itself. 25,000 years ago, beside a lake in Africa, a member of the Ishango community recorded the cycles of the Moon; in ancient Babylon, astronomers used their knowledge of the planets to advise their king about affairs of state and the price of barley. For 2,000 years the 'geocentric theory' provided the dominant explanation of the Universe, based on the evidence of everyday human observations.

The word 'planet' comes from the Greek word for 'wanderer', because originally any celestial objects that were not 'fixed' stars were regarded as planets. The Sun and Moon were thought to be planets and their motions could, with careful attention and the keeping of records, be predicted. It seemed reasonable to assume that the same was true of Mercury, Venus, Mars, Jupiter and Saturn, which were visible to the naked eye as bright lights roaming across the heavens. Many ancient observers of the skies thought that if they could understand and predict the motions of the seven planets they might uncover the deepest mysteries of the cosmos.

We have a tangible connection with one of the first known observers of the Moon, 25,000 years ago. He or she was a member of the Ishango people who, until they were wiped out by a volcanic eruption, lived, fished and farmed along the shores of what we now call Lake Edward, one of the sources of the Nile in central equatorial Africa. This person scratched markings in groups of 29 on the bone handle of a knife or chisel. These markings are of varying sizes that, according to anthropologist Alexander Marshack, represent the changing phases of the Moon throughout six months of the lunar cycle. Gaps in the markings seem to represent cloudy nights when the Moon was not seen.

We know nothing more about the maker of the Ishango bone: the earliest surviving lunar calendar. Perhaps the carver was a hunter or traveller recording a long journey, perhaps a woman keeping track of her menstrual cycle.

Across the continents the erection of monuments and the evidence of ancient myths and legends reveal the many different ways in which prehistoric men and women made sense of the cosmos and its relationship to their lives. It is unlikely that these peoples made precise astronomical observations, even at an elaborate primitive observatory like Stonehenge, but the keeping of calendars that tracked the solar and lunar cycles required the first civilizations to develop more complex patterns of thought and communication. Modern physicists use advanced mathematics to explain the operation of the Universe, but it was the observation of the heavens that first encouraged humankind to take the earliest steps towards mathematical thought.

1   **Ptolemaic universe** A pilgrim has travelled to the edge of the Earth and looks through the celestial spheres to see how they rotate. The illustration purports to be a German medieval woodcut, but is a fake, concocted in the early 20th century by Camille Flammarion, the French popularizer of astronomy, to illustrate the Ptolemaic universe.

2   **Ishango bone** The two sides of the Ishango bone from central Africa are scratched with markings in groups of 29, which record the phases of the Moon.

3   **The Moon and Saturn** The phases of the Moon and its patchy appearance (the 'man in the Moon' shapes) always suggested that the Moon was a sphere with topographical features, but it was not until the invention of the telescope that it became clear that the Moon was a world with mountains and plains, like the Earth. Before the telescope, the planet Saturn (the bright 'star' near the Moon in this photo) appeared only as a point in the night sky.

4

5

4 **Cuneiform tablet with observations of Venus** Cuneiform writing was made
by pressing a wedge-shaped stick into wet clay, which was then baked and
hardened. This tablet from Nineveh in northern Iraq lists observations of the
planet Venus made in the reign of Ammisaduqa, king of Babylon. His rule
is precisely dated by this tablet to between 1646 and 1626 BCE.

5 **Venus, Mars, Saturn and the Moon** Four 'planets' – Venus, Mars, Saturn
and the Moon – shine against a background of stars. This photo conveys
what the naked eye sees and there is little to suggest that the 'planets'
and stars are fundamentally different, although the progressive dimming
suggests that the Moon is closest to Earth and the stars furthest away.

The first known systematic observations of the planets
were recorded on Babylonian cuneiform tablets, starting from
about 1700 BCE. In the Neo-Assyrian period (911–612 BCE)
astronomers regularly recorded the motions of the planets
and gave astrological forecasts to the king. Astronomers of
the Achaemenid kingdom kept astronomical diaries of their
observations, which they used to make predictions about
affairs of state, the level of the river Euphrates, and the price
of goods such as barley, dates, mustard, sesame and wool.

Alexander the Great's conquests in the East in the 4th
century BCE brought the Babylonians' detailed astronomical
records to the attention of ancient Greece, where philosophers
such as Thales of Miletus, Pythagoras, Plato and Aristotle
had debated the nature of the Universe and used geometry
to explain planetary motions. The Babylonians' example
encouraged Greek astronomers to base their speculations
upon more exact observations of the stars and planets.

The Greeks assumed that the Earth, which was not a
planet, lay stationary at the centre of the Universe – after
all, the Earth doesn't rock about as if it is moving, nor do
the positions of the stars change as if we are viewing them
from a succession of positions along an orbit. This was the
'geocentric' theory of the planets and it is associated with
the name of the astronomer Claudius Ptolemaeus, more
often known as Ptolemy.

Ptolemy worked at or near Alexandria in Egypt during
the middle decades of the 2nd century CE. By 147 CE he had
developed the geocentric theory to a sophisticated level,
described in outline in a public inscription and presented in
full in a large treatise entitled the *Almagest* (from the title of
an Arabic translation, meaning 'the greatest').

The Ptolemaic theory of the Universe held that the Moon,
Mercury, Venus, the Sun, Mars, Jupiter and Saturn revolved in
a succession of concentric orbits, circling around the Earth as it
lay motionless at the Universe's centre. Although astronomers
had by this time introduced the idea of orbits, the picture that
was commonly in mind was that the planets were mounted
on a series of hollow crystal spheres. Outside the sphere of

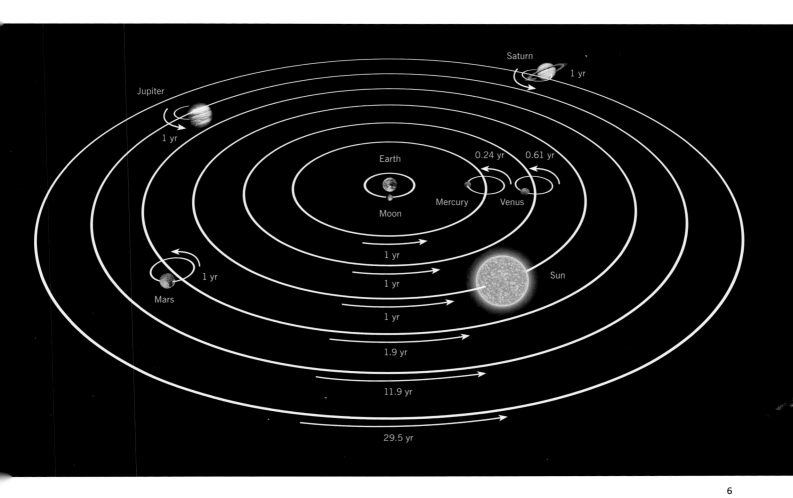

Saturn
1 yr

Jupiter
1 yr

0.24 yr   0.61 yr

Earth

Mercury   Venus

Moon

1 yr

1 yr

1 yr
Sun

Mars
1 yr

1.9 yr

11.9 yr

29.5 yr

6

the most distant planet, Saturn, was the sphere of the stars – because at any time of the night the stars seem to be studded on a hemisphere above us and this sphere appears to rotate around the Earth. (Navigators still use the convention of the celestial sphere to calculate the positions of the stars as seen from ships.) Beyond the sphere of the stars was an unseen sphere – the *primum mobile* ('first mover') – which was the primary mechanism that drove the movements of the Universe.

What were the engines that rotated the celestial spheres on their axes? The Christian astronomers who adopted Ptolemy's theories eventually offered their own answer: eight of Gabriel's angels pushed the spheres through their rotations.

In its essentials, the geocentric theory of the Universe is the common-sense view of its structure: it is what we feel and see. The true arrangement of the Universe would remain hidden from human perception for nearly 2,000 years.

6   **Ptolemaic system of the planets** In Ptolemy's theory of the Universe, the seven 'planets' circled on epicycles whose centres orbited the Earth (at the centre of the diagram). The rotation periods of the epicycles were chosen to fit astronomers' observations but reality soon proved more complex than this diagram, and further epicycles had to be added.

DAY 13 [20 SEPTEMBER]:

S unset to moonrise: 8°. There was a lunar eclipse. The totality of the Moon was covered at the moment when Jupiter set and Saturn rose. During totality the west wind blew, during clearing the east wind. During the eclipse, deaths and plague occurred. That month, the equivalent for 1 shekel of silver was: barley, [so many] kur; mustard, 3 kur; sesame, 1 pân, 5 minas. At that time, Jupiter was in Scorpio; Venus was in Leo, at the end of the month in Virgo; Saturn was in Pisces; Mercury and Mars, which had set, were not visible.
Cuneiform tablet, 331 BCE

Stars and Constellations

Our human link with the Ice Age

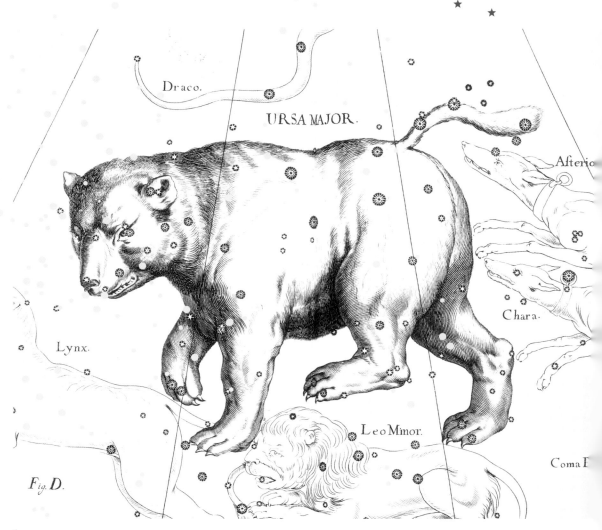

Draco.

URSA MAJOR.

Afterio

Lynx.

Chara.

Leo Minor.

Coma B

Fig. D.

1

Although the stars are random in position and brightness, they form shapes that observers can interpret symbolically, just as they read clouds, tea leaves or entrails. Even before recorded history, people saw mythical heroes, animals, birds and everyday objects in the patterns of the stars. The names of the ancient constellations are still used by astronomers, and are among the oldest surviving elements of human culture.

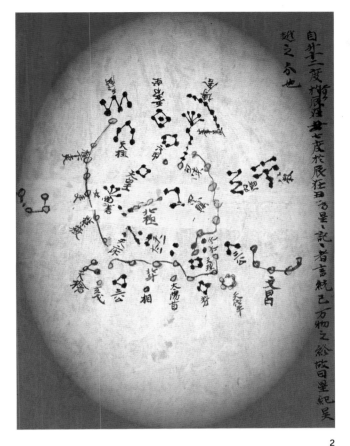

2

3

One of the most ancient constellations is Ursa Major. It was identified as a bear by northern peoples in both North America and Eurasia, and must therefore have pre-dated the disappearance around 15,000 years ago of the land-bridge joining Alaska and Siberia. The concept of the constellation as a bear then spread southward into the Middle East and the eastern Mediterranean, even though bears had disappeared from these lands as the glaciers retreated at the end of the last ice age. The Pointers of Ursa Major help to locate the North (pole) Star in the sky, and for this reason the Great Bear is often the first constellation that anyone in the northern hemisphere learns. Ursa Major symbolizes the northern regions and features on the flags of Alaska and the Cherokee nation.

45 constellations were described in the *Phaenomena*, a poem written in about 275 BCE by the Macedonian Greek Aratus of Soli, based on earlier work, now lost, by Eudoxus, a Greek astronomer and mathematician of the 4th century BCE. Eudoxus pored over old manuscripts in libraries in Egypt, which recorded original constellations dating back to the Babylonian civilizations of Mesopotamia. These manuscripts are now lost (some in the fires that destroyed the Library of Alexandria

1 **Ursa Major** Johannes Hevelius charted the stars of the constellation Ursa Major for a celestial atlas, *Uranographia* (1690), and embellished them with a picture of a grumpy bear.

2 **Chinese constellations** The circumpolar stars and constellations are mapped in the Dunhuang Star Chart, dating from about 720 CE. Chinese constellations are smaller than Western constellations and usually unrelated but the Seven Stars of the Northern Dipper (in Chinese: Beidou Qixing) are recognizably part of Ursa Major. This star chart is one of tens of thousands of scroll books discovered in 1908 in a library in the Mogao Buddhist caves near Dunhuang in north-west China, hidden in a room that had been bricked up a thousand years ago in anticipation of the imminent arrival of an invading army. It is the oldest surviving star chart on paper.

3 **Farnese Atlas** Sentenced by Zeus to hold up the celestial sphere, Atlas struggles under its weight in a 2nd-century Roman marble copy of a Hellenistic sculpture. The globe shows 41 of the constellations listed by Aratus and is the oldest surviving picture of classical Greek constellations; there is speculation that their depiction here was based on the lost star catalogue by Hipparchus (129 BCE).

by the 5th century CE), but the constellations they recorded survived in Greek literary culture. By the age of the poet Homer (*c.* 8th century BCE) the Babylonian constellations had become interwoven with Greek mythology; by the 3rd century BCE they had been completely replaced by Greek versions.

Most of the ancient constellations acquired their present Latin names in the 1st and 2nd centuries CE as Greek mythology became absorbed into Roman culture and Latin translations of Greek texts like *Phaenomena* appeared. The Greek-speaking astronomer Ptolemy, who lived in the Roman colony of Alexandria in Egypt in the 2nd century CE, described 48 constellations in his work called the *Almagest*. Ptolemy's description of the constellations formed the basis for the constellations of the present day.

However, many minor constellations have since been added to Ptolemy's list, filling the gaps between the Greco-Roman constellations. The Polish astronomer Johannes Hevelius named seven constellations in 1687. They included Lacerta, since, as Hevelius wryly reasoned, only a lizard could wriggle into the small space available, and the Lynx, because the eyes of a lynx were needed to see its faint stars. Some astronomers attempted to honour patrons by naming constellations for them, but only one survives: Hevelius's constellation Scutum was originally Scutum Sobiescianum, named after Poland's King John III Sobieski.

Other modern constellations never found widespread favour and were included on some charts but ignored on others. In 1922 the International Astronomical Union (IAU) took charge of the chaotic situation. Under the Belgian astronomer Eugène Delporte, the IAU standardized the constellations to the official modern system, abbreviating some over-elaborate names and rendering many constellations obsolete.

Alongside the official names of the constellations, there are common names in use in various languages. Gemini, for example, is known as the Twins in English, Gémeaux in French, Zwillinge in German and Gemelli in Italian. There are also names for asterisms (sub-constellations), such as the Plough, the Big Dipper and Charles Wain (all of them actually naming the same part of Ursa Major) and the Hyades (a star cluster in the constellation Taurus). The Summer Triangle of the northern sky (an equilateral triangle formed by the stars Altair, Deneb and Vega) is an asterism that spans several constellations. Some Australian aboriginal peoples saw a constellation in the Southern Cross that was not made of stars, but of dark clouds in the shape of an emu (**51**).

The zodiacal constellations are the major constellations that the Sun, Moon and planets cross as they travel across the sky. The zodiacal signs emerged in Babylonian astronomy during the 5th century BCE and travelled to Greece, Egypt, Rome, and India. They originally numbered six, all of them animals – hence the name 'zodiac', a Greek word meaning 'figured like animals'. The present system of zodiacal signs consists of twelve constellations, each 30° long. The continuing popularity of consulting one's 'star signs' in many daily newspapers attests to the lasting influence of the ancient constellations on the public imagination, even though their prophetic powers are scientific nonsense.

4 **Cassiopeia** The constellation, which is across Polaris from the Plough, as depicted in *Urania's Mirror* (1825), a boxed set of 32 hand-coloured cards showing 'all the stars visible in the British Empire'. The images of the stars are pricked through with holes, so that it is possible to see the constellations by holding the cards in front of a candle flame. The author was given as an anonymous 'young lady', but Peter Hingley, librarian of the Royal Astronomical Society, has discovered that the author was really the Reverend Richard Rouse Bloxam, a master at Rugby School.

THE COLLECTIVE UNCONSCIOUS

The whole of mythology could be taken as a sort of projection of the collective unconscious. We can see this most clearly if we look at the heavenly constellations, whose originally chaotic forms are organized through the projection of images. This explains the influence of the stars as asserted by astrologers. These influences are nothing but unconscious, introspective perceptions of the collective unconscious.

Carl Jung, *The Structure of the Psyche*, 1927

5 Zodiacal figures A celestial map centred on the ecliptic pole, with the constellation figures of the zodiac around the ecliptic circle. The German humanist Peter Apian constructed 21 *volvelles* (printed pages of moving images, like a child's pop-up book) for his *Astronomicum Caesareum* (1540), including rotating celestial maps like this one.

03. The Milky Way

Path of the gods, souls and pilgrims

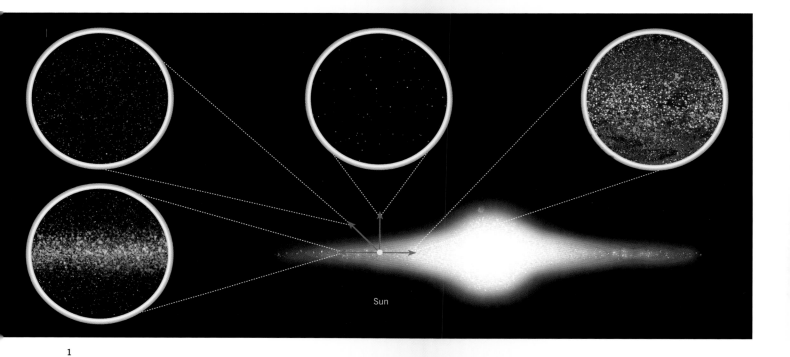

Sun

1

● Our Sun is part of a disc-shaped collection of stars called the
Milky Way Galaxy, or just 'the Galaxy'. The Sun lies in the plane
of the disc. When we look at the sky along the direction of this plane,
we see large numbers of stars, massed into a filmy band. This is the Milky
Way, which has the appearance of a milk-like stream of light, although
it is not actually a band or a stream, but a disc of stars seen edge-on.

1   The Milky Way As we view our Galaxy in different directions we see more or fewer stars. When we view the main disc of the Galaxy along its plane of symmetry, the Milky Way appears as a band.

2   *The Origin of the Milky Way* (Jacopo Tintoretto, *c.* 1575) Wishing to immortalize his son Hercules, born to the mortal Alcmene, Jupiter held the infant to suckle the breasts of the goddess Juno as she slept. Some milk spurted upwards to form the Milky Way.

3   The Milky Way A remarkably natural view of the stars as they are seen by the naked eye was made in a large drawing completed in 1955 by astronomer Knut Lundmark and engineers Martin Keskŭla and Tatjana Keskŭla at the Lund Observatory, Sweden. The whole sky is shown as an oval shape, with the plane of our Milky Way Galaxy along the central axis. 7,000 individual stars are plotted as white dots and the Milky Way is spray-painted with brighter star clouds and dark dust lanes that accurately represent its shape and texture.

2

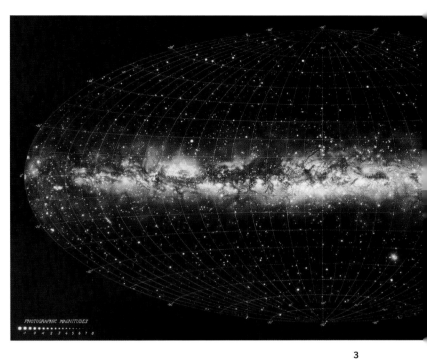

3

The earliest surviving written description of the Milky Way is in Ptolemy's *Almagest*. 'The Milky Way is not simply a circle,' he wrote, 'but a zone having almost the colour of milk, whence its name. It is not regular and ordered but different in width, colour, density and position; in one part it is double.' Of course, the very word 'milky' suggested the literal mythological explanation, a favourite scene of painters, Tintoretto among others, that the Milky Way was the milk of the goddess Juno. The Greek scientist and philosopher Aristotle made the first scientific discussion about the Milky Way in his *Meteorologica* (written 350 BCE) and many Western ideas about the origin or interpretation of the Milky Way can be traced back to this source, in which he discussed and classified all the possible theories.

Because we live centrally in the plane of the disc of our galaxy, the direction of greatest number of stars in the Milky Way stretches in a great circle around the sky. The Milky Way thus has the appearance of an arch, bridge or road across the night sky, conspicuous from dark locations both in summer and in winter evenings. Near the polar regions, the summer nights are short and twilit, so Swedes, for example, can scarcely see the Milky Way during the summer and are only able to view it during the long nights of the winter season. Swedes call the Milky Way *Vintergatan* ('Winter Street').

The Milky Way is bifurcated in some of its sections, giving it the appearance of a meandering river split by dark elongated islets. This is caused by a crinkled thin layer of dust, which from Earth is seen edge-on, silhouetted against the Milky Way and hiding an irregular zone of light originating from the broader disc of the stars beyond (**51**). Because of this visual effect, in Arab lands the Milky Way was known as *al-Nahr*, 'the River', but the name never became adopted into Western languages because of the potential for confusion with another long meandering constellation with the name of a river, Eridanus.

The Milky Way has also been named after specific roads. Country names from England for the Milky Way were

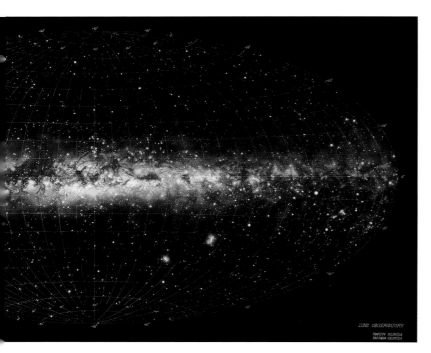

LUND OBSERVATORY
MARJÚN KEIKOLA
TATJANA KEIKOLA

4

Watling Street and Walsingham Way. Watling Street is the Roman road built from Chester to London and on to Dover. Walsingham Way is the road from London to the Virgin's shrine in Norfolk, and local people thought that the Milky Way pointed to the shrine. In Spain the Milky Way is sometimes called El Camino de Santiago. This is the pilgrim's track to Santiago de Compostela in northern Spain. Both pilgrims' ways were thronged with pilgrims, as the Milky Way is with stars.

In many cultures the Milky Way symbolized the journey of the soul into the afterlife. Coupling the belief that the stars represent souls with the idea that the Milky Way was a road, the Romans described the Milky Way as the path to the Seats of the Heroes, *Heroum Sedes*, travelled by the departing souls of illustrious men.

The similarity of the pale appearance of the Milky Way to grey ash produced another class of explanations: that it was the scorched path of the Sun or, in classical mythology, the disastrous route travelled by Phaeton, son of the sun-god

Helios, when he lost control of his father's chariot, which carried the Sun across the sky. Jupiter averted this potential disaster by throwing a thunderbolt, and causing Phaeton to hurtle in flames into the river Eridanus. The final theory mentioned by Aristotle was that the Milky Way was some sort of manufacturing imperfection, splitting or joining, like the seam of a metal casting or the sewed seam around a leather-covered ball.

The true explanation for the Milky Way's 'milk-like' appearance is that it is made up of many stars, which are too faint and close to each other to be visible individually. This was first conjectured in the 5th century BCE by Democritus and proved by Galileo with his telescope in 1610. The first picture to show the Milky Way realistically as a collection of stars was *Flight into Egypt* (1609) by Adam Elsheimer, now in the Alte Pinakothek, Munich.

5

4 *Flight into Egypt* (**Adam Elsheimer, 1609**) In a small night-scene, Joseph,
torch in hand, leads Mary and the infant Jesus along a lake-shore on an
ass. The full Moon overhead does not completely swamp the Milky Way,
stretching up to the left. Its stars are painted individually using the single
hair of a paintbrush, the first representation of the Milky Way as a a
band of stars.

5 **Centre of the Milky Way** A wide-angle picture shows the Sagittarius star
clouds towards the centre of our Galaxy. They are sliced by rifts of dust,
which hide the stars beyond.

WINTER STREET

S ilent with star-dust, yonder it lies,
The Winter Street, so fair and so white;
Winding along through the boundless skies.
Edith Matilda Thomas

04.

The Shape of the Earth

Our planet, a flattened sphere

The ancient understanding that the Earth was round rather than flat was only the beginning. Over the past 2,000 years, a clock pendulum, a librarian's horseback journey to Alexandria and a wayward satellite have helped us to establish the exact size and shape of the Earth. It turns out that our planet is not a perfect sphere, but resembles a squashed, dimpled golf ball.

3

From antiquity every educated person knew that the Earth was approximately spherical. The question of whether Christopher Columbus would fall off the edge of the world if he sailed from Spain westwards was founded in ignorance. The real doubts centred on whether he would survive the dangers of the journey (weather, sailing hazards, sea monsters) and whether he would be able to discover an alternative route to the East Indies. As history tells us, Columbus survived the ocean crossing and in 1492 landed in what he called the West Indies, actually part of the Americas.

Ancient astronomers argued that the Earth was spherical because it always cast a circular shadow on the Moon during a lunar eclipse. It was well known, too, that a lookout at the top of a ship's mast would see land before his fellow sailors on deck, because he was able to see over the curvature of the Earth. Greek philosophers of the Pythagorean school in the 6th century BCE disseminated these standard arguments for a spherical Earth throughout the educated world.

1 **Spherical Earth** As a ship leaves shore and sails over the horizon, its hull disappears first, while the top of its mast can still be seen. A woodcut illustration from a 1550 edition of Sacrobosco's textbook *Tractus de sphaera*, written about 1230, and the most widely used astronomical text of the middle ages, depicted this well-known evidence for a round Earth.

2 **Curvature of the Earth from space** Photographed by astronauts flying at altitudes above 100 km, the horizon shows the Earth is spherical.

3 **Map of the world** A diagrammatic map of one side of the globe, illustrating the *Prokheiroi Kanones* ('Handy Tables'), compiled by Ptolemy for astronomical and geographical calculations. Of the known world (upper half of the map), Egypt is represented by the rectangle and the Indian Ocean by the semicircle. Hades and its rivers occupy the southern hemisphere.

4 **Lunar eclipse** The shadow of the Earth appears circular when projected on to the Moon during a lunar eclipse, showing that the Earth is spherical.

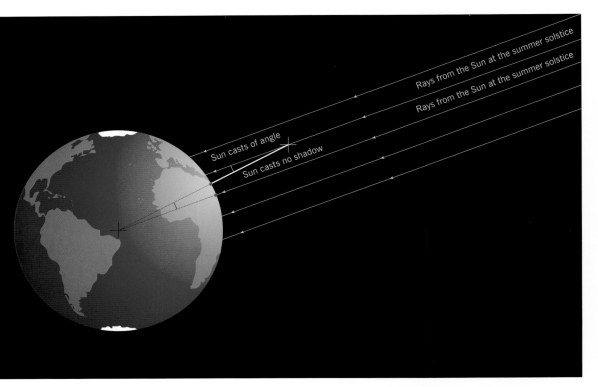

5

6

In the 3rd century BCE, Eratosthenes, the librarian of Alexandria, determined the size of the Earth. He had heard that at Syene in Upper Egypt (present-day Aswan) the Sun was directly overhead at noon on the day of the summer solstice: the Sun's rays reached the bottom of a deep well. He determined the length of the shadow of a vertical post at Alexandria on the same day and found that the angle of the Sun was 1/50 of a circle to the south of the zenith. It is said that he determined the distance between Syene and Alexandria by driving a carriage between the two cities and counting the revolutions of the wheels. He multiplied this distance – 5,000 stadia – by 50 to calculate that the circumference of the Earth was 250,000 stadia. The modern equivalent of a stadium is not well established, but Eratosthenes's figure is thought to be the equivalent of about 45,000 km, remarkably close to the modern measurement of the Earth's circumference as 40,000 km.

However, by the 17th century it had become clear that the Earth was not entirely spherical. The first evidence came from a clock pendulum. In 1671 the Paris Academy of Sciences sent Jean Richer to Cayenne in Guyana, South America, on the equator. He was to observe the close approach of Mars to the Earth in 1672, in order to establish the distance between them and thus the scale of the Solar System. To accomplish

5 **Size of the Earth** Eratosthenes estimated the circumference of the Earth by measuring the length and the angle of the arc between Alexandria and Syene.

6 **Eratosthenes** The head librarian in Alexandria, Erastothenes was sometimes nicknamed 'Pentathis', meaning a champion in pentathlon athletic events, and also 'Beta', the second letter in the Greek alphabet. These were compliments for a librarian with broad interests, but also back-handed criticisms of someone who tried his hand at everything and was outstanding at nothing.

7 **The Geoid** Areas of the sea have been coloured to represent places of different height above and below the average shape of the Earth. Additionally the globe has been distorted in shape to represent the same differences, exaggerated relative to the size of the Earth by hundreds of times.

8 **Académie Royale des Sciences de Paris** In the 17th century, the Royal Academy of Sciences in Paris carried out numerous scientific investigations into a wide variety of programmes, as shown in a composite illustration.

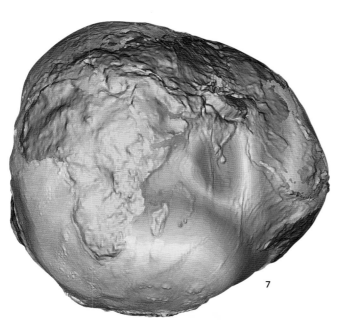

7

this, Richer needed an accurate clock. He took a clock with a pendulum that had beaten seconds correctly in Paris, but in Cayenne the same clock ran slow and lost two and a half minutes every day. To make it beat seconds accurately in Cayenne, Richer discovered that he had to shorten the pendulum by about 3 mm. The reason for this was a mystery.

In 1687 Isaac Newton offered a solution to Richer's discovery that a pendulum beat slower at the equator than in France. The Earth is not exactly spherical, but bulges at the equator and is squashed at the poles. It rotates once every 24 hours and the resulting centrifugal force raises the region along the equator; this also reduces gravity at the equator, which is why Richer's pendulum beat more slowly in Cayenne.

During the 18th century Newton's explanation was confirmed by a massive body of work, chiefly organised by the Academy of Sciences of Paris. The magnitude of the task is represented by the long list of scientists involved: the four astronomers of the Cassini dynasty, all successive directors of the Paris Observatory (Jean-Dominique Cassini, his son Jacques Cassini, his grandson César-François Cassini de Thury and his great-grandson, also called Jean-Dominique Cassini); the astronomers Jean-Baptiste-Joseph Delambre and Pierre Méchain; the French mathematicians Pierre Maupertuis, Pierre Bouguer and Louis Godin; the explorer Charles-Marie La Condamine; and the Swedish astronomer Anders Celsius, whose name is remembered in the centigrade temperature scale. They carefully measured the three-dimensional shape of France, and some participated in adventures to the equator in Ecuador and to the Arctic Circle in Lapland to measure the curvature of the Earth. They discovered that the Earth was indeed a flattened sphere. The modern value for the flattening is 1/298.25 – there is a difference of more than 21 km between the Earth's radius at the equator (6,378 km) and its radius at the pole (6,357 km).

The shape of the Earth affects the motion of orbiting satellites: the lumps of extra material protroding from its spherical surface pull satellites off course. Some satellites, like the LAGEOS, or Laser Geodynamics Satellites, are simple spheres covered with reflectors so that their position can be measured to millimetres when laser pulses are reflected from them. Others, like JASON-1 and TOPEX/Poseidon, carried radar to measure the wrinkles on the surfaces of the Earth's oceans to an accuracy of centimetres. These satellites have shown that the Earth's ocean surface is dimpled like a golf ball, with wind-driven currents piling water up into mounds against the continental shores, and warmer sea areas standing higher, in the same way that mercury rises in a thermometer. Overall the height of Earth's ocean surface is lower by about 150 m in the north Indian Ocean (off the south coast of India) than in the western Pacific Ocean (off New Guinea); travelling eastwards from the sea off New Guinea to the sea off California is to go downhill by 90 m. In the northern Atlantic Ocean, the sea off Florida is 130 m lower than the sea off Iceland.

Additionally, the shape of Earth changes over the seasons and the years, as the mass of water covering the Earth shifts in position. Global-scale climate changes (such as the El Niño phenomenon) have melted sub-polar glaciers and changed the currents in the Southern, Pacific and Indian Oceans, causing the bulge in the Earth's equator to grow larger and mass to move away from the poles.

9 **Ocean topography** There are undulations of about 200 m from peak to peak in the surface of the sea. These are caused by a variety of factors, such as strong currents piling water up against a continental shore. The undulations have been measured by radar from satellites like Topex/ Poseidon, and are used to predict tides and changes of sea level caused by global warming.

10 **Causes of the irregular shape of the Earth** Below the surface of the Earth, the circulation of the liquid core causes plate movements that crack the continents and raise mountains. Volcanoes grow where magma wells up from the interior. The rotation of the Earth flattens the polar regions. An ice cap grows at the South Pole and flows outwards to the edge of Antarctica, while depressed land areas released from the weight of melting glaciers spring upwards. Water is moved from area to area at different seasons of the year. Ocean currents and winds distort the sea and push on obstructions.

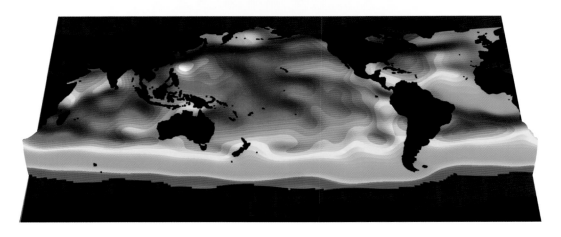

9

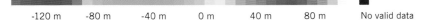

-120 m   -80 m   -40 m   0 m   40 m   80 m   No valid data

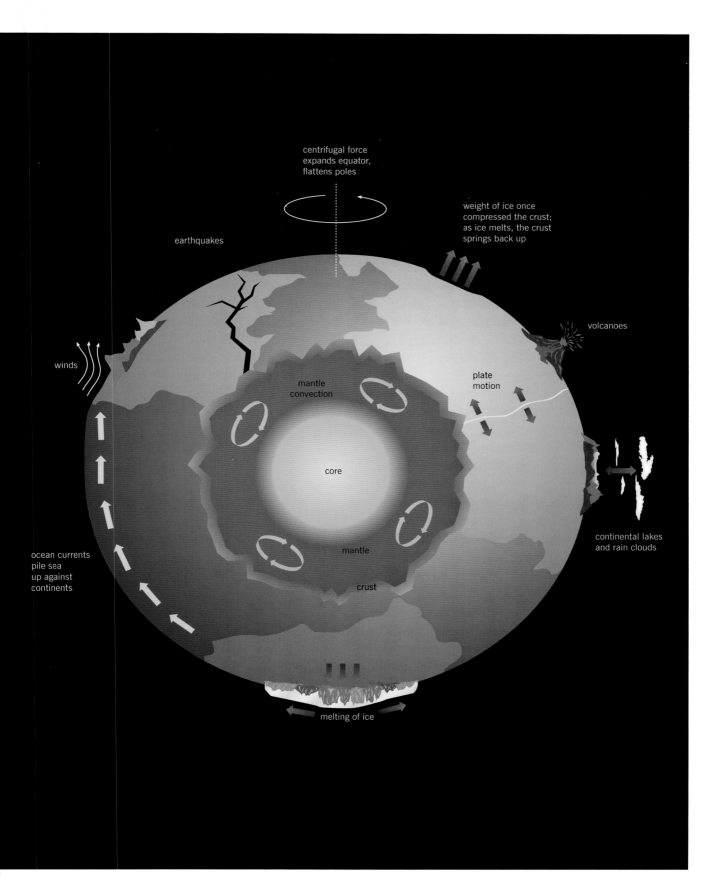

centrifugal force
expands equator,
flattens poles

earthquakes

weight of ice once
compressed the crust;
as ice melts, the crust
springs back up

winds

volcanoes

plate
motion

mantle
convection

core

mantle

crust

continental lakes
and rain clouds

ocean currents
pile sea
up against
continents

melting of ice

The Southern Constellations

Hidden stars revealed by the tilting Earth

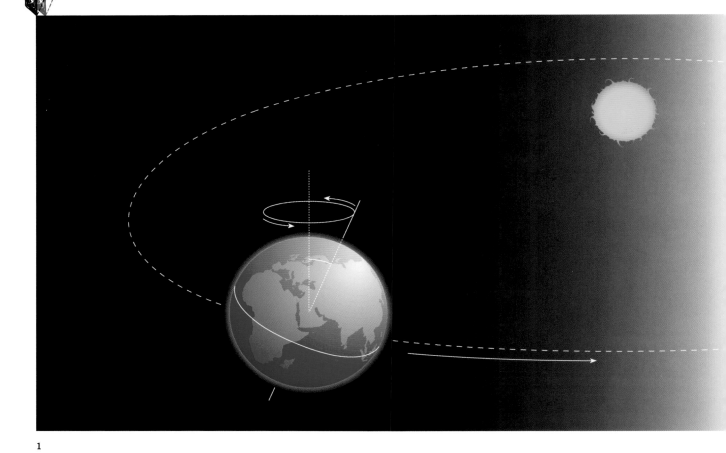

1

How did stars that cannot be seen from Europe and the Mediterranean come to have Greek and Roman names? Although the traditional constellations were named thousands of years ago, their visibility has changed dramatically since antiquity. In some parts of the world new stars have risen in the sky while ancient constellations have disappeared below the horizon. These changes are caused by a 26,000-year cycle of 'wobbles' in the Earth's rotation.

2

3

1 **Precession** The Earth's axis of rotation leans at an angle of 23.5° to its orbital plane, but over a period of 26,000 years the direction of the axis in space moves in a cone, the phenomenon of 'precession'.

2 **Star trails over Finland** On the Arctic Circle, star trails run nearly parallel to the horizon because the north celestial pole is almost directly overhead.

3 **Star trails over Hawaii** At latitudes close to the equator, the stars rise and set nearly perpendicular to the horizon, rotating about the north celestial pole, which lies near to the horizon rather than directly overhead.

4 **Precessing top** In his *Popular Astronomy* (1894) Camille Flammarion showed how a spinning top precesses, its rotational axis tracing out the surface of a cone under the gravitational influence of the Earth, just as the Earth precesses under the gravitational influence of the Moon and Sun.

4

If you look up at the stars from the North Pole, only half the sky is visible, the half that is centred around the north celestial pole immediately above your head, the point at the extension of the Earth's axis of rotation. The other half of the sky is perpetually below the horizon, even though the Earth rotates. But if you are at the Equator, you will be able to see every part of the sky in turn at different times of the year. At intermediate European latitudes, where most of the classical constellations were named after figures from Greco-Roman mythology, half the southern sky is perpetually out of sight. Because the constellations in this part of the southern sky could not be seen from Europe or the Mediterranean, they were not named by Europeans until sailors began to explore the southern seas in the 15th century.

There are two peculiar exceptions to this rule. In a part of the sky that lies south of the bright star Fomalhaut and is, in fact, readily visible from Europe, there are no constellation figures that have mythological names dating back to classical times. On the opposite side of the celestial globe is an area of the southern sky, which, although it never rises above the horizon even in southern Europe, contains a substantial part of a constellation that was known to Greek and Roman classical scholars even though they were never able to see its stars. This constellation is called Argo, the name of the ship in which, according to classical mythology, Jason

5

6

sailed with the Argonauts in search of the Golden Fleece.
The situation seems very puzzling: stars that were not
visible from Europe came to have European classical
names, while visible stars were for some reason left
unlabelled. The solution to the mystery is precession.

The traditional constellations are centred on Polaris. The
North Star lies near to the centre of the 24-hour rotation of the
stars, but it has not always been 'in pole position'. The axis of
the rotation of the Earth shifts, cyclically, in space, wobbling
like the stem of a spinning top and pointing to different stars
from time to time. This wobbling is called 'precession', and
its effect is to twist the axis of the Earth's diurnal rotation in
a circle on the sky. The Earth's axis slowly follows this circle
over a period of 26,000 years. At this speed the changes in
the constellations are not readily noticeable from generation to
generation, but may be perceived from civilization to civilization.

In the several thousand years that have passed since
the constellation figures were first recorded, precession
has twisted the Earth's axis from its original position
to the present North Pole. This has the effect of hiding
part of the sky mapped by classical astronomers, while
revealing another area, which contains stars that could not
be seen from Europe and the Mediterranean in classical
times, and was therefore not mapped until much later.

A further effect of the twist is that constellation figures
(such as Orion) that used to stand erect in the sky, with their
heads facing the North Pole, now lean at an angle. These
changes started to be apparent to astronomers as early as

the 2nd century BCE, when the Greek astronomer Hipparchus
noticed discrepancies between his own observations of the
sky and the constellations that had been recorded by his
predecessors Eudoxus and Aratus two centuries earlier.

Precession also affects the part of the sky in which the
Sun appears at the spring equinox. For roughly 2,000 years
before the start of the Common Era, the sun appeared near
the constellation Aries at the spring equinox, shifting gradually
into Pisces towards the beginning of the Common Era. Soon
Aquarius will become the constellation of the spring equinox
(some say that it already has), and this will be (or has been)
the 'dawning of the Age of Aquarius'. An 'age' in this sense lasts
about 2,000 years (the period of precession, 26,000 years,
divided among twelve signs of the zodiac).

Using the known speed of precession and working
backwards, it is possible to calculate that all the classical
constellation figures were visible and stood erect in the sky
around 2800 BCE (plus or minus 300 years) as seen from
latitude 36°. This corresponds to the time and place where
the constellations were formulated: during the height of
the great civilizations of the Tigris-Euphrates valley, which
straddles this latitude.

While most of the constellations of the northern sky are
named for figures from classical mythology, many southern
constellations commemorate modern inventions, having
been named by 17th- and 18th-century astronomers during
expeditions to southern lands. The 18th-century French
astronomer Nicolas Louis de Lacaille carried out an extended

7

5 The first printed celestial maps This map of the stars was made by the German engraver Albrecht Dürer (1515). His now iconic constellation figures were accurately placed, but he pictured the sky as if on a sphere seen from the outside, as God might look at it.

6 Dürer's map of the southern constellations Dürer's map of the southern constellations shows the remarkable 'hole' in the region never visible from the north, where no constellations had been described. The hole is offset from the present position of the south celestial pole because of the change of orientation of the Earth's rotation axis through precession.

7 Abbé Nicolas Louis de Lacaille Lacaille mapped the 'foreign, strange-eyed' southern stars and formed them into constellations, including the Chemical Furnace, belching smoke and flames in the lower left hand corner of Fig 8.

8 Cellarius's map of the southern constellations The Dutch-German mathematician and cosmographer Andreas Cellarius was the author of the *Harmonia macrocosmica* (1660), a large, spectacular astronomical atlas.

astronomical expedition to the Cape of Good Hope. He invented constellations celebrating what were at the time modern inventions such as Fornax (the Chemical Furnace) and Horologium (the Pendulum Clock). He broke down one large southern constellation, Argo Navis, into more conveniently sized component parts including the Keel, the Poop Deck and the Sails.

The constellation of Crux, the Cross, which lies near to the south celestial pole, is of particular significance in Christian cultures because of its cruciform shape. Known in Europe as a separate constellation only since the 16th century (Amerigo Vespucci claimed to be the first European to see its stars on his third voyage of 1501) in modern times the Southern Cross has become a sentimental symbol of southern-hemisphere patriotism: it appears on the flags of Australia, Papua New Guinea, New Zealand, Samoa and Brazil. The four stars of the Southern Cross were in fact known to Ptolemy as part of the constellation Centaurus but were 'lost' to European eyes as

the Earth tilted away from the constellation due to precession. The Southern Cross ceased to be visible from the British Isles at the time that the building of Stonehenge was started (about 3000 BCE) and from northern Mediterranean shores at the time of the collapse of the Roman Empire (about 500 CE).

Because the southern stars were imagined but unseen by Europeans, and associated with exploration and long voyages in the strange lands overhung by the southern skies, they are often used symbolically in literature and art to convey a sense of not belonging, of adventure and 19th-century imperialism. For example, in Thomas Hardy's poem 'Drummer Hodge', the drummer-boy is killed in the Boer War and lies buried in the Karoo desert in the Cape, in a grave on an isolated *kopje* in the veldt, under foreign, 'strange-eyed' constellations.

9 **The southern Milky Way over Paranal** The dark patch (centre) is the Coal Sack, and on its upper edge is the Southern Cross.

10 **Brazilian flag** The Southern Cross is the constellation of five stars at centre, below the letters 'E PR'.

11 **Navigating a ship** Navigators observe the stars with sighting instruments from the stern of a galleon in André Theret's *Cosmographie* (1575). To avoid disaster at sea, it was essential for navigators to have exact knowledge of the positions of the constellations throughout the year. The unfamiliar southern stars were therefore especially unsettling and disorientating for European sea-explorers in the Age of Discovery.

THE DOMAIN OF SCIENCE

The sky is more the domain of science than of poetry. It is the stars as not known to science that I would know, the stars which the lonely traveller knows.
Henry David Thoreau, *Journal,* **1853**

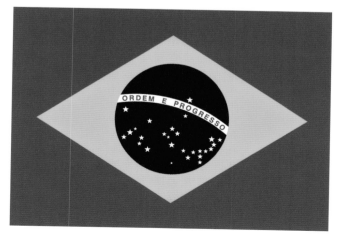

The Sun

At the centre of the Solar System

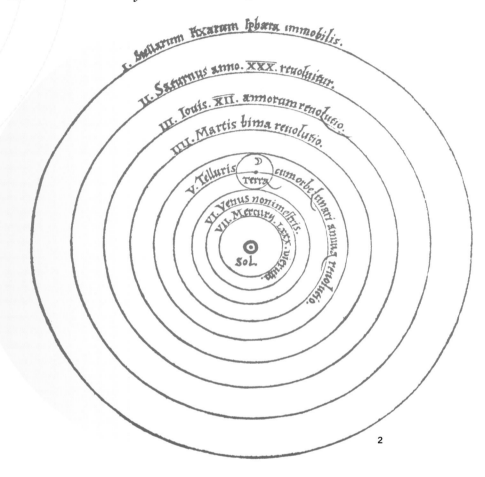

In the first half of the 16th century, the Polish cleric Nicolaus Copernicus outlined a hypothesis to replace the 1,500-year-old Ptolemaic system, proposing that all the planets, save the Moon, revolved around the Sun. Few scientific discoveries have demanded such a fundamental transformation of human thought. Yet Copernicus's revolutionary theory was actually the product of centuries of careful astronomical observation that had gradually exposed the failure of the Ptolemaic system to predict the movement of the planets.

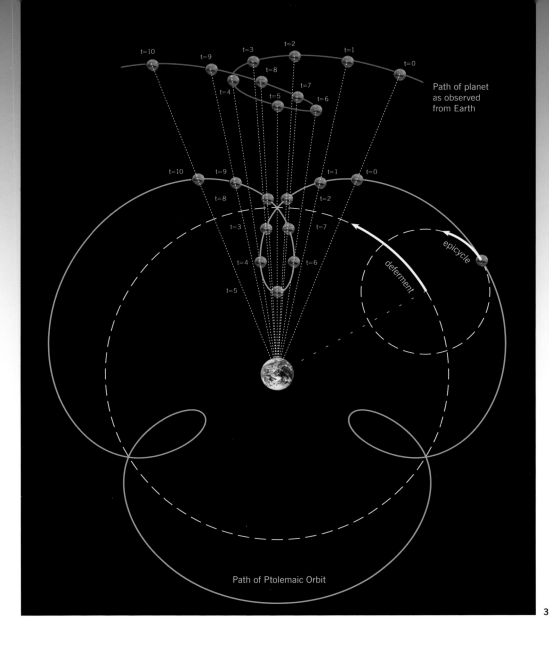

t=10 t=9 t=3 t=2 t=1 t=0

t=8 t=7

t=4 t=5 t=6

Path of planet
as observed
from Earth

t=10 t=9 t=1 t=0

t=8 t=2

t=3 t=7

epicycle

deferent

t=4 t=6

t=5

Path of Ptolemaic Orbit

3

Since the 2nd century CE, the geocentric theory, called the Ptolemaic system, had held that the Earth was stationary at the centre of the Universe (01). The theory survived unchallenged for over 1,500 years as it did not contradict literal readings of the Bible or the immediate evidence of the five senses.

Yet by the 16th century the Ptolemaic theory was beginning to suffer a progressive loss of confidence. It had become clear that the geocentric model required constant additions and changes to make it correspond with astronomers' actual observations. Astronomers were beginning to feel that these successive add-ons were arbitrary and did not give a decisive overall explanation for the observed behaviour of the heavenly bodies. The most serious problem was in the addition of epicycles to explain why some planets did not appear to orbit the Earth as predicted by the Ptolemaic theory.

1 **Nicolaus Copernicus** Astronomy was not Copernicus's profession, but a personal interest he pursued while practising medicine and working as an ecclesiastical administrator and government adviser in Polish Prussia. His friends persuaded him to publish his theories, and delivered the first printed copy of his treatise into his hands as he lay on his deathbed.

2 **The Copernican Universe** Copernicus's diagram of the Universe, from his *De revolutionibus orbium coelestium* ('On the Revolutions of the Celestial Spheres') is remarkably similar to modern illustrations of the Solar System.

3 **Epicycles** The Ptolemaic system envisaged each planet orbiting the Earth in a series of consecutive small orbits, or 'epicycles' (shown in the two-o'clock position on the diagram) within a larger orbit (designated by the broken circular line around the Earth). This produced a looping orbit (represented by the clover-shaped outermost line), which made the planet seem to move backwards in a loop when observed from the Earth, as illustrated at the top of the diagram.

Because the Earth has a shorter orbit than the outer planets (Mars, Jupiter, Saturn, Uranus and Neptune), it overtakes them from time to time as it revolves around the Sun. When viewed from the Earth on these occasions, the outer planets seem to halt in their orbits and move backwards in a loop. This is called 'retrograde motion'. As first proposed by Appolonius of Perga (*c.* 200 BCE) and Hipparchus (*c.* 130 BCE), Ptolemy added epicycles to his system to account for this loop.

An epicycle was envisaged as a kind of revolving wheel that carried the planet on its outer rim, while itself revolving in an orbit around the Earth. The combination of the two motions was thought to produce the retrograde loop. In modern engineering, the concept of epicycles survives in the name of so-called 'planetary gears', in which a small gear wheel rotates outside a central one, all enclosed within a hollow toothed chamber.

Despite this attempt to refine the existing Ptolemaic theory, discrepancies between the predicted and observed orbits of the planets continued to build up over time. This was particularly a problem with the planet Mars, which moves unusually quickly through an especially eccentric orbit, and whose orbit seen from Earth is therefore complex. To predict the movement of Mars accurately, astronomers were obliged to add ever more epicycles and to assign these epicyclical 'wheels' a variety of different sizes and spin rates.

Although the theory of epicycles eventually did allow astronomers to calculate the movements of the planets with relative accuracy – no inconsiderable achievement – epicycles did not seem to be grounded in real observation. All they did was make the calculations work. Moreover, the theory never seemed to be reaching a stage where it was able to provide a satisfactory unified explanation for the motions of the planets. When further epicycles had to be added, it seemed that there was no end to the arbitrariness of the theory – 'wheels within

4

WISDOM AND FOOLISHNESS

People give ear to an upstart astrologer who strove to show that the Earth revolves, not the heavens of the firmament, the Sun and the Moon ... This fool wishes to reverse the entire scheme of astronomy; but sacred Scripture tells us that Joshua commanded the Sun to stand still, not the Earth.

Martin Luther, *Table Talk,* **1540**

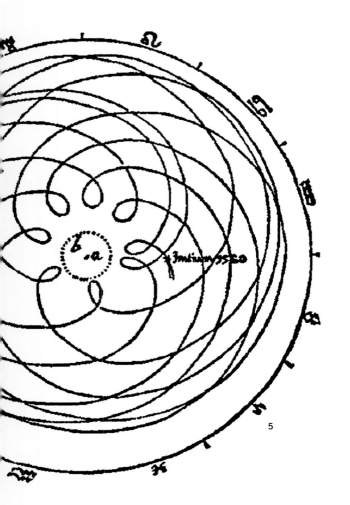

wheels' is an expression describing complex social machinery, referencing both astrologers' systems of epicycles, and the biblical expression in Ezekiel 1:16.

The situation began to change in 1543, when the Polish cleric Nicolaus Copernicus published his *De revolutionibus orbium coelestium* ('On the Revolutions of the Celestial Spheres'), proposing a radical new hypothesis to replace the Ptolemaic theory: all the planets, save the Moon, revolved around the Sun in a series of concentric orbits. He produced a remarkably accurate diagram of the Solar System, which fixed the order of the planets from the Sun according to their different orbital speeds, on the assumption that the slowest planet was the furthest from the Sun. His theory also simplified the calculations of planetary motions, explaining retrograde motion as a visual effect created by the orbital movement of the Earth, although he still formulated the calculations in terms of a (smaller) number of epicycles.

These were astonishing thoughts – too astonishing for many, and Copernicus's hypothesis remained just that for more than 60 years, until it was given convincing form by the brilliant and eccentric German astronomer Johannes Kepler. Kepler based his calculations on observations by his tutor Tycho Brahe, a 16th-century Danish astronomer. A strong Protestant, Brahe did not believe that the Earth moved, because that appeared to contradict biblical texts. Brahe had proposed that the Moon and Sun orbited the Earth, but that all the other planets revolved around the Sun. Nevertheless, the accuracy of Brahe's observations provided the data that allowed his pupil to complete the simplification of the Copernican theory, which, ironically, was anathema to Brahe's literalist religious beliefs.

In simplifying Copernicus's theory, Kepler produced the first recognizably modern plan of the Solar System. He abolished the concept of epicycles altogether, and in their place offered

4 **Johannes Kepler** Tycho Brahe's pupil used his tutor's planetary observations to prove that the planets orbit the Sun in ellipses, not circles or epicycles. The portrait was the frontispiece of Kepler's *Complete Works*. The inscription erroneously describes the German astronomer as an Austrian mathematician.

5 **'On the Orbit of the Planet Mars'** In 1619 Kepler published a diagram of the Solar System, drawn as if the Earth was stationary, to demonstrate how ridiculously complicated was the system of epicycles necessary to account for the orbit of Mars between 1560 and 1596.

SCORN OF INCOMPETENCE

If there should chance to be any mathematicians who, ignorant in mathematics yet pretending to skill in that science, should dare, upon the authority of some passage of Scripture wrested to their purpose, to condemn and censure my hypothesis, I value them not, and scorn their inconsiderate judgment.

Nicolas Copernicus, preface to *De revolutionibus orbium coelestium*, 1543

6

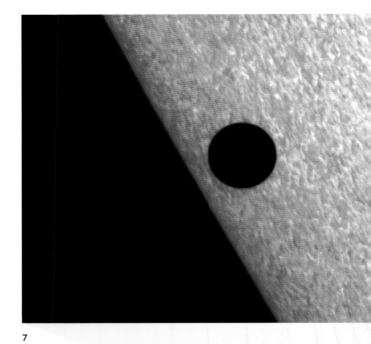

7

his own description of the motions of the planets. He calculated tables to estimate where the planets would be in the future (the 'Rudolphine Tables', published in 1627), and was able to predict accurately that Mercury would be aligned with the Sun in 1631 so exactly that it would pass in front of the Sun's disc. This 'transit of Mercury' was witnessed by the French astronomer Pierre Gassendi. Eight years later, using Kepler's tables and theories, the English cleric Jeremiah Horrocks calculated a transit of Venus and, with the merchant William Crabtree, actually saw the transit with his own eyes, just before sunset on 24 November 1639. They were both ecstatic with their discovery: Horrocks wrote that, 'rapt in contemplation, [Crabtree] stood for some time motionless, scarcely trusting his own senses, through excess of joy; for we astronomers are of a womanish disposition and are overjoyed with trifles'.

These observations were convincing proof that Kepler's calculations of the planets were the most accurate ever produced. How did he do it? Kepler showed that the orbits of the planets about the Sun, particularly the orbit of Mars, were simple ellipses. He also demonstrated that

there were basic relationships between the motions of the planets and the sizes of their orbits – his 'laws' of planetary motion (**29**). These interrelationships removed a lot of the arbitrariness in the arrangement of the planets.

Kepler attempted an explanation of these relationships by proposing that there existed a 'magnetic' virtue or force between the Sun and the planets, including the Earth and Moon. In so doing, he prepared the way for the identification of the force of gravity by Isaac Newton (**29**).

The theory of epicycles had seemed arbitrary to astronomers, but the implications of the Copernican system were nothing short of startling. Far from lying stationary at the centre of the Universe, the Earth is moving. Our planet spins on its axis; it orbits around the Sun, and it wobbles as it orbits. Despite the lack of sensation its gyrations produce in us, the Earth's motions are dizzying as a dance.

8

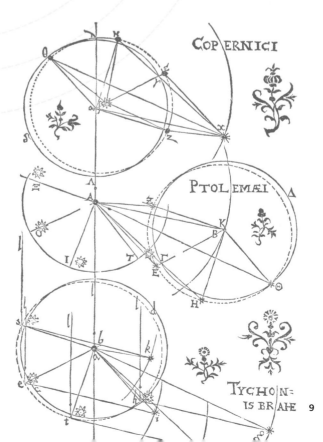

6 **Pierre Gassendi** The French philosopher and astronomer was a friend of Galileo and a supporter of Kepler, whose theories he sought to verify.

7 **Transit of Mercury** Mercury passed in front of the Sun on 8 November 2006, and was imaged from space by the Japanese Hinode solar observatory.

8 **William Crabtree** (*Crabtree Watching the Transit of Venus AD 1639*, Ford Madox Brown, 1903). The awestruck draper views the transit of Venus projected into the attic of his draper's shop through darkened shutters and curtained windows, while his wife struggles to keep the children in order.

9 **Orbit of Mars** In his *Astronomia nova* ('New Astronomy', 1609), Kepler compared the orbit of Mars in the heliocentric Copernican planetary system, the geocentric Ptolemaic system, and Brahe's system (a compromise between the two) – all using variations on the idea that celestial orbits were circular. On the basis of Brahe's own data, Kepler showed that it was more accurate to use the Copernican theory to calculate the position of Mars, but only when its orbit was assumed to be an ellipse, not a circle.

Discoveries in the
Solar System

07. Comets

Disasters, sun-grazers and the 'Lady's Comet'

Unlike stars and planets, comets appear without warning and move rapidly and erratically across the sky. In early cultures, comets were regarded as harbingers of doom because of this behaviour, but in the centuries following the Enlightenment, this same unpredictability has become an irresistible attraction for comet-hunters. One of earliest and most famous comet-hunters was an 18th-century Englishwoman, Caroline Herschel, who laboured under extreme conditions to discover fourteen comets and was rewarded with a royal stipend for her pioneering efforts.

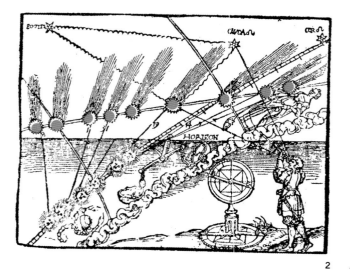

2

Comets are small, dark bodies in the Solar System, which are hard to see when far from the Sun. They have a solid part, the 'nucleus,' which is composed of both ices and solid material. When a comet comes close to the Sun, its nucleus melts, vaporizes and becomes 'active'. It develops a bright, dusty atmosphere, the 'coma', and 'tails' of gas and dust, which reflect more sunlight. Coupled with the speed of the comet as it approaches the Sun, this means that comets can spring into sight quickly and unexpectedly.

For a long time, comets were interpreted superstitiously, because, in an age when the cyclical movements of the stars were thought to control events on Earth, comets were sporadic and unpredictable. In his 1665 book *De Cometis*, John Gadbury warned that comets were 'threatening the world with Famine, Plague and War: To Princes, Death! To Kingdoms, many Crosses; To all Estates, inevitable Losses! To Herdsmen, Rot; to Plowmen, hapless Seasons; To Sailors, Storms, To Cities, Civil Treasons!'

Anyone who is in the right place at the right time can discover a comet, which will usually be named for its discoverer. The most successful comet-hunters are of course those who devote lots of time to systematic searches: one highly successful Japanese comet-hunter is said to have dedicated the rest of his life to finding comets to fulfill a deathbed vow to his dying father. The usual method is to sweep a telescope or binoculars from side to side to examine a section of the sky, looking for objects that are fuzzy. Nebulae, galaxies and star clusters can be eliminated by reference to a catalogue or atlas. Comet-hunter Charles Messier compiled a list of such objects that might be mistaken for a comet; this became known as the Messier Catalogue (1774–81), and is still in use. The clinching difference is that a comet moves in the sky, whereas nebulae and galaxies are stationary.

Between 1996 and 2008 armchair astronomers discovered over 1,500 comets using the Solar and Heliospheric Observatory (SOHO) satellite; it is still on station gazing constantly at the Sun, so more will be discovered. Comets show in SOHO's camera as they graze the surface of the Sun; some of them approach the Sun but are not seen to leave (i.e. they melt). Most are members of the Kreutz sungrazing family. In 1888 Heinrich Kreutz discovered that a number of comets have similar orbits passing near the Sun. They come from a single comet that shattered into many fragments, which have further broken into small pieces about 10 m in size. Armchair astronomers scan online SOHO pictures daily to discover new fragments before they disappear.

1 **Disasters caused by the Comet of 363** CE The comet allegedly caused thunder, lightning and earthquakes, foretold the death in battle of the Roman emperor Flavius Claudius Julianus during a war with the Persians, and provoked the destruction by fire of the library in Antioch – much of which is crammed unconvincingly into this tiny engraving from Stanislaus Lubienietzki's book *Theatrum cometicum* ('The Theatre of Comets', 1668). The comet reappeared over 1,600 years later, in 1969, as Comet Bennett (C/1969Y1).

2 **Halley's Comet** Peter Apian observed Halley's Comet on its return in 1531 and noticed that its tail always pointed away from the Sun. This illustration is from Apian's 1532 treatise on the comet.

3 **Sungrazer comet** The space satellite SOHO stares persistently at the Sun, using a telescope with a central obstruction (the blank orange area) that blocks out the Sun's bright disc, revealing its tenuous atmosphere, the stars beyond and, in the lower left of this image, a comet melting as it plunges close to the warmth of the Sun.

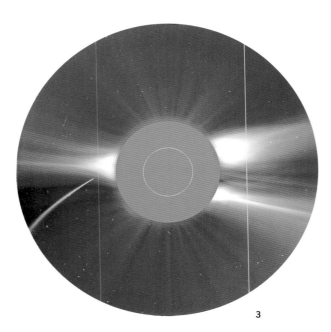

3

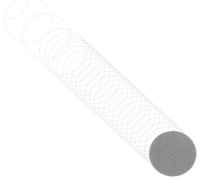

4 **Caroline Herschel** The only known portrait of Caroline as a young woman before she left Hanover in 1772, aged sixteen, to join her brother William in England. The silhouette conceals the scars and damaged eye left from an attack of smallpox when she was four years old, but it belies the advice that she had received from her father 'against all thoughts of marrying, seeing as I was neither handsome nor rich'.

5 **Caroline Herschel** Caroline Herschel resting in her bed, aged 97. A long-time friend of Caroline's, Mme Beckedorff, commented that the drawing 'did not do justice to her intelligent countenance; the features are too strong, not feminine enough, and the expression too fierce'.

6 **Comet Herschel** Caroline described and sketched her comet at the time of its discovery and made a fair copy in her notebook.

4   5

THE FIRST 'LADY'S COMET'

Caroline Herschel, the sister of William Herschel, who discovered Uranus in 1781, was the first woman to discover a comet, in 1786. She went on to find a total of fourteen. This account of her first discovery starts after William has discovered Uranus, has been been honoured by King George III, and appointed as the Royal Astronomer:

Much of my Brother's time was taken up with going, when the evenings were clear, to the Queen's Lodge to show the King objects through the 7-foot telescope. But after midnight he could return to our house, to observe double stars and compile his catalogue of nebulae.

I had the comfort to see that my Brother was satisfied with my endeavours in assisting him when he wanted another person, either to run to the Clocks, writing down a memorandum, recording the time of an observation, fetching and carrying instruments, or measuring the ground with poles, of which something of the kind would occur every moment.

My brother began his series of sweeps over the sky even before the new 20-foot instrument was finished, and my feelings were not very comfortable when every moment I was alarmed by a crash or fall, knowing William to be elevated 15 or 16 feet on a temporary cross-beam instead of a safe gallery. The ladders had not even braces at the bottom, and one night in a high wind he had hardly touched the ground before the whole apparatus came down. That my fears of dangers and accidents were not wholly imaginary I had an unlucky sample on the night of New Year's Eve, 1784. A few stars became visible and in the greatest hurry all was got ready for observing. My Brother at the front of the telescope directed me to make a small alteration in the lateral motion. This was done by machinery at each end of which was an iron hook such as butchers use. Having to run in the dark on ground a foot deep in melting snow, I fell on one of these hooks, which entered my right leg above the knee. I was hooked. William and the workmen were instantly with me but could not lift me without leaving near two ounces of my flesh behind. I was obliged to be my own surgeon by tying a kerchief about it. At the end of six weeks I had some fears about my poor limb and had Dr Lind's opinion, who on seeing the wound said it was going on well.

I had the comfort to know that my Brother was no loser through the accident, for the remainder of the night on which it happened was cloudy, and the nights following afforded only a few short clear intervals. It was a full two weeks before there was the necessity to expose myself for a whole night to the severity of the season.

h 50' I saw the object in the center of fig I
like a star out of focus
while the others were
perfectly clear. the
prec. star is very faint
but the weather is
hazy, and in a clearer
night undoubtedly some
more will be visible

I

33' They make now a
perfect fee figure
this figure is right

II

'10' III

I think the situation
is now like in Fig 3
but it is so hazy
that I could only imagine
I saw the second star &

the preceding I could not see
The comet is ab.t half way between
54 Ursa maj. and some stars
I found after looking over the
at leisure, to be 14. 15. & 16 Com
by the obtuse angle it make
those stars in H. I conclude
to be ab.t ° above the parall
the 15 of Coma.

Aug.t 2 1786.

10 h 9' I saw the red star in the center
the following faint one, in the fan

fig I   fig II

as last night, the comet was

6

It was at Slough that my brother built his greatest telescope, 40 feet long. He had two grants from the King of £2000 each, and £200 a year afterwards for the expenses: the ropes, the painting, the keep and the clothing of the men to attend at night. A whole troop of labourers were engaged with the iron tool to grind the mirror and there were no less than 30 to 40 at work for upwards of three months, felling and rooting up trees, digging, laying bricks for the foundations.

I generally chose my situation by my brother's side, almost entirely attached to my writing desk. I seldom had an opportunity to use my own newly acquired telescope. But when my Brother was obliged to deliver a 10-foot telescope as a present from the King to the Observatory of Göttingen, I began with bustling work, first on William's account and then my own. I cleaned all the brasswork of the telescopes, and put curtains before the shelves to prevent the settling of dust. I calculated my Brother's earlier observations of nebulae. And, then, when the sky cleared, I made sweeps with my own 2-foot telescope.

On the 1st of August 1786 I found an object very much resembling in colour and brightness the 27th nebula of Messier's catalogue, with the difference, however, of being round. I suspected it to be a comet, but, a haziness coming on, it was not possible to entirely satisfy myself as to its motion

until the following evening, and thus to confirm it as my first comet discovery. My brother William was on his return commanded to show it to the King, who said that it was very small and had nothing striking in its appearance. Miss Fanny Burney acclaimed my comet as the first 'Lady's Comet'. It gave great pleasure to the ladies of the Court. I heard that Princess Augusta was in particular desirous that the lady guests should view it, calling them from the card-table.

Partly as a result of this, and in specific consequence of a recommendation made to the King by Sir Joseph Banks, President of the Royal Society, a salary of 50 pounds per year was settled on me as an assistant to my Brother. In October 1787 I received the first quarterly instalment of 12 pounds 10 shillings. It was the first money in all my lifetime that, at the age of 37 years, I ever thought myself at liberty to spend to my own liking.

Caroline Herschel

This is a composite and edited pastiche compiled from family sources, including Caroline's memoirs.

The Satellites of Jupiter

Galileo shatters the crystal spheres

1

2

The belief that the heavens orbited the Earth on a series of crystal spheres had survived for 2,000 years, but shortly after the invention of the telescope it was shattered in an instant. A few years after Kepler and Copernicus had used their calculations of planetary movements to develop the theory that the Earth and the planets orbited the Sun, Galileo Galilei saw the irrefutable evidence with his own eyes: mountains on the Moon, and four mysterious stars near Jupiter.

1   **The Moon** Galileo's illustrations of the Moon, from his *Sidereus nuncius* ('The Starry Messenger', 1610). He saw craters, 'seas' and mountain chains, including some mountains standing in shadow with their peaks illuminated by the Sun.

2   **Jupiter's satellites** Galileo noted the changing disposition of four (sometimes only two or three) small stars around Jupiter from night to night, suddenly realizing that the 'stars' were four satellites in orbit around the planet.

3   **Hans Lippershey** Born in Wesel, Germany, Lippershey moved to Middelburg, in the Netherlands, where he set up as a spectacle-maker and became a citizen in 1602. His application for a patent on the invention of the telescope was heard in October 1608.

4   **Galileo Galilei** It was Galileo (and Kepler) of whom Isaac Newton was speaking when he said that 'If I have seen further, it is by standing upon the shoulders of giants.'

5   **Galilean telescopes** Two telescopes attributed to Galileo, mounted on an ornamental display stand in the Museum of the History of Science in Florence.

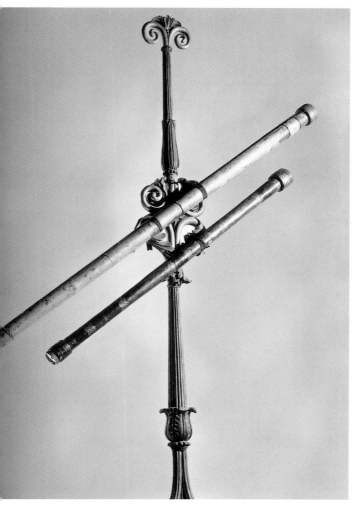

In 1608 Hans Lippershey (or Lipperhey), a German-born optician living in Middelburg, Zeeland, applied for a patent from the Dutch government for 'a certain device by means of which all things at a very great distance can be seen as if they were nearby, by looking through glasses which he claims to be a new invention'. One story is that Lippershey discovered the principle of the telescope when two children were playing with lenses in his shop and looked through two at the same time, one held up behind the other, exclaiming with surprise at what they saw.

In Venice later that year, the learned monk Pietro Sarpi, recently retired from high office in the Venetian government, heard about the patent application; a few months later Sarpi was visited by a protégé, Galileo Galilei, and they discussed Lippershey's invention. Galileo was professor of mathematics at the University of Padua in the Venetian republic and he realized the importance of the invention to Venice, a maritime power. He had been seeking a salary increase, so in 1609 he made a prototype of a telescope that, through Sarpi, he brought to the attention of the Venetian authorities. Using his invention, Galileo was able to describe ships approaching the port of Venice before they could be seen at all with the unaided eye. The value of the telescope was clear and his salary was doubled, although there were strings attached, such as he would never get a salary increase ever again and could never move from Padua. This resulted in him looking elsewhere for a new position, a prize that the momentous astronomical discoveries he was about to

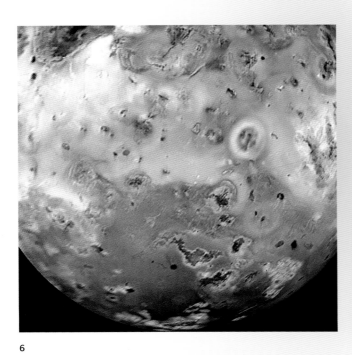

6

7

6–9 Jupiter's satellites In the space age, Jupiter's satellites have been seen as individual worlds, each with its own different character. **6 Io** is pock-marked with volcanoes and scoured by lava flows. **7 Europa** is completely covered with cracked floes of ice. **8 Ganymede** is made of a mixture, half and half, of rock and ice, its surface plastic, mobile and wrinkled under tectonic forces. **9 Callisto** is heavily cratered, like Ganymede, and indeed, both are very like our own Moon.

make with his telescope would eventually win for him. In the winter of 1609–10 Galileo observed the Moon and saw bright spots in the un-illuminated part of its surface that gradually grew in size and merged with the illuminated area as the month progressed. He correctly interpreted the bright spots as mountain tops that had caught the first rays of the Sun, moving into full sunlight as the Moon rotated. He measured the heights of lunar mountains using their shadows, calculating one at 6 km high. He saw that some mountains were arranged in straight mountain ranges and some in circles, surrounding craters. He discovered that the Moon was not a smooth, perfect sphere as taught by Aristotle and Ptolemy; rather, its surface was 'rough and uneven, and just like the surface of the Earth itself.'

In England, Thomas Harriott, a mathematician and cartographer, used a telescope to observe the Moon and drew it in August 1609, several months before Galileo; he also discovered sunspots. He never published his astronomical work and thus remains largely unknown for this achievement. He is however remembered as one of the founders of the Virginia colonies, having visited Roanoke Island in 1585–86 and written an influential report about its agricultural and mineralogical potential, and its Algonquian inhabitants.

On 7 January 1610, Galileo drafted a letter to an unknown recipient describing a further momentous discovery that he had made the previous night: 'Besides my observations of the Moon, I have observed the following in other stars. First that many fixed stars are seen with the telescope, which are not otherwise discerned; and only this evening, I have seen Jupiter accompanied by three fixed stars, totally invisible by their smallness…'

At first Galileo did not think there was anything remarkable about these stars: a triplet arranged in a straight line through Jupiter, two on one side and one on the other. But, according to his observation journal, when he came to look at Jupiter again on the 8th of January the three stars were all on the other side of Jupiter. Presumably Jupiter had moved from its previous position. The 9th was cloudy, and on the 10th and 11th there were only two stars, on one side of Jupiter, with the third conjoined with Jupiter (or so Galileo speculated). On the 12th the three stars were again arranged differently: two on one side of Jupiter, one on the other. 'It appears that around Jupiter there are three moving stars invisible to everyone up to this time', he reported. On the 13th, he realized that there were in fact four little stars; he saw them all again on the 15th.

8

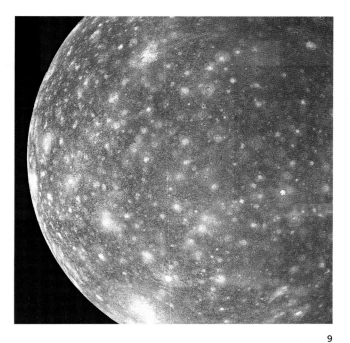

9

Galileo seems originally to have thought that the stars were moving back and forth in a straight line. But if this was the case, how did they pass one another? Suddenly he realized that the four 'stars' were in actually in orbit around Jupiter. In an instant, the four tiny stars had disproved the 2,000-year-old Ptolemaic theory that every celestial body orbited around the Earth.

The orbital motion of the satellites around Jupiter was very like the orbital motion of the planets around the Sun as expressed in the Copernican theory, and also very like the motion of the Moon around the Earth. It was clear that the Earth was just like the celestial bodies and that it was not the centre of the Universe.

Galileo wrote up his discoveries in January and February 1610 in a book called *Sidereus nuncius* ('The Starry Messenger'). He dedicated the book to Cosimo II de' Medici. The Medici family was a powerful and wealthy family that ruled Florence from the 13th to the 17th centuries. Galileo had tutored Cosimo in mathematics and in 1609, at the age of nineteen, the young man succeeded to the title of Grand Duke of Tuscany on the death of his father, Ferdinand. To flatter the family, Galileo named the satellites of Jupiter 'the Medicean stars', although this term did not stick. As a reward and as a mark of esteem, in 1610 Cosimo appointed Galileo as his philosopher and mathematician for life, and made him chief mathematician at the University of Pisa.

CAN SUCH PLANETS EXIST?

O n nights such as these, all over Italy telescopes are being turned towards the Heavens. Jupiter's moons will not make milk any cheaper. But they have never been seen before and they are there. From that the man in the street draws the conclusion that there may be other things to see if only he opens his eyes. It is not the movements of a few distant stars that make all Italy prick up its ears but the news that opinions hitherto held inviolable have now begun to totter.
Bertolt Brecht, *The Life of Galileo*, 1937–43

09. The Phases of Venus

Revealing the shape of the Copernican system

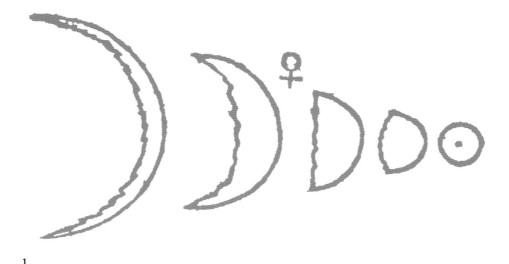

1

Having discovered that the 'Medicean Stars' orbited Jupiter rather than the Earth, Galileo turned his telescope towards the planet Venus. He discovered that Venus had phases similar to the Moon, which proved that Venus was orbiting the Sun and that Copernicus's theory of the Solar System was correct. Galileo's discoveries were at first acclaimed by astronomers, but then perceived as a threat to the biblical worldview, earning him persecution from the Church during his lifetime. History now recognizes Galileo as one of the world's great scientists.

1   **Galileo's observations of Venus** With his telescope, Galileo saw both the changing phase and changing size of Venus. He even noted the softer and more uneven character of the terminator (the line between the clouds on the sunlit and dark sides of the planet), compared with its limb (the edge of the planet sillhouetted against space).

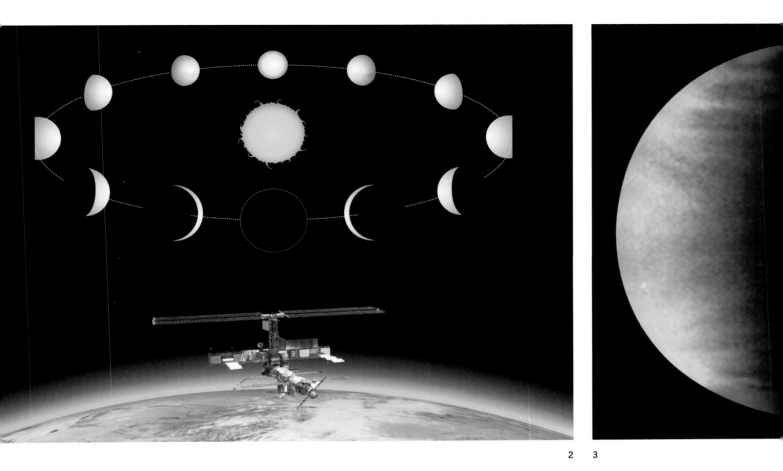

2   3

In 1610, in Pisa, after discovering the satellites of Jupiter with his new telescopes (**08**), Galileo was finally able to observe the planet Venus; when he had first assembled his telescope, Venus had been too near to the Sun to be examined closely. He saw that Venus had phases like the Moon. When the planet was at its greatest distance from the Sun, it looked like a half-moon, with the bright side facing in the direction of the Sun. As it approached the Sun again, its phase either increased towards full circular illumination, or narrowed to a thin crescent. Galileo reported his discovery using a Latin anagram, which he sent to Kepler. The coded sentence (in not very elegant Latin) was 'Haec immatura a me jam frustra leguntur o.y.' It can be translated as 'Things not ripe for disclosure are read by me', but its letters also can be re-arranged to read 'Cynthiae figurae æmulatur mater amorum'. This means 'The shapes of Cynthia [the Moon] are emulated by the mother of loves [Venus]'. The anagram was a device used in the 17th century to establish priority for a discovery; if the announcement was made straightforwardly and was promulgated at the slow pace of communication of the time, someone who got the announcement earlier than anyone else could falsely claim the discovery as his own, and pretend surprise when the announcement became more widely disseminated. The terse nature of the anagram also bought time and room for manoeuvre for the discoverer as a hedge against an incomplete or misinterpreted result.

Galileo's discovery of the phases of Venus showed immediately that Venus travelled around the Sun in an orbit that lay inside the orbit of the Earth. When Venus was on the far side of the Sun, its face was fully illuminated like the full Moon. When Venus was moving between the Earth and Sun, its unilluminated rear partially faced the Earth and it showed only as a thin crescent of light. This geometry fitted the arrangement that Copernicus had hypothesized in his model of the Solar System. It could not be reconciled with the Ptolemaic theory that Venus orbited the Earth in a crystal sphere between the Earth and the Sun.

2   Phases of Venus The changing phase and size of Venus are natural consequences of its orbit around the Sun, inside the orbit of the Earth.

3   Venus Due to the opaque cloud that covers the planet, the Hubble Space Telescope is unable to see Venus in much more detail than could Galileo.

Galileo's observations also confirmed another of Copernicus's predictions: that the size of Venus would appear to change as the planet came closer to Earth and then retreated. Galileo saw that Venus did indeed change size – it appeared smallest when it was showing a full face and beyond the Sun at its most distant, and four times larger when it was crescent-shaped and at its closest approach to Earth.

As Galileo himself realized, his discovery of the phases of Venus in 1610 was crucial in confirming Copernicus's theory of the Solar System. In principle, Galileo's discovery was still consistent with Tycho Brahe's compromise between the Copernican theory and the Ptolemaic theory. Brahe had proposed in 1583 that the planets orbit the Sun, which itself orbits the stationary Earth (**06**). However, Galileo dismissed Brahe's theory (also called the 'Tychonic theory') because it assumed that the Sun caused the planets to move, while leaving the Earth stationary and unaffected, even though some planets were evidently bigger than the Earth.

Galileo knew that his discovery would not be readily accepted. On New Year's Day in 1611 he sent the Tuscan ambassador in Prague the solution to the Venus anagram, explaining that the phases meant that the planet must orbit around the Sun: '…something indeed believed by the Pythagoreans, Copernicus, Kepler and myself, but not proved as it is now. Hence Kepler and other Copernicans may glory in their successful theories, although as a result we will be thought to be fools by most bookish philosophers, who will regard us as men of little understanding or common sense.'

The consequences for Galileo were actually much worse than this. His discoveries were acclaimed by astronomers and sophisticated churchmen like Cardinal Maffeo Barberini (later Pope Urban VIII), but he was denounced to the

GALILEO'S ABJURATION

'**I**, Galileo, son of the late Vincenzio Galilei, Florentine, aged 70 years, arraigned personally before this tribunal and kneeling before you…swear that I have always believed, do believe, and by God's help will in the future believe all that is held, preached, and taught by the Holy Catholic and Apostolic Church.…I have been pronounced by the Holy Office to be vehemently suspected of heresy, that is to say, of having held and believed that the Sun is the centre of the world and immovable and that the Earth is not the centre and moves.…I abjure, curse, detest the aforesaid errors and heresies…and I swear that in future I will never again say or assert, verbally or in writing, anything that might furnish occasion for a similar suspicion regarding me.'
Galileo Galilei, 1633

Inquisition as a heretic by Tommaso Caccini, a Dominican friar, and advised by Cardinal Bellarmine to treat the Copernican theory only as a hypothesis. The telescope was derided as a device capable of making things appear in the sky that were not actually there. (Galileo said that he would pay 10,000 *scudi* to anyone who could make a telescope that showed satellites around one planet but not the others.) In 1616 he was officially ordered not to advocate or teach Copernican astronomy except as a hypothesis, but nevertheless continued to publish his scientific work, and evidently pushed the boundaries of church tolerance too far. He was put on trial in Rome for teaching, contrary to Scripture, that the Earth moved and in 1633 was convicted, forced to recant, banned from further publications on his scientific discoveries and placed under house arrest until his death in 1642.

4 The Tychonic planetary system Tycho Brahe constructed a model of the Solar System, illustrated here by Andreas Cellarius (*Harmonia macrocosmica*, 1660), that was a compromise between the Ptolemaic system (the Sun and the Moon orbited around the Earth, which was stationary at the centre of the Solar System) and the Copernican system (the other planets orbited around the Sun).

5 *Galileo Facing the Roman Inquisition* **(Cristiano Banti, 1857)** In this romanticized artist's portrayal of his examination, Galileo appears stronger and more defiant than accounts suggest he actually was. A frail old man, shown the instruments of torture that would be used if he persisted against the authority of the Church, Galileo understandably collapsed under the intense pressure and, kneeling rather than standing proudly, agreed that he was wrong to hold the opinion that the Earth moves.

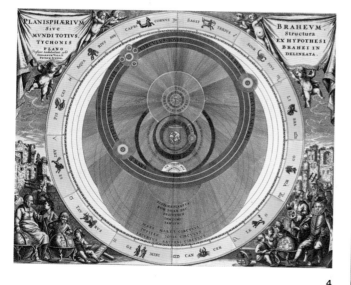

4

AUTHORITY AND REASON
In questions of science the authority of a thousand is not worth the humble reasoning of a single individual.
Attributed to Galileo by François Arago in his
Eulogy to Galileo, 1874

5

10

Uranus

William Herschel discovers the first new planet

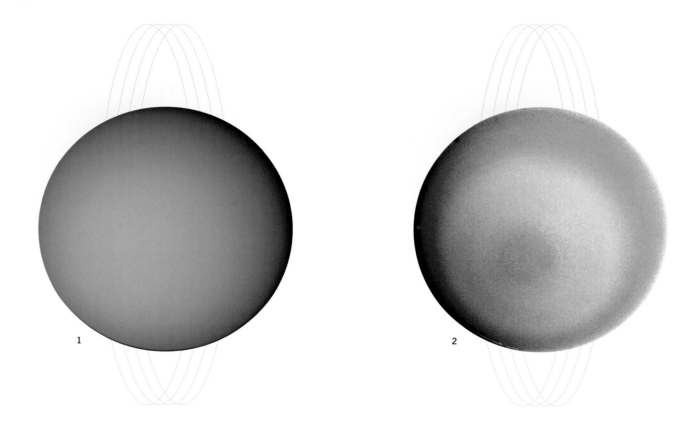

1

2

The only planets that were known to the astronomers of antiquity were those that could easily be seen with the naked eye: Mercury, Venus, Mars, Jupiter and Saturn. This changed in the 18th century when, armed with a powerful telescope that he had made himself, a music teacher from Bath discovered Uranus, the first new planet identified in recorded history, and the third-largest in the Solar System.

3

1 **Uranus** The planet is covered with opaque and all-but-featureless cloud.

2 **Uranus** Sophisticated image-processing techniques bring out slight differences in colour in the cloud layers as they circle the polar regions, and reveal here that its pole is pointing towards us as the planet rolls along in its orbit.

3 **William Herschel** The original portrait by John Russell was painted in 1794 and shows William aged 56, at the height of his fame, having discovered Uranus thirteen years before.

4 **A Herschelian telescope** William Herschel constructed and sold many telescopes of the same model, which was simple to wheel out of the house onto a paved area and turn to the chosen direction. The telescope's position was adjusted up or down with screws, and its eyepiece was at head height.

4

William Herschel was born in 1738 in Hanover, at a time when it was a British possession. He followed his father into a military career as a bandsman and fought alongside him as part of the British army at the battle of Hastenbeck. Attacked by the French army, Herschel escaped to England, eventually settling in Bath, where he enjoyed great success as a music teacher to society ladies and as a concert artist. He was eventually joined in Bath by his sister, Caroline, who not only kept house for him and tried to defend him against predatory widows but also accompanied him, singing, in his concerts.

William's father had been interested in astronomy and William also studied the subject, forming an ambition to see the heavens with his own eyes. He made telescopes, casting and grinding the mirrors himself in the basement of his house, designing and forming the telescope tubes from wood and tin, and erecting the telescopes on the garden lawns of his various houses. Herschel then set out systematically to 'review' the entire sky with his telescope, inspecting every star and even the spaces between them. On 13 March 1781 he discovered a 'curious either nebulous star or perhaps a comet'. Tracking it over the next hours and days, he found that it moved and could not be a star. Over the next few weeks, the curious object was observed by astronomers at Greenwich and Oxford, and proved to be in a near-circular orbit beyond Saturn, rather than in the highly eccentric orbit crossing the Solar System that would be expected of a comet. Furthermore, there was no trace of a comet's fuzzy coma or tail, and the object had a circular disc like a planet. It was indeed a planet, the first discovered since antiquity. At sixth magnitude in brightness, Uranus is marginally visible with the unaided eye and easily seen in modest telescopes. It had been mistakenly recorded as a star several times before Herschel's discovery.

Herschel was invited to London to tell King George III of the new planet. He was asked to erect a telescope at Windsor Castle in order to show astronomical sights to the royal household, and rewarded with a patronage appointment as the King's Royal Astronomer and a stipend – his sister Caroline, who assisted him with his observations and discovered the first comet found by a woman, was later given a stipend, though of only half as much (**07**). In gratitude for this royal patronage, William Herschel suggested that the new planet should be named after the era of its discovery 'in the reign of King George III' as 'Georgium Sidus' – the Georgian star (or planet). This name was used for a time in England, but never in any other country, and it was the mythological name Uranus suggested by the German astronomer Johann Bode that eventually stuck.

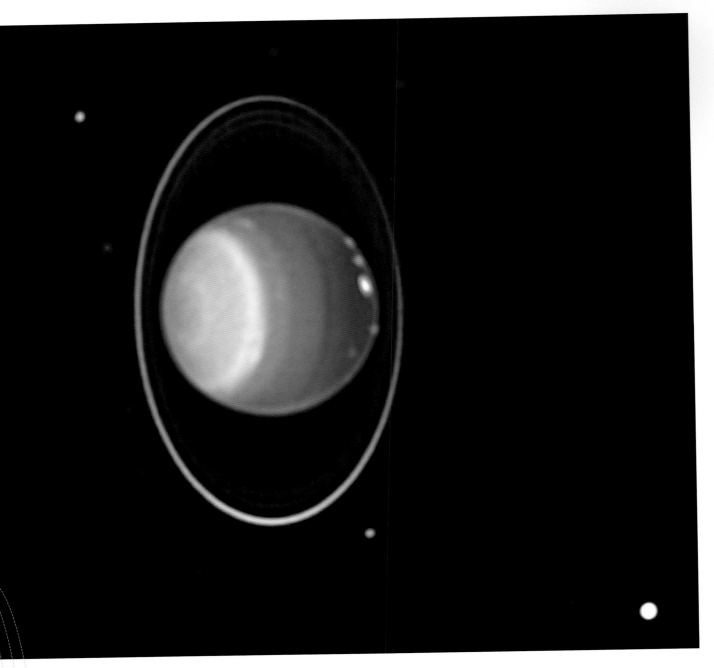

5

5 **Uranus with its rings and satellites** This 1998 colour view by the Hubble
Space Telescope was imaged with infrared radiation and shows a band of
bright orange-coloured high clouds moving quickly (at more than 500 km/
hr) around the planet's equator. Green and blue regions mark relatively clear
areas where it is possible to see deeper into the planet's atmosphere.

6 **Discovery of Uranus** William Herschel's notes for Tuesday, 13 March 1781
record that he looked at the region around the star Pollux, and noticed (fifth
line below the date) a 'curious either nebulous star or perhaps a comet'.
William's sister, Caroline, crossed the page out with two vertical lines after
she had transcribed a fair copy into a logbook.

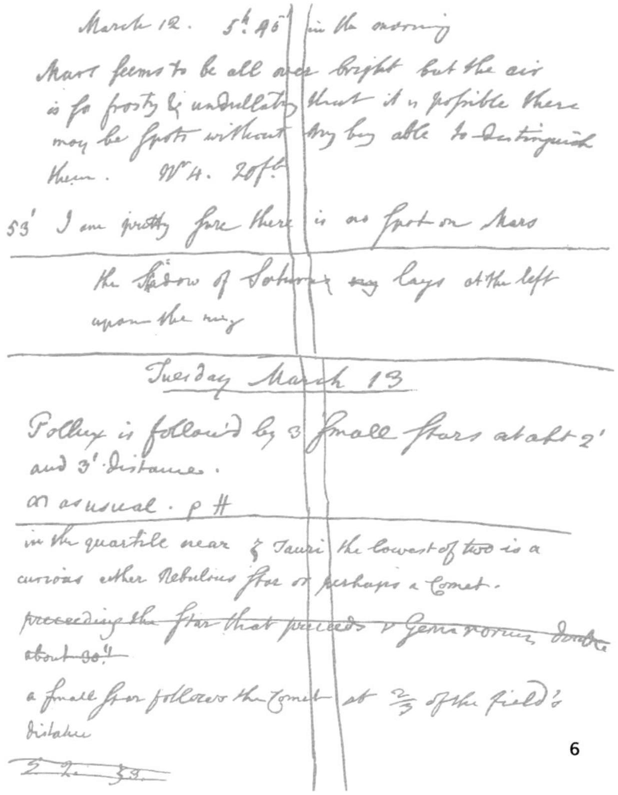

[Handwritten notes in left column:]

March 12. 5ʰ 45′ in the morning
Mars seems to be all over bright but the air
is so frosty & undulating that it is possible there
may be spots without my being able to distinguish
them.   N⁴⁴. 20ft

53′ I am pretty sure there is no spot on Mars

the shadow of Saturn say lays at the left
upon the ring

Tuesday March 13

Pollux is follow'd by 3 small stars at aft 2′
and 3′ distance.

on as usual. p ♂

in the quartile near ζ Tauri the lowest of two is a
curious either Nebulous star or perhaps a Comet.

preceeding the star that precedes ν Gemi norum, double
about 80″

a small star follows the comet at ⅔ of the field's
distance

6

Uranus is the third-largest of the planets in the Solar System, and ranks fourth in mass. Its characteristic blue-green appearance is due to a high layer of methane ice clouds. It has a ring system, discovered in 1977 when starlight was unexpectedly blocked by the rings as high-speed photometers observed a star that was being occulted (passed in front of) by Uranus (**27**). The rings were first imaged in 1986 by the Voyager 2 spacecraft.

Uranus has five bright satellites and more than a dozen fainter ones. The satellites provided the first evidence of the unusual tilt of Uranus. They orbit around Uranus's equator and show that its pole is tilted by more than 90°. One theory is that Uranus originally had a more normal tilt, but that it was struck by an Earth-sized planetesimal that knocked it over. Because its rotational axis stays fixed in direction while the planet orbits around the Sun, its polar regions are sometimes pointing directly towards the Sun and sometimes at right angles to it. By contrast, the Earth's polar regions are never presented directly to the Sun, although they do tilt a little because the Earth's axis is tilted by 23.5° to its orbit, generating the pattern of the Earth's seasons (**19**). The differences in solar heating during Uranus's orbital 'year' are much more extreme than those on Earth, and its seasons are peculiar: for example, during its summer time, when the Sun is always above its horizon, one of Uranus's polar regions will be hotter than the equatorial regions ever are.

THE THRILL OF DISCOVERY

Then felt I like some watcher of the skies
   When a new planet swims into his ken;
Or like stout Cortez when with eagle eyes
He star'd at the Pacific.
John Keats, 'On First Looking into Chapman's Homer', 1817

DISCOVERY OF URANUS

It has generally been supposed that it was a lucky accident that brought the planet to my view; this is an evident mistake. In the regular manner that I examined every star of the heavens, not only of that brightness but many far inferior, it was that night its turn to be discovered. I had gradually perused the Great Volume of the Author of Nature, and was now come to the page which contained a seventh Planet. Had business prevented me that evening I must have found it the next, and the goodness of my telescope was such that I perceived its visible planetary disc as soon as I looked at it. And by the application of my micrometer I determined its motion in a few hours.
William Herschel

11. Neptune

The planet discovered by the pen

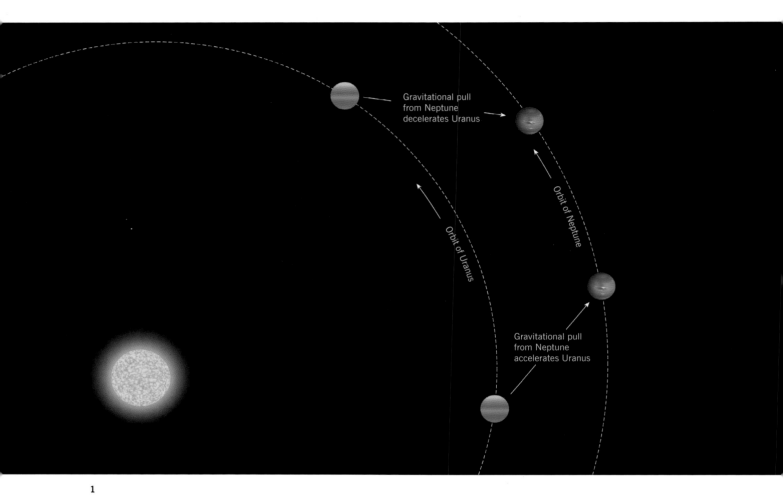

Gravitational pull
from Neptune
decelerates Uranus

Orbit of Neptune

Orbit of Uranus

Gravitational pull
from Neptune
accelerates Uranus

1

While Uranus was discovered with a telescope, Neptune was discovered with a pen and paper – twice. Two very different astronomers – the Frenchman Urbain Le Verrier and the Englishman John Couch Adams – independently tried to explain why Uranus was being pulled off course in its orbit. Both sets of calculations pointed to a previously unknown planet, making Neptune the first planet to be located through mathematics rather than by direct observation of the heavens.

1 **Discovery of Neptune** As Uranus overtakes Neptune on the inside orbit, Neptune at first pulls Uranus along ahead of its predicted position but then retards Uranus, pulling it back behind schedule. This was the clue that indicated the existence of Neptune.

2 **Urbain Le Verrier** As a rather beautiful young man, Le Verrier came to work at the Paris Observatory and discovered Neptune by mathematics at the age of 34. It is hard to reconcile this portrait with the description of him, after he had become director of the observatory at the age of 43, as 'if not the most detestable man in France, the most detested'.

3 **John Couch Adams** Even before he had graduated from university, Adams planned to calculate the position of the planet perturbing the orbit of Uranus. The portrait suggests his 'faint and forgettable personality', which, coupled with his obsessive perfectionism in making lengthy and repetitive calculations, has encouraged speculation that he had Asperger syndrome.

4 **Neptune** Given the planet's position in such a cold region of the Solar System, and its known low temperature, it was a surprise when Voyager 2 discovered in 1989 that Neptune had active clouds in its atmosphere, such as the Great Dark Spot.

After the discovery of Uranus by William Herschel, several previous observations of the planet were found, which had been made and recorded by astronomers who had not recognized it as a planet. By 1830 these records enabled an accurate orbit to be calculated for Uranus. At the same time it became clear that Uranus was departing from this orbit.

The Director of the Paris Observatory, François Arago, suggested to Urbain Le Verrier that a previously unseen planet was pulling the planet off track. In 1845 Le Verrier calculated its expected position. He used Bode's Law (10, 12) to assume the distance of Neptune from the Sun, although the accuracy of Bode's Law actually breaks down after Uranus. Le Verrier sent his prediction in 1846 to Johann Galle, an astronomer at the Berlin Observatory. Galle, together with his assistant Heinrich Louis d'Arrest, began a search on the same night (23 September 1846) that they received the letter.

At d'Arrest's suggestion, Galle used the latest star chart of the area, which had only just been produced. Within 30 minutes they had identified a star that was not on the map. They confirmed that it was the new planet on the following night by observing its motion relative to the other stars. Galle wrote to Le Verrier, saying, 'Monsieur, the planet of which you indicated the position really exists'. Le Verrier replied, 'I thank you for the alacrity with which you applied my instructions. We are thereby, thanks to you, definitely in possession of a new world'.

Meanwhile, in England, the Scottish mathematician Mary Somerville had earlier suggested to a young Cambridge student, John Couch Adams, that an unknown planet was affecting the orbit of Uranus. With a letter of introduction from James Challis in Cambridge, who had been impressed by Adams's preliminary calculations regarding the position of the hypothetical planet, Adams applied to George Airy, the Astronomer Royal, for research assistance. However, due to his youth, humble background and reticent manner – and perhaps also due to Airy's unapproachable character – Adams twice failed to secure an interview with Airy. Airy told none of his colleagues about Adams's prediction, dismissing the

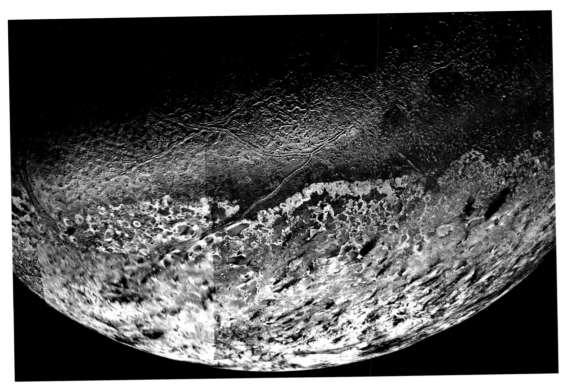

5

5 **Triton** Neptune's largest satellite (viewed by Voyager 2) is so cold that it has a surface of nitrogen ice, streaked by dark dust deposited by huge geyser-like plumes. The origin of the so-called cantaloupe terrain (top) is unknown.

6 **Johann Gottfried Galle** Immediately upon receipt of a letter from Le Verrier, Galle persuaded his reluctant superior, Johann Encke, to permit him to search for the predicted planet, and found it over the next two nights.

7 **Heinrich Louis d'Arrest** As a student who happened to be at the telescope at the time, d'Arrest enthusiastically helped Galle to look for Neptune, suggesting that they should search for a 'star' that was not on recent star charts, rather than for a 'star' with a perceptible disc, as Galle was unsuccessfully attempting to do.

8 **George Airy's transit circle at the Royal Observatory in Greenwich**

discovery of a new planet as not the job of the Greenwich Observatory. He passed the project back to Challis, who began a somewhat half-hearted search for Neptune on the basis of Adams's calculations. Challis did not have the same up-to-date charts as Galle, and had to compare successive observations of the same area of the sky to see if there was a star that moved. Hobbled by a rigid institutional hierarchy and outdated materials, Adams and Challis were overtaken by the brisk efficiency of Le Verrier and Galle, although Adams's calculations, which he shared only with Challis and Airy, had accurately predicted the position of Neptune eight months before Le Verrier's findings were published. When Airy and Challis were forced to justify their delay in following up Adams's calculations, Adams is said to have reacted with characteristic modesty, offering his congratulations

to Le Verrier without a trace of bitterness: as the president of the Royal Astronomical Society it was he who handed its gold medal to Le Verrier in 1875. Eventually both Adams and Le Verrier would be credited jointly with the discovery of Neptune.

After Neptune had been located, William Lassell, a wealthy Liverpool brewer, was one of the first to inspect the new planet with a large telescope. According to family tradition, Lassell had missed out on being the discoverer of Neptune, a letter communicating Adams's predicted position having been accidentally destroyed by an overzealous maid clearing away and burning rubbish. When the coordinates of 'Le Verrier's planet' were published in *The Times*, Lassell rushed to examine its position in the sky with his telescope. He saw that Neptune had a distinct disc, confirming its identification as a planet. During the same observation, he also discovered Neptune's satellite, Triton. Triton orbits Neptune backwards and at a steep angle, suggesting that it was not formed together with Neptune, but was a passing planet that Neptune captured.

The naming of Neptune generated as much controversy as its discovery. In keeping with the custom for naming celestial features after Roman mythological characters, Galle suggested that the new planet should be called 'Janus', and Challis proposed 'Oceanus'. Le Verrier, eager to put his own stamp on 'the planet exterior to Uranus', first proposed 'Neptune' and later 'Le Verrier', but in the ensuing dispute over who should be credited with the planet's discovery, the latter suggestion attracted fierce opposition outside France. By the end of the discovery year the more neutral 'Neptune', the name of the Roman god of the sea, had become the accepted usage. Neptune is the outermost of the four giant planets, forming a pair with Uranus. It rotates quickly and is slightly flat (or 'oblate'). Curiously it emits more than twice as much energy as it receives from the distant Sun. The excess energy is generated by the cooling of Neptune's hot interior. Neptune and its approximately half-dozen satellites were explored by the Voyager 2 fly-by in August 1989, which revealed Neptune's 'Great Dark Spot', thought to be a hole in the methane atmosphere large enough to fit the entire Earth. Voyager also obtained images of Triton that showed a bright south polar cap. North of the polar cap, a rugged terrain, reminiscent of the skin of a cantaloupe melon, is cross-cut by a pattern of intercepting ridges, left from some past tectonic event.

6

7

8

DISCOVERY OF NEPTUNE BY PEN

The method pursued by M. Le Verrier totally differs from all previous attempts of geometers and astronomers. The latter have sometimes accidentally found a movable point, a planet, in the field of their telescopes. M. Le Verrier perceived the new body without the necessity of casting a single look towards the heavens. He saw it at the end of his pen; he determined, by the mere force of calculation, the place and approximate magnitude of a body situated far beyond the hitherto known limits of our planetary system, of a body the distance of which from the Sun exceeds 2,800 millions of miles, and which, seen in our powerful telescopes, barely exhibits a sensible disc.
François Arago, *Comptes rendus*, 1846

12. Asteroids

Remnants of the early Solar System

A mysterious gap between Mars and Jupiter led astronomers to the discovery of asteroids. A few are very small planets, as was thought when they were first identified in the 19th century, and some are fragments from the collisions of larger asteroids, but many asteroids are actually scrap material left over from the formation of the planets in the early Solar System. They are random in size and shape, and most follow orbits in a belt between Mars and Jupiter.

1

2

3

1 **Asteroid track** An asteroid streaks across a long-exposure image made by the Hubble Space Telescope (HST) of a distant galaxy. The exposure was separated into thirteen separate exposures, which break the trail into thirteen segments. The trail is curved into arcs because the HST was orbiting around the Earth during the exposures; the curvature of the trail is due to the parallax of the asteroid as seen from the extremities of its orbit. The asteroid was 270 million km from Earth at the time of observation.

2 **Giuseppe Piazzi** confirms the position of Ceres, the first known asteroid, on his star charts. Ceres has been pointed out to him by Urania, the muse of Astronomy, in this portrait of 1811 by Francesco La Farina.

3 **The Osservatorio di Palermo** was founded by Piazzi in 1790, and is shown here as it looked in 1804.

In the 16th century, Johannes Kepler noticed that there was a rather large gap in the arrangement of the planets in the Solar System. Setting the Sun–Earth distance at 10 units, the distances from the Sun to the known planets were as follows:

Sun–Mercury	3.9 units
Sun–Venus	7.2 units
Sun–Earth	10 units
Sun–Mars	15 units
Sun–Jupiter	52 units
Sun–Saturn	96 units

Kepler saw that the distances between the planets roughly doubled at each step. Yet Jupiter was nearly four times the distance of Mars from the Sun – almost twice the expected amount. Kepler suspected that there might be an undiscovered planet in the gap. Isaac Newton noticed the same phenomenon and suggested that perhaps Providence had put Jupiter and Saturn at an extra-large distance from the Sun to minimize their disruptive effects on the inner planets of the Solar System.

In 1766 Johann Daniel Titius, a professor of physics at the University of Wittenberg, discovered a more accurate formula for estimating the distances of the planets from the Sun. His formula was popularized by a German astronomer, Johann Bode, and consequently became known as Bode's Law (more correctly, but only sometimes, called the Titius–Bode Law). The formula is still regarded as an interesting fact, although in the intervening 200 years no one has been able to explain why the planets in the Solar System are spaced apart so regularly.

For all planets except Mercury, the formula for Bode's Law works like this:

$$A = 4 + 3 \times 2^n$$

A represents the distance from the planet in question to the Sun (measured in units equivalent to 1/10 the distance between the Earth and the Sun). Mercury has a fixed value of A = 4.

n represents the consecutive order of the planets after Mercury in the Solar System, beginning with Venus at 0: 0, 1, 2, 3…

Applying Bode's Law to all the known planets gave the following results:

	Actual distance (as measured)	*n*	*Distance according to Bode's Law*	
Sun–Mercury	3.9 units		4	= 4 units
Sun–Venus	7.2 units	0	4+3	= 7 units
Sun–Earth	10 units	1	4+6	= 10 units
Sun–Mars	15 units	2	4+12	= 16 units
?		3	4+24	= 28 units
Sun–Jupiter	52 units	4	4+48	= 52 units
Sun–Saturn	96 units	5	4+96	= 100 units
Sun–Uranus	192 units	6	4+192	= 196 units

4

5

If you compare the actual distances of the planets with the figures calculated by Bode's Law, you can see that the law produces a good (but not exact) estimate for all the planets known to astronomers up to the 18th century (Mercury–Saturn).

There are, however, some problems. First of all, there is the funny way that the distance to Mercury is calculated. The value for Mercury is really just put in to make the formula look like it works. Secondly, there is an obvious gap at $n=3$ between Mars and Jupiter. Like Kepler, Bode thought that there must be an undiscovered planet in the gap: 'Can one believe that the Creator of the Universe has left this position empty? Certainly not!' When William Herschel discovered Uranus in 1781 (**10**), belief in the validity of Bode's Law strengthened, because it fitted the formula so well. But the gap became even more significant. It was an obvious challenge to discover another unseen planet.

The court astronomer of the Duchy of Saxe-Gotha, Germany, Baron Franz von Zach, took up the search for the new planet in September 1800 and organized a team of two dozen astronomers to share the work. They became known as the 'Celestial Police'. The first day of the new century (properly reckoned), 1 January 1801, brought success even before the Celestial Police could get to work. A member of

the team (who did not even know that his efforts had been volunteered!) discovered a new planet. He was a Sicilian monk, Father Giuseppe Piazzi, who came across a moving object as he constructed a star catalogue with a telescope in Palermo. As his observations progressed, it became clear that the new object was not a comet. It had a nearly circular orbit in the right zone between Mars and Jupiter, with a distance of 28 units as measured by Bode's Law. Piazzi named the planet Ceres, after the Roman goddess of the harvest and the patron goddess of Sicily, but subsequently lost sight of the object, the sequence of his observations interrupted by illness. However, the German mathematician Carl Friedrich Gauss discovered a new method (called 'the method of least squares') for computing orbits, which enabled Piazzi's planet to be located again.

When William Herschel examined the new planet with his large telescope, he could see no disc. The planet must be small – a 'minor planet' – and Herschel used a new word to describe it: 'asteroid', meaning 'an almost star-like object'. But then, to everyone's surprise, in 1802 another minor planet was discovered by Heinrich Wilhelm Matthäus Olbers, who was a doctor in Bremen by day, and an amateur astronomer by night (**61**). Olbers located a second asteroid in 1807. A fellow German astronomer, Karl Ludwig Harding,

6

4   **The Palermo Circle Telescope** This telescope was used by Piazzi to measure star positions. While engaged in this work, he came across a moving 'star', which proved to be the asteroid Ceres.

5   **Descent of NEAR onto Eros** Four images taken in 2001 by a down-pointing camera as the spacecraft approached the surface of the asteroid from altitudes of 1150, 700, 250 and 130 m, respectively. During transmission of the last picture back to Earth, the spacecraft impacted on the surface at 6 km/hr. The pebbles visible on the surface of the asteroid in the last picture are a few centimetres across.

6   **The asteroid Ida and its asteroid-moon Dactyl** The small image on the right is Dactyl, shown on the same scale as the crater-pitted asteroid Ida on the left.

had discovered yet another asteroid in 1804. Like Ceres, these three asteroids were named after classical goddesses: Pallas, Vesta and Juno. A further 100 new asteroids had been discovered by 1868, 200 by 1879 and 300 by 1890 – as astrophotography became more sensitive so many asteroids began to spoil photographs of the stars that by the late 19th century they were dismissed as 'vermin of the skies'.

Compared to the planets of the Solar System, most asteroids are very small: the total mass of all objects in the asteroid belt is only about 1/1,000 of the Earth's mass. The first four asteroids to be discovered were the larger ones; of these, Ceres is the largest at 950 km in diameter. The 100 largest are above 140 km across and 1,000 others are above 30 km. Most of the asteroids that are well-observed by astronomers orbit in the region known as the 'asteroid belt' between Mars and Jupiter (21–33 units away from the Sun, as measured by Bode's Law).

Some asteroids are fragments left from the catastrophic collision of one asteroid with another. Most asteroids are thought to be the remnants of the planetesimals (small proto-planets) from which the planets formed: the strong gravity of Jupiter stopped the planetesimals in the asteroid belt from settling together and congealing into a planet.

Many asteroids have been kicked out of the asteroid belt by near-miss collisions. Some fall in towards the Sun and, if they come near the Earth, become a natural hazard both for astronauts and for life on the surface of our planet (**20**). A few pass close enough to the Earth that they can be imaged by radar, although none of these have come close enough to pose a danger to the Earth's inhabitants. Several asteroids actually have been visited and imaged at close range by spacecraft, in recognition of their scientific importance as remnants of the early Solar System.

The Galileo spacecraft visited Gaspra in October 1992 and Ida in August 1993. It discovered that Ida had a satellite, which was later named Dactyl – the first asteroid satellite discovered. The Near Earth Asteroid Rendezvous (NEAR) mission targeted Mathilde in June 1997, and then Eros, which the mission's Shoemaker spacecraft orbited for a year until it was skilfully landed on Eros's surface on 14 February 2001. NEAR's inspection of Eros showed that its surface is rubble-strewn with boulders, which come mostly from the single meteor impact that created its largest crater, 7.6 km in diameter. This crater was subsequently named Shoemaker after CalTech planetary scientist Eugene Shoemaker, like the NEAR mission spacecraft itself.

13. Pluto

A planet deliberately sought, but not a planet and discovered by accident

1

2

The discovery of Uranus, the asteroids and Neptune made scientists suspect that there might be even more planets in the Solar System. This suggestion was reinforced by the fact that Neptune mysteriously appeared to be drifting off its calculated orbit. Could a large planet be hiding in the darkest and most distant reaches of the Solar System?

Mindful of Le Verrier's and Adams's discovery of Neptune as the source of a gravitational attraction that pulled Uranus off course, two American astronomers – Percival Lowell in 1908 and William Pickering in 1919 – suggested that an undiscovered planet was perturbing Neptune. In 1911 the Indian astronomer Venkatesh Ketakar published a calculation that posited details of the orbits of two hypothetical planets beyond Neptune, a model based on the assumption that the gravity generated by each of the three bodies affected the orbit of the others.

Lowell devoted ten years to his search. He was a member of a wealthy, influential Boston family, who had trained at Harvard as a mathematician. From the age of 38 he devoted himself to astronomy, moving in 1894 to Flagstaff, Arizona, where the clear skies were favourable for astronomical observation. He built a private observatory to study the planet Mars and to search for the trans-Neptunian planet, which he termed Planet X. Meanwhile, Pickering unsuccessfully searched the photographic archive of the Mt Wilson Observatory for images that might show the mysterious new planet.

How would you find a distant planet? The stars are fixed in position relative to each other, but planets are in orbit and change position quickly. Lowell and his assistants repeatedly photographed the sky in the search regions to check for star-like images that moved. Using this method, Lowell discovered 515 asteroids but no Planet X.

Lowell Observatory remained in operation after Lowell's death in 1916 and recommenced the search for Planet X in 1927 under its new director, Vesto Melvin Slipher. In December 1929, he hired an amateur astronomer, Clyde Tombaugh, as an assistant to take pairs of photographs two weeks apart. The pairs were put side by side in a viewing device called a 'blink comparator'. Its operator rapidly shifts a mirror back and forth to view each photograph alternately. Any part of an image that has moved between the exposures leaps from one position to another, and is readily identifiable. At first Slipher and his brother worked the comparator but Tombaugh produced pictures faster than the Sliphers could process them and the brothers became bored. They delegated the task of inspecting the pictures to Tombaugh. At the age of 24, he discovered Planet X in February 1930.

There are prediscovery photographs of Pluto dating back to 1914, including two images taken before Lowell's death. Pluto was much fainter than had been expected and was therefore overlooked in these early pictures. Ironically, it seems that Pluto was not massive enough to have

4

1 **The Rape of Persephone by Pluto (Gian Lorenzo Bernini, 1622)** This statue shows the Roman god of the underworld abducting Persephone, daughter of Ceres, and carrying her off to rule with him in cold, barren Hades. Zeus interceded to permit her to return to Earth in springtime to carpet the ground with flowers.

2 **The New Horizons spacecraft** The mission will approach Pluto and its moon Charon in 2015.

3 **Clyde Tombaugh** searches for Pluto with his blink comparator at the Lowell Observatory in 1930.

4 **Percival Lowell** Seen here observing Mars with the Lowell Observatory refracting telescope, Lowell inspired the search for Planet X. His initials were incorporated in the name Pluto.

caused the deviation of Neptune from its predicted orbit. The discrepancies were the result of errors in estimating the masses of the other planets, and using more accurate modern values it has been shown that there is no actual deviation in Neptune's orbit. Thus the reason that Tombaugh discovered Pluto was not that he searched where the planet was calculated to be, but that he diligently searched at all.

In powerful telescopes Pluto shows an unusually large and strangely broadened image, but the planet is actually much smaller than it looks. Astronomer Jim Christy of the US Naval Observatory discovered the reason for Pluto's odd appearance in 1978, noticing that in especially clear photographs of Pluto taken at regular intervals, the elongation seemed to rotate around Pluto. This turned out to be because Pluto has a moon that is nearly as large and as bright as the planet itself (the moon is half of Pluto's diameter). The moon is distinctly separate from Pluto but still unusually close for at satellite, at a distance of only seventeen times its radius. The moon was named Charon, after the mythological ferryman to Hades.

The existence of Charon was confirmed between 1985 and 1990 when the orbital plane of Pluto and Charon became visible edge-on from Earth, producing the expected series of mutual eclipses as Pluto and Charon passed in front of each other in turn. Since the right alignment occurs for only two five-year intervals in Pluto's 248-year orbit, it was lucky that

6

January 23, 1930   January 29, 1930

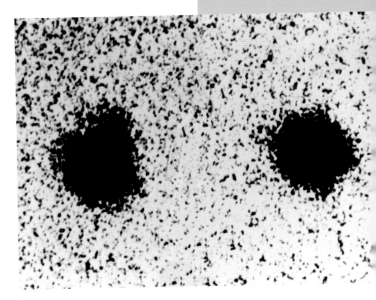

7

5

THE NAMING OF PLUTO

Tombaugh and the Sliphers received much advice on the name of the new planet. Lowell's widow suggested that the planet should be named after herself. The astronomers at the Lowell Observatory took a dim view of this, given that she had been trying for ten years to get her hands on Lowell's endowment for the observatory. The name Pluto was proposed by Venetia Burney, an eleven-year-old schoolgirl living in Oxford. She made the suggestion to her grandfather, a librarian at the university, after he read a newspaper article to her about the planet that raised the issue of its name. He passed the suggestion to the Oxford astronomer Herbert Hall Turner, who passed it on to his American colleagues. Venetia was interested in classical mythology and suggested the name of the Roman god of the underworld because Planet X was presumably dark and cold. The name found unanimous favour in a vote of the Lowell Observatory astronomers and was announced in May 1930. A further point in its favour was that the name started with Percival Lowell's initials, and neatly got round the historical disinclination of astronomers to name planets after people.

this happened so soon after Charon's discovery. Any lingering doubts about the existence of the moon were removed when the Hubble Space Telescope (HST) imaged Pluto and Charon side-by-side in 1994. In 2005 the HST discovered two more moons orbiting around Pluto.

It is likely that Pluto and Charon formed as separate objects in a near-circular, low-inclination orbit beyond Neptune. Neptune perturbed their orbits and they got fed into the same eccentric, steeply inclined orbit, controlled by Neptune. Eventually Pluto and Charon collided and formed the present binary system.

Pluto is an unusual planet, so unusual that in 2006 it was removed from the status of 'planet' by the International Astronomical Union and termed a 'dwarf planet'. It is a member of the Kuiper Belt (**14**). Pluto has strongly contrasting surface features, although it is so distant that surface features are difficult to see, even with the HST. Its surface is probably covered in part by highly reflective ice.

Pluto has not yet been visited by a spacecraft for a close look. A voyage through space to Pluto takes a decade. The New Horizons spacecraft was launched on 19 January 2006 and is scheduled to reach Pluto on 14 July 2015; it will then continue further into the Kuiper Belt region (**14**).

5   **Venetia Burney** As a schoolgirl in 1930, she named Pluto.

6   **Discovery of Pluto** Pluto's image shifts against the background of stars in the two discovery photographs taken six days apart.

7   **Discovery of Charon** These photographic images of Pluto are grainy but show a distinct bump off to one side, which rotates through about 90° from one of these photographs to the next, a full revolution every six days. Jim Christy discovered Charon in 1978 using these images.

8   **Pluto and its three moons** Pluto (A) and its largest moon, Charon (B), are the two brightest images in a HST picture obtained in 2005. In the picture are two additional small moons discovered by the HST: Nix (C) and Hydra (D), which are two to three times farther from Pluto than Charon.

9   **Pluto from Nix** The artist's concept shows the nighttime view from the surface of Nix. Pluto is the large disc, centre right, and Charon is the smaller disc, extreme right.

The Kuiper Belt

The frontier of the Solar System

1

Following the discoveries of Neptune and Pluto, astronomers began to suspect that the Solar System was much larger than previously thought, and that its frontier extended well beyond the orbits of the known planets. This theory, which originated in the 1950s, was proved forty years later with the discovery of the Kuiper Belt, a band of asteroids and comets that lies beyond Neptune and is made up of material left over from the formation of the Solar System. Many of the comets that we see from Earth are objects that have travelled in from the Kuiper Belt.

2

Like Pluto and Neptune, the Kuiper Belt was 'discovered' in theory long before it was actually seen. American professional astronomer Gerard Kuiper (rhymes with 'viper') is conventionally credited with proposing the existence of the Belt, which he put forward at a symposium in 1951, hence its name. However, another American astronomer, Frederick Leonard, had mentioned the idea in print as early as 1930, in a publication for amateur astronomers that was largely ignored by professionals.

Irish amateur astronomer Kenneth Edgeworth laid out the idea more clearly in scientific papers that he wrote in the late 1940s. Edgeworth was a soldier and engineer who won the Military Cross in the First World War and worked in the post and telegraph office in the Sudan, taking up his earlier interest in astronomy when he retired. He wrote review accounts for amateur and professional journals about the origin of the stars and the Solar System. In 1949 he published an account of the formation process of stars, in which he speculated that the formation of the Sun might have left a debris field beyond Pluto. In its first 40 years Edgeworth's article was referenced only a few times but it is much better known today and the 'Kuiper Belt' is therefore sometimes called the 'Edgeworth–Kuiper Belt'.

1 **Kuiper Belt Object 1998 WW31** This binary Kuiper Belt Object was discovered in 1998 beyond the orbit of Neptune (artist's concept). It has a moon, like Pluto. The two bodies are very close in size and mass and orbit one another with a period of 570 days.

2 **Gerard Kuiper** One of the dominant figures in 20th-century American planetary studies, Kuiper pioneered infrared astronomy and founded the Lunar and Planetary Laboratory in Arizona.

3 **Discovery of Sedna** The Kuiper Belt Object was discovered moving across three images taken in 2004 with Palomar Observatory's Schmidt Telescope.

THE FIRST SIGN OF A DISCOVERY

The most exciting phrase to hear in science, the one that heralds new discoveries, is not 'Eureka!' [I found it!] but 'That's funny ...'
Isaac Asimov (attr.)

These astronomers argued that there was no reason why the Solar System should end abruptly at Neptune or Pluto. In its modern form, the basic argument would be that the solar nebula, from which the planets were formed, extended well beyond Neptune. But planets as large as Neptune could not form further out in the Solar System. Because of the low gravity in the outermost reaches of the Solar System, everything moves so slowly that collisions in which smaller bodies stuck together to form bigger ones are very infrequent, and only small bodies can accumulate. Even some of the original small component bodies survive. Because it is so cold in those distant regions, these small bodies would be ice-rich asteroids, or comets.

In fact, some of the comets that we see from Earth have come from this population. Pioneering mathematical work by American astronomer Paul Joss in 1970 and Spanish astronomer Julio Fernández in 1980 proved that this was theoretically possible. In 1988 Canadian astronomers Martin Duncan, Tom Quinn and Scott Tremaine used comprehensive computer simulations to show that when individual bodies in the Kuiper Belt encountered each other in near-misses,

their orbits might indeed be disturbed dramatically. Many of the Kuiper Belt Objects are flung out of the Solar System into interstellar space but some fall inward towards the Sun and become trapped into short-period orbits by the massive planets that they encounter as they fall. If this happens, the Kuiper Belt Objects become comets, repeatedly passing by the intense heat of the Sun and progressively melting, until eventually they break up, the residual dust causing meteor showers.

As they studied the properties of the short-period comets, Duncan, Quinn and Tremaine concluded that all of these comets had originated from a single flat disc of material, because their orbits around the Sun were confined to a single plane. The plane of the comets' orbits was close to the plane of the Earth's orbit, which suggested that the disc of comet and asteroid material had originated during the formation of the Solar System. This disc is the Kuiper Belt, with Neptune at its inner edge.

In 1987, British-born astronomer David Jewitt was becoming increasingly puzzled by emptiness of the outer Solar System. 'It was just freaky.' He persuaded Jane Luu,

6

4   **Eris (formerly Xena) and its satellite, Dysnomia (formerly Gabrielle) (artist's concept)** Eris is a trans-Neptunian object as large as Pluto (2,400 km in diameter) and has a very eccentric orbit with a period of 557 years. As of 2008, it lies nearly 100 times further from the Sun than the Earth and is the most distant object known in the Solar System.

5   **Kuiper Belt Object** 2000 FV53 was discovered in this Hubble Space Telescope image, in the centre, a quarter way up from the lower edge. The vast majority of Kuiper Belt objects will only ever appear as little dots. The whole concept of the Kuiper Belt has therefore been built up by astronomers who appreciated the significance of slim evidence like this.

6   **Dave Jewitt** Clowning with a lens, Jewitt looks for Kuiper Belt Objects.

5

a colleague at the University of Hawaii, to help him find what was beyond Neptune and Pluto, because, as he told her, 'If we don't, nobody will.' They set out on a search using the same technique Clyde Tombaugh had used to find Pluto, repeatedly imaging an area of the sky to find moving objects in the Solar System, though using digital electronic detectors called Charge Coupled Devices (CCDs) instead of photography. In August 1992, after a five-year search, they found the first member of the Kuiper Belt, which was catalogued as '1992 QB1'. This was the archetypical Kuiper Belt Object, and gives its 'QB1' catalogue number to similar objects in the Kuiper Belt, which are collectively called Cubewanos.

The small planets or comets in the Kuiper Belt are called Kuiper Belt Objects, or perhaps more neutrally Trans-Neptunian Objects or TNOs. Over a thousand Kuiper Belt Objects have been discovered, but there are probably at least 70,000 of them with diameters larger than 100 km. The Cubewanos, which account for most of these, have nearly circular orbits that lie in the same plane as the major planets. They are the original planetesimals and have always orbited this way. But some of the objects have orbits that have been strongly disturbed. A few are in large, eccentric orbits, perhaps put there by a long-ago interaction with Neptune or disturbed by a passing star.

About a quarter of the known Kuiper Belt Objects are in resonance with Neptune. This means is that the object makes a certain number of orbits around the Sun for every orbit completed by Neptune. Pluto does this, making three orbits for every two that Neptune makes, so it is in the '3 to 2' resonance. A large number of Kuiper Belt Objects are also in the same '3 to 2' resonance, and are consequently called Plutinos. Because it orbits together with a large group of Kuiper Belt Objects, and is physically distinct from the other planets of the Solar System, Pluto is considered the largest known object in the Kuiper Belt, not a planet. This is the stance that informed the decision of the International Astronomical Union in 2006 to revoke Pluto's status as a planet in 2006 (**13**).

15. Meteors and meteorites

The sky is falling!

Meteors are what are popularly called 'shooting stars'. An interplanetary piece of rock or dust, called a meteoroid, may fall to the Earth if its orbit crosses the Earth's orbit. As the meteor enters the Earth's atmosphere, its surface melts because of friction with the air, causing its outer layers to vaporize as glowing gas. Small meteoroids completely disintegrate in the air. If the meteoroid is large or robust, it can survive and reach the ground as a rock or iron lump, which is called a meteorite and may hit the ground so hard that it makes a crater.

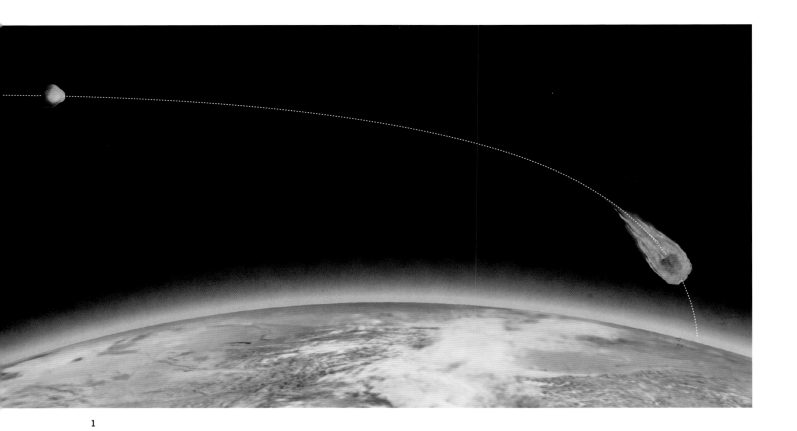

2

1 **Meteor** A 'meteoroid' orbits in space, becoming a 'meteor' as it plunges in a fiery trail through the atmosphere. With some of the rock worn away ('ablated') by atmospheric friction, the residual stone falls to Earth as a 'meteorite'.

2 **Wold meteorite memorial** A brick pillar marks the point on the hillside where the Wold meteorite just missed young John Shipley.

3 **Ernst Chladni** Best known for his work on acoustics, especially of musical instruments, Chladni is also considered the father of meteoritics.

MEMORIAL TO THE DEATH AND EARLIER NARROW ESCAPE OF A METEORITE IMPACT SURVIVOR

Erected to the memory of John Shipley
who departed this life
May 17th 1829
Aged 51 years

All you that do behold my stone
O: think how quickly I was gone:
death does not always warning give
therefore be careful how you live.
John Shipley's headstone in Wold Newton churchyard

3

Meteors are relatively common but meteorites are much rarer. For centuries, people have recorded unusual stones that fell from the sky, sometimes regarding them as sacred. Some meteorites have been seen to fall, accompanied by the sound and flash of an atmospheric explosion. In 1492 a huge triangular stone noisily made a metre-deep hole in a wheat field outside the small town of Ensisheim, Alsace, witnessed by a young boy. Because the Emperor Maximilian decided that the fall was a good omen, the 'Thunderstone of Ensisheim' is preserved in the Regency Palace there. One of the best-documented early falls occurred near the village of Wold Newton near Scarborough in England, where in 1795 a seventeen-year-old ploughman, John Shipley, saw and heard a 25 kg meteorite impact on the ground 8 m away and was showered by earth from the resulting 50 cm-deep crater. The meteorite is now in the Natural History Museum, London and a pillar marks the spot where it fell. In 1911 a meteorite fall in Egypt killed a dog (so it is said) and in 1992 another damaged a Chevrolet Malibu car in Peekskill, NY.

The Greek philosopher Aristotle (4th century BCE) explained meteors as wholly atmospheric phenomena. He thought that everything was composed of different proportions of four basic elements: 'earth,' 'water,' 'air,' and 'fire,' the latter being something like 'inflammable material'. In his book *Meteorologica* he suggested that thin streams of a mixture of 'fire' and 'air' rose to the top of the atmosphere. Ignited by the rotating motion of heavenly bodies turning around the stationary Earth, the exhalations burst into flame, like sparks off a grinding machine, making a 'shooting star'. Variations of this explanation persisted until the end of the 18th century. Aristotle dismissed meteorites as an unrelated phenomena caused by bits of volcanoes that had been launched into the sky by distant explosions.

Aristotle's view was that stones could not originate from the sky, and for over two millennia learned people dismissed accounts of meteor strike as peasants' fables. However, at the beginning of the 19th century Aristotle's view was suddenly replaced by a new paradigm in the face of overwhelming scientific evidence.

In 1794 the German physicist Ernst Chladni laid out the connection between meteors and meteorites. He studied a 700-kg iron meteorite that had been found in Siberia. Its surface was blackened and had been melted, and its composition was similar to other meteorites that had been found in widely distributed areas around the world. These meteorites were mainly iron but had been found in places that had no iron deposits. Chladni concluded that meteorites

4

5

must therefore come from a single source and that this source must cover the whole Earth – they had fallen from space.

Chladni's hypothesis was confirmed by the French scholar Jean-Baptiste Biot, who was sent by the French Academy of Sciences to investigate reports of many stones falling from the sky at L'Aigle, Basse-Normandie, in 1803. Biot had firmly believed in the Aristotelian explanation for meteorites, but two pieces of evidence changed his mind. One was the number of reports by respected people who had actually witnessed the 'fall of a rain of stones thrown by the meteor'. The other was the sudden appearance of stones across the area which had no similarity with any kind of mineral or human artefact from the region.

The best place to discover meteorites is Antarctica. They are relatively easy to find on the white surface of ice, as the nearest terrestrial rock is 3,000 feet underneath it. The first Antarctic meteorite was discovered in 1912 by a member of Douglas Mawson's Australian expedition. In 1969, Japanese glaciologists discovered nine meteorites within 3 km of each other; the meteorites were of five

different types and therefore not fragments from the same fall. This find emphasized the importance of Antarctica as a place to discover meteorites. The Japanese National Institute of Polar Research and the University of Pittsburgh set up expeditions to Antarctica in the mid-1970s, which led to the establishment of the Japanese Antarctic Meteorite Research Center in Tokyo and the US Antarctic Search for Meteorites programme (ANSMET), now led by Scott Sandford. Tens of thousands of meteorites have since been collected.

Meteorites are valuable sources of information about the makeup of other planets in the Solar System. Roberta (Robbie) Score of ANSMET team discovered ALH84001 (its number signifies that it was found in the Alan Hills icefield in North Victoria Land in 1984). It is from the planet Mars, as shown when the gases trapped inside it were compared with measurements of the Martian atmosphere made by the Viking lander. ALH84001 was ejected from the surface of Mars by the impact of a comet or asteroid. It shows mineral deposits that have, very controversially, been interpreted as from Martian microbes (**65**).

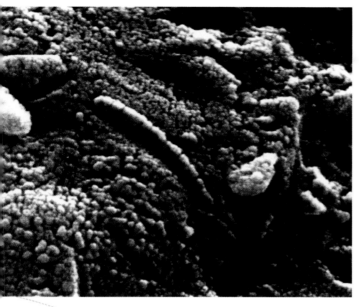

6

7

4 **Antarctic meteorite** At its moment of discovery on the surface of the ice, this meteorite is photographed with an identification number.

5 **ALH84001** An Antarctic meteorite that came from Mars.

6 **ET or not ET?** A high magnification image of part of ALH84001 shows a worm-like structure originally interpreted as a fossil nanobacterium from Mars, but later by other scientists as terrestrial contamination or inorganic deposits.

7 **Kaaba** Muslims at a religious rite centred on a meteorite once owned by Mohammed, which is kept in the black cubic building.

SACRED METEORITES

The stone at Delphi in the Temple of Apollo, known as the *Omphalos* or 'Navel of the World' was a meteorite said to have been thrown to Earth by the god Chronos when he created the Universe. In the Masjid al-Haram mosque in Mecca, the *Hadschar al-Aswad* is a sacred 'Black Stone' kept in the Kaaba, a cubic building, the axis of the Islamic world. Although the stone has never been examined scientifically, it is said to be a meteorite, given to Abraham by the archangel Gabriel and at one time possessed by the prophet Mohammed. The 14-tonne Willamette iron meteorite, now on display in the American Museum of Natural History, New York, was originally used in pre-hunting rituals by the people of the Clackamas tribes in Oregon to harden their weapons.

16. Meteor Showers

In the middle of the night, stars fell like rain

1

2

On any night it is possible to see sporadic meteors, but regularly, on certain days of the year, meteors come in showers. The first record of a meteor shower dates from 16 March 687 BCE, in the Chou dynasty: 'In the middle of the night, stars fell like rain.' Mistaken for the apocalyptic collapse of the heavens in ancient times, and for V-2 rockets during the Second World War, meteor showers are caused by clouds of dust and rock that have regular orbits and are closely associated with comets.

1 **Meteor radiant** Meteoroids travelling on parallel orbits in space streak through the atmosphere and appear in perspective to radiate from a vanishing point termed the 'radiant', just as parallel railway lines and wires do.

2 **Meteor** Its trail gets brighter as a dust particle encounters denser atmosphere but fades abruptly as the meteor evaporates and dissipates to atoms.

3 **Leonids** In 1833 the Leonid meteor shower fell like rain.

The meteors in a shower have a common origin, usually in a comet (or in one case an asteroid). Meteoroids released from the parent body spread along its orbit and form a meteoroid stream, which might intersect the Earth's orbit. When the Earth passes through the stream, lots of the meteoroids shower into the Earth's atmosphere, travelling in parallel. Seen in perspective from the Earth's surface, the meteors appear to radiate from the same point, just as parallel railway tracks do. The radiant (that is, the vanishing point of the paths of the individual meteors, observed from Earth) lies in a given constellation or near a given star.

There might be half a dozen sporadic meteors per hour on a normal night, but anything up to tens of thousands of meteors per hour during a shower. The intensity of an annual shower varies from year to year because the meteoroids travel around their orbit in clumps, with the main clump closely associated with the parent comet or asteroid, and in a given year the Earth might or might not pass through a clump. Also, the meteors' orbit may move or split into substreams, so the Earth might pass closer or further from the centre of the stream or between two substreams.

In mid-November of each year occurs the Leonid meteor shower (known by this name because its radiant is in the constellation Leo). The Leonid meteor shower of November 1833 was very spectacular, peaking at 1,000 meteors per minute. It was best seen from North America and was recorded

4

4 **The Opening of the Fifth and Sixth Seals** (from Albrecht Dürer's *The Apocalypse of St. John*, 1498) On the Day of Judgment, through the clouds, between the eclipsed Sun and Moon, a shower of stars falls onto pope, emperor and common folk alike '...the stars of Heaven fell unto Earth as the fig tree casts its unripe fruits when shaken by a great wind' (Revelation 6:13).

5 **Fireball** A large meteoroid suffers several explosions as it traverses the atmosphere, and leaves a bursting meteor trail.

6 **Meteor and the Milky Way** A fireball graces a long exposure photograph of the Milky Way as it apparently plunges to the ground in the distance (but most likely is burnt up in the lower layers of the atmosphere over the horizon, beyond the cloud).

5

by Plains Indians. The calendars ('winter counts') kept by the Sioux tribes name each year after a notable event and 1833–34 was called 'stars-all-falling-down year', with the comment: 'They feared the Great Spirit had lost control over his creation.'

This Leonid shower of 1833 was analysed by Denison Olmsted and Catlin Twining of what became Yale University, who discovered the radiant of the shower and realized that the radiant was actually the orbital path of the meteor stream. Hubert Newton, also of Yale, later calculated that the orbital period of the meteoroids was 33 years and identified appearances of the shower dating back to 902 CE. After Temple's Comet was discovered in 1866, it became clear that it was the parent body of the Leonids, because its orbit was identical to the orbit of the meteors.

In 1836, Adolphe Quetelet, a Belgian astronomer and statistician, discovered a second shower, the Perseids, which occurred in mid-August each year. The 1834 shower had been seen by John Locke, the headmaster of a girls' school, who published a letter about it in the Cincinnati Daily Gazette. His account, which also mentioned that he had discovered the shower's radiant in Perseus, went unnoticed by astronomers, most of whom did not read this newspaper.

Quetelet predicted the next display of meteors in August 1837. Edward Herrick, a bookseller–librarian in New Haven, Connecticut, observed the shower on 9 August and identified seven previous occasions when August meteors had been seen, ranging from 1029 in Egypt to 1833 in England. He found a reference to a European superstition that the burning tears of St Lawrence are seen in the sky on the night of 10 August, this day being the anniversary of his martyrdom in Rome in 258 CE. Evidently the Perseid meteor shower had been known for centuries.

In 1867 Italian astronomer Giovanni Schiaparelli discovered that the Perseid meteors came from a stream whose orbit was the same as the bright comet of 1862, Comet Swift–Tuttle, named after the two American astronomers who had discovered it.

Meteor showers can only be seen at night (unless the meteor is exceptionally bright). If a shower occurs at a time of the year when the Sun lies near to the direction of the radiant, the shower remains invisible, because you can't see meteors in daylight. However, in 1944 radio engineer James Hey discovered that meteors generate radar echoes, which can detected by day as well as by night. In September of

that year, during the Second World War, V-2 missiles were being launched on London from Germany. As a member of the British Army Operational Research Group, Hey developed a radar system to detect incoming missiles in the hope that the civilian population could be given warning of an attack to minimize casualties. Hey's system indeed detected incoming V-2s but also gave many false alarms, detecting launches of rockets when none were reported by spies and predicting attacks that never materialized.

After the war had finished, Hey set out to discover what had caused the false echoes. He discovered that his spurious radar echoes occurred when a meteor trail passed through the radar beam. The radar reflection was from the ionized air produced in the meteor trail. Final proof came from an organized campaign, during which Hey and colleagues at other stations coordinated to look for radar echoes during the Giacobinid meteor shower of October 1946, a shower associated with Comet Giacobini–Zinner. They saw 10,000 radar echoes per hour rather than the usual two or three. Using radar echoes, Hey and his team quickly discovered several new daytime meteor showers known as the Arietids, the Zeta (ζ) Perseids and the Beta (β) Taurids.

ABRAHAM LINCOLN AND THE LEONIDS.

According to Walt Whitman, President Abraham Lincoln saw the Leonids in 1833:

In the gloomiest period of the war, he had a call from a large delegation of bank presidents. In the talk after business was settled, one of the big Dons asked Mr. Lincoln if his confidence in the permanency of the Union was not beginning to be shaken – whereupon the homely President told a little story: 'When I was a young man in Illinois, I boarded for a time with a deacon of the Presbyterian church. One night [in 1833] I was roused from my sleep by a rap at the door, and I heard the deacon's voice exclaiming "Arise, Abraham, the day of judgment has come!" I sprang from my bed and rushed to the window, and saw the stars falling in great showers! But looking back of them in the heavens I saw all the grand old constellations with which I was so well acquainted, fixed and true in their places. Gentlemen, the world did not come to an end then, nor will the Union now.' Walt Whitman, *Notes Left Over*, 'A Lincoln Reminiscence', 1892

17

The Earth's Magnetosphere
Our defence against the Sun

The Earth is a giant magnet. Its liquid iron core generates a magnetic field around the planet that extends outward as far as the Moon. This magnetic field not only causes compass needles to point north, but also shields the Earth from lethal doses of solar radiation. Without its magnetic field, the surface of the Earth would resemble the desolate landscape of Mars.

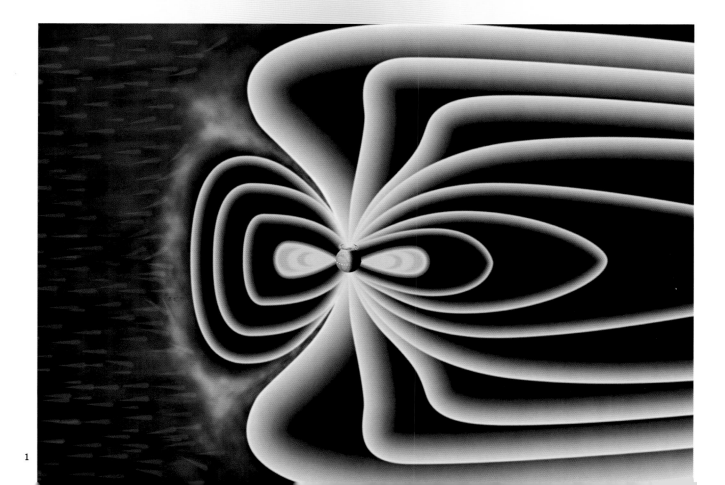

2

1 Magnetosphere The magnetic field lines from the Earth's core are swept downstream by the solar wind (from the left) but there are closed 'bottles' within the magnetic field that contain the Van Allen Belts. Solar particles descend to the auroral ovals near the Earth's surface where there are gaps in the magnetic field.

2 Aurora Curtains of green light (and other colours) are generated by the impact of charged particles in the air.

3 James van Allen is in the centre, between Wernher von Braun (right), who led the American rocket programme after the Second World War, and William Pickering (left), first director of the Jet Propulsion Laboratory; they hold aloft a full-size model of America's first satellite, Explorer 1, with which van Allen discovered the radiation belts that bear his name.

The Earth's magnetic field is caused by the circulatory motions of the liquid iron core of the Earth, which generates currents and a magnetic field much in the same way that dynamos and generators do. The region that is subject to the magnetic field's direct influence is called the magnetosphere (the term was coined by Cornell University scientist Tommy Gold in 1959) and extends into space towards the Moon. The magnetosphere was the first major scientific discovery of the space age.

For centuries, sailors in many cultures knew of the lodestone, which indicated the direction north. If freely suspended – for example, floating on cork on the surface of water – some minerals like magnetite can be used as a compass to aid navigation if the shore or the stars are not visible. The magnetic properties of the lodestone can be transferred to an iron needle for greater clarity of direction.

In 1576, Robert Norman, a ship's instrument maker, noticed that in London a magnetized needle not only turned to point north, but also tended to dip down below the horizontal by about 70°. In 1600, William Gilbert realized that this was because the needle was following lines of magnetic force that converged down towards the surface of the Earth, suggesting that the Earth is itself a magnet whose

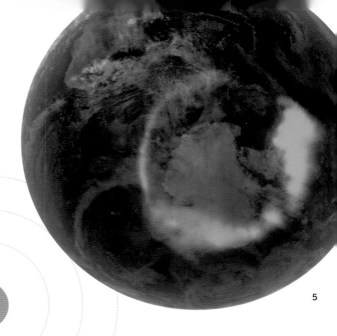

5

4

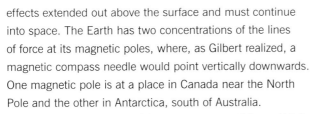

4 **Coronal mass ejection** A magnetic spasm on the surface of the Sun (inset within the central circular obstruction of the main picture) triggers an explosion in the corona like a sneeze, and a cloud of particles from the Sun's atmosphere travels out into space.

5 **Auroral oval** Seen from space, above the South Pole, aurorae are active within an oval area.

6 **Aurorae** The aurorae occur mostly in the upper layers of the atmosphere but especially active areas extend downward, closer to the surface of the Earth.

effects extended out above the surface and must continue into space. The Earth has two concentrations of the lines of force at its magnetic poles, where, as Gilbert realized, a magnetic compass needle would point vertically downwards. One magnetic pole is at a place in Canada near the North Pole and the other in Antarctica, south of Australia.

Between 1698 and 1700, the astronomer Edmund Halley combined his own magnetic survey of the Atlantic Ocean with other people's measurements, producing the first map of the world showing the direction in which a magnet pointed at any given location. Magnets always point a little off the true north because the north magnetic pole is not identical with the north pole of the Earth's axis of rotation. There were repeated surveys in the centuries since then, but in 1957–58 there was a major coordinated global effort to study geomagnetism called the International Geophysical Year (IGY). Its scientific role was usurped when it became the public arena for the military objectives of the Cold War between the USA and the USSR.

American scientist James van Allen became a key figure in the execution of the IGY. He had worked on high-altitude experiments, at first using V-2 rockets, then developing small so-called 'sounding rockets' and a hybrid vehicle called a rockoon, a rocket taken to altitude and launched from a balloon. He used all these to 'sound' the upper atmosphere.

On 4 October 1957 the USSR reached beyond the confines of the Earth and launched Sputnik 1, the first artificial satellite to orbit the Earth. This event is reckoned as the start of the 'Space Race' between the USSR and the USA. Early in 1958, the USA responded by launching its first space probes, Explorer 1 and Explorer 3, both carrying experimental equipment with which van Allen measured the density of charged particles in space.

Charged particles from the Sun and from outer space are caught by the Earth's magnetic field and funnelled through the magnetosphere into the magnetic polar regions, as discovered by the Norwegian physicist Kristian Birkeland in 1895. Birkeland put a magnetized iron sphere, a *terrella* ('little Earth'), in a vacuum chamber, and aimed a beam of electrons towards it. He saw that the electrons were steered by the magnetic field to the *terrella*'s magnetic poles. This experiment replicated in miniature the phenomenon of the aurora (or Northern and Southern Lights). When the electrons strike the Earth's atmosphere they produce a colourful shimmering aurora that can readily be seen at polar latitudes. Birkeland's *terrella* was an experimental way to simulate the magnetosphere, which is nowadays carried out by numerical simulations in computers.

Van Allen discovered that the Earth was encircled by a doughnut-shaped region of charged particle radiation. This

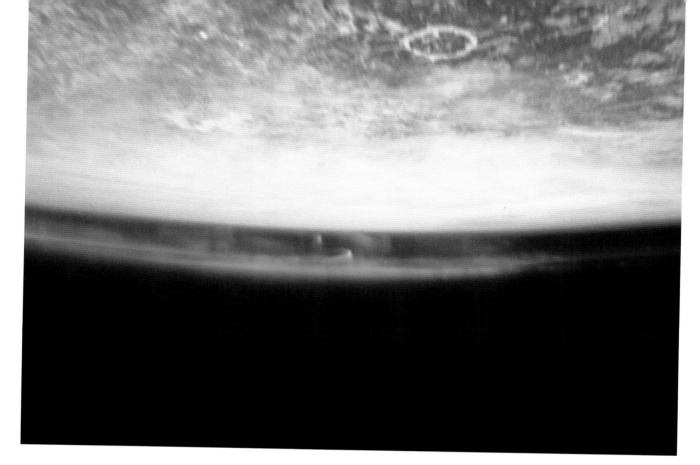

radiation was trapped within 'magnetic bottles' that were enclosed by the Earth's magnetosphere. The doughnut-shaped region of radiation was named the Van Allen Belt in honour of its discoverer. A second, outer belt was identified in 1958 by the Pioneer 3 lunar probe (which failed to get to the Moon but nevertheless reached an altitude of 101,000 km) and also detected by the Sputnik 2 and 3 satellites. The Van Allen Belts were the first major scientific discovery to be made as a result of space exploration.

The Van Allen Belts reach from the top of the Earth's atmosphere out to a distance of about 3.5 times the diameter of the Earth. They are major components of a system of electric currents and high radiation regions that have been mapped by a succession of space satellites. The satellites sent into the high radiation regions have to be specially robust because the particles deleteriously affect the environmental conditions for space flight. The particles also affect the transmission of power and electrical signals at the Earth's surface. All these effects are known as 'space weather'. They are caused by storms produced by the Sun and are associated with sunspots.

The first indications that the Sun causes changes in the magnetosphere were noticed in the 19th century. In 1843, Heinrich Schwabe, a German apothecary and amateur astronomer, discovered that sunspots came and went on a ten-year cycle (now more accurately reckoned at eleven years). The explorer Alexander von Humboldt popularized Schwabe's work. Consequently, when an army officer named Edward Sabine discovered that magnetic storms (violent vibrations of a compass needle) were more frequent at intervals of ten years, he suggested that this was related to the solar cycle. Sabine's theory was dramatically confirmed when English amateur astronomer Richard Carrington discovered a large flare on the Sun in 1859, which was immediately followed by violent a magnetic storm and aurora.

Solar wind – the second major discovery of the space programme – is the actual mechanism in the Sun that affects the Earth's magnetosphere. This is a constantly flowing but erratic stream of charged particles that emanates from the Sun and impinges on the Earth – or rather onto the Earth's magnetic field. The behaviour of comet tails – which always point away from the Sun regardless of the direction in which the comet is travelling – gave early indications that something was flowing outwards to sweep the tails away, like the loose ends of a scarf blowing in the wind. Physicists Eugene Parker, Ludwig Biermann and Hannes Alfvén independently came to the conclusion in the 1950s that there existed a relentless outward flow of charged particles from the Sun: the solar wind. The solar wind was confirmed

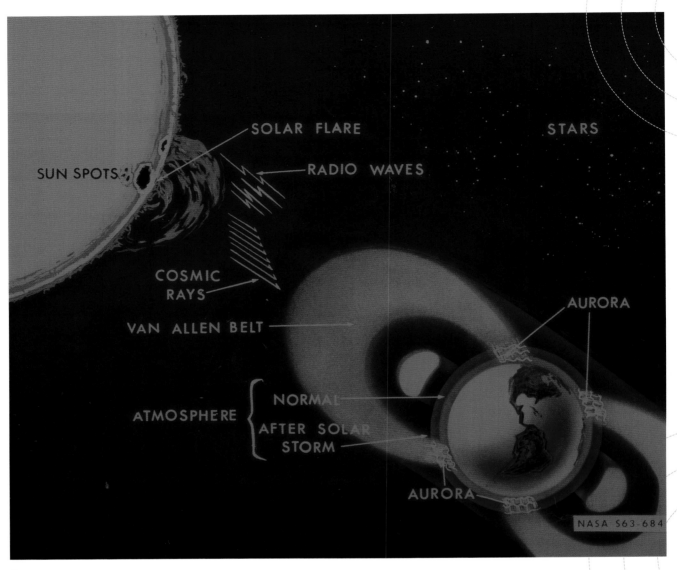

7

7 **Magnetosphere** The major components of the magnetosphere as explained
by a NASA press release contemporaneous with the early Explorer
missions, *c.* 1960.

8 **The solar wind** The SOHO satellite recorded an explosion ('coronal mass
ejection') on the Sun in 1997. Minutes later the burst of solar particle
radiation ripped through the electronic detectors on the satellite and
peppered them with streaks of interference. Other particles in the burst
would have gone on to enter the Earth's magnetosphere.

9 **Astronaut** During a solar storm, such as those illustrated in Figs. 4 and 8,
an astronaut on a space walk (like these, repairing space equipment) would
be safer back inside the Space Station, better protected from radiation.

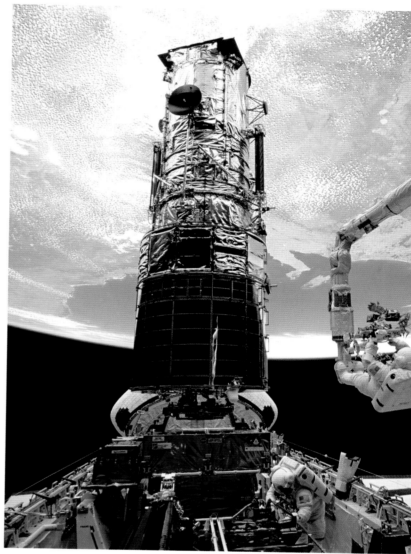

97/11/06 14:19
97/11/06 14:32
97/11/06 14:26
97/11/06 14:12

8

9

by the Soviet lunar probes Luna 1, 2 and 3 as they transited from the Earth to the Moon in 1959 and by the American Mariner 2 probe as it travelled to Venus in 1962.

Except at the poles, the magnetic field of the Earth defends the Earth's atmosphere and surface from the solar wind. If the magnetosphere did not exist, the Earth's atmosphere would be blown off by the solar wind and its surface would be exposed to lethal radiation. This seems to be what happened to the planet Mars when its magnetic field died away to almost nothing. By contrast, the Earth's magnetosphere is unusually large because the iron core of the Earth is unusually large, the amalgamation of two iron cores in the collision that formed the Moon (21).

The magnetic field of the Earth is always shifting, sustained by the circulatory motions in the Earth's liquid iron core, which are erratic and oscillate. This causes the magnetic field of the Earth to drift and tilt, and is the reason why, as astronomer Henry Gellibrand discovered in 1635, the north magnetic pole in Canada is slowly drifting westward towards Alaska. The Earth's magnetic field also changes strength, and even reverses polarity (the north magnetic pole changes to the south pole and vice versa) every 250,000 years or so. It is not known how quickly this happens, or whether for some periods of time the Earth's magnetic field becomes so weak that the magnetosphere is turned off. If so, the atmosphere and the surface of the Earth would be exposed to the solar wind. Nor is it known what happens temporarily to the natural environment at such a time.

18

Comets

Sandbanks or dirty snowballs?

1

Although comets are no longer superstitiously associated with bad fortune, the mystery of where they come from and what they are made of continues to intrigue astronomers. Recent images from space probes show that comets are made of ice and fine dust. It is possible that some of the water in our oceans and perhaps even the molecular seeds of early life were brought to Earth by comets.

1 Comet McNaught This photograph, taken in 2007, showed ripples in the comet's magnificent tail caused by periodic releases of dust along its curving orbit and the pressure of solar radiation on the dust particles.

2 The Great Comet of 1680 Kirch's Comet was the first comet discovered by telescope. With this diagram, Isaac Newton showed that its orbit was a parabola and conformed to Kepler's laws – it was the comet that proved Newton's law of gravitation.

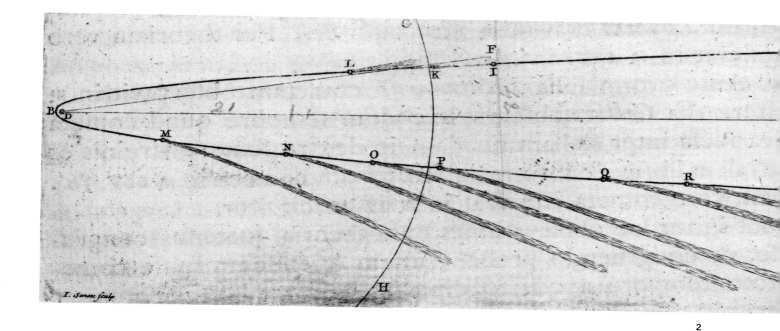

2

Comets are small bodies that orbit in the Solar System, like planets (**07**). However, unlike planets, whose orbits are nearly circular and confined largely to one plane (the 'ecliptic'), comets' orbits are highly eccentric and may be inclined upward or downward at any angle. Aristotle considered comets to be atmospheric phenomena (**15**) but in 1577 Tycho Brahe discovered by measuring the parallax of the comet that it was located beyond the Moon, and astronomical in origin.

Isaac Newton discovered that the comet of 1680 was following a parabolic orbit around the Sun, visiting the Sun only once as far as he could tell. But his colleague Edmond Halley discovered that the orbits of three comets, which appeared in 1531, 1607 and 1682, were identical. Halley suggested that all three comets were actually the same object, revisiting every 76 years, and predicted the next appearance in 1758 or early 1759. This happened after his death, when what is now called Halley's Comet was rediscovered by Johann Georg Palitzch, a farmer from Saxony and an amateur astronomer.

Comets have an icy 'nucleus,' which is up to several tens of kilometres in size. When a comet nears the Sun, the ice melts and the comet develops a bright, dusty atmosphere, called the 'coma', and 'tails' of gas and dust. The tails of comets point away from the Sun. The dust of a comet tail is pushed away from the comet by solar radiation pressure, as discovered in 1900–1 by the Swedish chemist Svante Arrhenius and the German physicist Karl Schwarzschild. The dust trails of some comets (Halley's Comet, for one) are associated with meteor showers (**16**). Comets also have a second, faint, straight tail of gas, which is pushed back by the solar wind (**17**).

There are two competing models explaining the composition of comet nuclei. The evidence of meteor showers suggested that the nucleus is a loose assembly of dust grains, easily dislodged, for instance, by solar radiation pressure; this is known as the 'sandbank model'. A number of physicists and chemists, such as the Belgian Pol Swings and the Russian Boris Yu Levin, developed theories in the 1940s to show just how weakly the dry grains were held together, but these studies also indicated that 'sandbanks' would quickly disintegrate and that periodic comets would not survive as many passages around the Sun as they actually do. In the early 1950s, American astronomer Fred Whipple proposed the 'dirty snowball' model. He suggested that a comet's nucleus is made of dust grains cemented together by ices, such as water, ammonia and methane. The ices sublimate

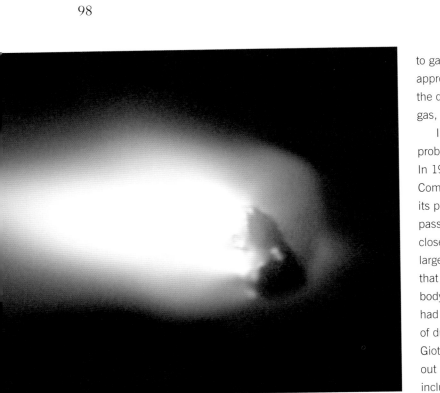

to gas and are released from the comet as it warms on approaching the Sun. The vaporization of the ices lets loose the dust grains, which are dragged from the nucleus by the gas, and the comet re-freezes once it has left the Sun behind.

In 1985, NASA's ICE (International Cometary Explorer) probe explored the plasma tail of Comet Giacobini–Zinner. In 1986, two Russian Vega spacecraft surveyed Halley's Comet from a distance and two Japanese probes investigated its plasma tail. The European Space Agency probe, Giotto, passed within 600 km of the nucleus of Halley's Comet, so close that the spacecraft was knocked out of alignment by a large particle from the comet's dust cloud. Giotto discovered that the comet nucleus had a coal-black, potato-shaped body about 15 km long and 7–10 km wide. The nucleus had a 'fluffy' porous texture, and was covered with a crust of dust, caked by a tar made of complex organic molecules. Giotto discovered seven jets on Halley's Comet, throwing out 3 tonnes of material per second. Some of it was gas, including vaporized tarry substances, gases of carbon

3

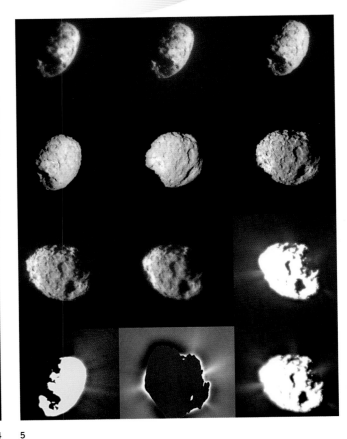

4   5

compounds that make comets some of the smelliest places in the Solar System. Most of the ejected grains of dust were no larger than specks of cigarette smoke. The largest grain detected weighed only 40 mg, but the large particle that damaged the spacecraft was perhaps as heavy as 1 gm.

Giotto confirmed the presence of ice in the comet's nucleus, which proved that Halley's Comet is not a 'sandbank'. However, the comet is not really a 'dirty snowball' either, since dirt is dominant, not ice. The structural arrangement of the comet is dictated by the physical properties of dust rather than ice.

It is thought that comets formed when the Sun formed, 4.5 billion years ago, from interstellar ices condensing onto grains of interstellar dust. They were originally planetesimals that congealed from scraps of dust and gas in the presolar nebula (**50**). They then remained almost unaltered in two cold, outer regions of the Solar System, until they fell towards the Sun, ultimately doomed to melt like snowmen when the Sun rises. Short-period comets come from the Kuiper Belt, which is located in the outer Solar System beyond the orbit

of Neptune (**14**). The source of the long-period or sporadic comets is thought to be the Oort Cloud, a spherical reservoir of comets surrounding the Solar System. Dutch astronomer Jan Oort discovered in 1950 that many long-period comets fall towards the Sun from a distance of between 20,000 and 200,000 times the distance from the Sun to Earth. Comets that formed inside Neptune's orbit were ejected into distant orbits during encounters with giant planets, and formed the Oort Cloud and the Kuiper Belt. Occasional encounters with each other, or with passing stars or giant clouds of interstellar material, re-inject some comets from the Oort Cloud and Kuiper Belt back into the inner Solar System. It is likely that, early in the history of the Solar System, there were frequent collisions of comets with Earth. Some of our ocean water may have been brought to Earth by comets. In addition to water, complex organic molecules (and especially 'prebiotic' organic molecules) could also have acted as seeds for the development of life on Earth (**20**, **65**).

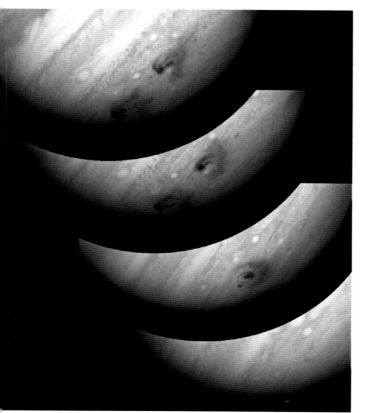

3 **Halley's Comet** Viewed by the Giotto space probe from a distance of 1,200 km, the comet's nucleus was shown to be coal-black and in the shape of a potato, with fountains of gas and dusty material.

4 **Comet (artist's impression)** Fountains of dust and gas spray from pits on the sunlit and sun-warmed side of the comet's nucleus, as ice vaporizes and bursts up through the tarry surface.

5 **Comet Wild 2** A series of pictures of the comet's nucleus, made by the Stardust space probe in 2004. The first nine pictures show the comet as viewed successively by Stardust. The final three pictures are deliberately overexposed (revealing the faint background stars to aid in navigating the spacecraft) and show the atmosphere of the comet, which is produced by jets of gas and dust.

6 **Comet impact on Jupiter** Comet Shoemaker–Levy 9 plunged into Jupiter in 1994, and caused an upwelling of red chemicals from inside the planet's atmosphere that made a heavy stain at the impact point and a lighter crescent-shaped spray on the cloud tops. Jupiter's weather gradually caused the chemicals to disperse.

19

The Earth's Climate, the Seasons and the Weather

Astronomical cycles

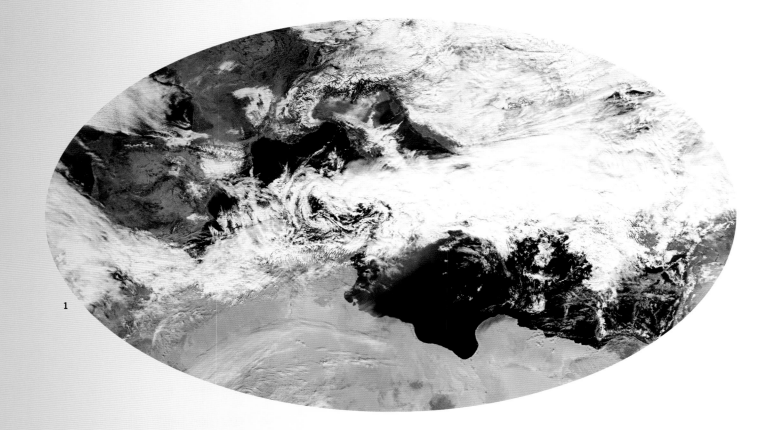

1

As the Sun warms the atmosphere, the land and the oceans, it sets in motion the cycles that generate the Earth's weather. But what triggers the catastrophic chill of an Ice Age? A Serbian maths teacher, Milutin Milanković, suspected that Ice Ages were caused by cyclical changes in the Earth's orbit. Deep below the ocean floor and the Antarctic ice, the proof of Milanković's theory was waiting to be uncovered.

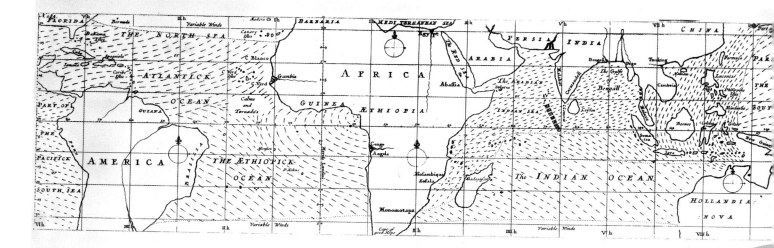

The Sun warms the Earth most directly in the regions around the equator. There, sunlight radiates through the atmosphere to warm the ground and surface of the sea. As they warm, the layer of air at the base of the atmosphere becomes less dense and rises (a process called 'convection'). Cold air is drawn in underneath the rising warm air. At high altitude, the warm air cools and flows away from the equator, eventually beginning to sink again at about latitude 30° and returning to the equator at ground level. In 1686, Edmund Halley discovered that this cycle was the engine for the major wind systems of the world.

Although this convection cycle creates the wind pattern, the wind does not blow just north and south from the equator. In the days of sailing ships, sea travel was heavily dependent on the 'trade winds', which flow strongly and consistently from the east in the areas around the equator. Why do the trade winds blow at right angles to what you would expect? George Hadley, an English lawyer and amateur meteorologist, explained the phenomenon in 1735; the closed cell of air cycling from the equator to latitude 30° and back is consequently called a 'Hadley cell'. Hadley realized the important effect of the rotation of the Earth on wind motion. As air blowing over the Earth's surface moves from a higher latitude region, where the wind's eastward velocity is lower, into a region of higher eastward velocity (at the equator), the wind picks up a westward motion.

Hadley's intuitive explanation was given a sound physical basis in 1835 by the French physicist Gaspard-Gustave de Coriolis, who applied Newton's fundamental mathematical theories to the problem. Because of the rotation of the Earth, winds deflect to the right in the northern hemisphere and to the left in the southern. This is called the Coriolis Effect. In 1856 Hadley cells and the Coriolis Effect were brought together in a

1   **Earth and its climate** The Orbview-2 satellite captured this view of a severe winter storm that swept across Europe on 16 December 2001. The image shows the heavy clouds across the Mediterranean Sea; snowfall is visible in some areas and the strong wind has picked up dust in North Africa.

2   **Halley's chart of the tropical winds** The chart, published in 1686, shows the direction of the prevailing winds in the tropical regions, including the trade winds, which in the larger areas of the sea, far from the disturbing influence of the land, blow consistently towards the west and the equator.

3   **Edmond Halley** As a friend of Isaac Newton, he helped him develop the law of gravitation. He was the second Astronomer Royal, in charge of the Royal Observatory at Greenwich.

4   **Global temperature cycles** This is the record of the temperature over the last 420,000 years measured from the deepest ice core yet drilled, 3,623 m deep, at the Russian Vostok station at the centre of Antarctica. There is a 100,000-year periodicity in the temperature.

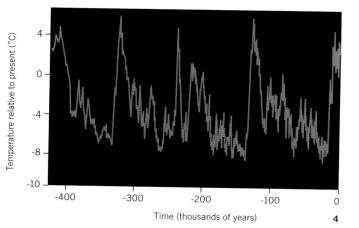

unified theory by the American meteorologist William Ferrel.

The position and the strength of the winds determine the weather, which therefore ultimately depends on the strength of solar radiation. The warmest zone on the Earth is in general the equator, but more precisely the sub-solar latitude (that is, the area of the Earth that lies directly 'below' the Sun). Because the Earth is tilted at 23.5° to its orbital plane around the Sun, the sub-solar latitude changes position throughout the year, moving from the Tropic of Cancer at 23.5° N in June to the Tropic of Capricorn at 23.5° S in December. This produces the annual cycle of the seasons, which are influenced locally by factors such as latitude, altitude, terrain, vegetation and proximity to oceans, and may be characterized by particular weather phenomena such as a rainy season or a persistent wind like the mistral.

Additionally, the eccentricity of the Earth's orbit around the Sun has a small but noticeable effect on the annual weather cycle. The Earth does not orbit the Sun in a perfect circle, but in a slight ellipse, which causes the distance between the Earth and the Sun to fluctuate at different times of the year. The Earth is closest to the Sun in the first week of January and furthest in the first week of July. This magnifies the effect of the Earth's tilt on the seasons, so in December and January the summer solar radiation is stronger in the southern hemisphere than it is in June and July in the northern hemisphere, which is why summers tend to be hotter in the southern hemisphere.

If the Earth's tilt and its eccentric orbit stayed the same forever, these annual cycles of the seasons would remain the same. However, long-term cycles of change in the Earth's position and orientation cause changes in the seasons over time. The Earth's axis does not always point in the same direction, but precesses in a cone over a period of 26,000 years, which causes the seasons to shift within that time-scale (**05**). Moreover, the tilt does not remain constant at 23.5° but nods between 21.5° and 24.5° over a period of 41,000 years. The larger the tilt, the greater the variation of the seasons. The eccentricity of the Earth's orbit (which produces the hotter summers in the southern hemisphere) is also variable, changing between almost 0 and 7%, on a time scale of 100,000 years. At present, the Earth is midway through that cycle, with a difference of 3.4% between January and July.

The total effect of all these orbital cycles on the weather is complicated. Throughout the 1920s and 1930s the Serbian civil engineer and geophysicist Milutin Milanković, in his second career as a mathematics teacher, devoted himself to studying their effects on climate change. For this reason they

ASTRONOMICAL JOURNAL.

No. 109.

VOL. V.   ALBANY, 1858, JANUARY 20.   NO. 13

THE INFLUENCE OF THE EARTH'S ROTATION UPON THE RELATIVE MOTION OF BODIES NEAR ITS SURFACE.

BY W. FERREL.

5

6

are called Milanković (or Milankovitch) cycles. He attributed the Ice Ages (periods when there are extensive ice sheets, like the one in Antarctica) to these cycles, discovering that recent cold periods have occurred approximately every 100,000 years, when all the Earth's different orbital cycles coincided to produce maximum cooling. His discovery was verified after his death by analysis of ocean sediments and Antarctic ice cores, which show isotopic variations in different layers that were caused by temperature differences when the layers were deposited. The present Ice Age (defined by glaciologists as a period in which there are extensive ice sheets, like the

7

5 **'The Influence of the Earth's Rotation'** William Ferrel put forward an original, rigorous explanation of the direction of the winds and the Coriolis Effect in this paper.

6 **Ice core storage** The US National Ice Core Laboratory houses an archive of ice cores in a building in Denver, Colorado. It has over 14,000 m of ice cores from 34 drill sites in Greenland, Antarctica, and high mountain glaciers in the Western United States. The facility is maintained at a temperature of -35 °C, with four levels of backups and safety systems.

7 **Milutin Milanković** Serbian mathematician, civil engineer and geophysicist, who identified the connection between Earth's orbit and the global climate.

8 **Melting glacier** Pine Island Glacier flows into the Amundsen Sea, off Antarctica, contributing to the greatest flow of ice to the sea in the world. Its flow rate has about doubled over the last 30 years as global warming melts the underside of the ice shelves and lubricates the flow. The picture shows a crack across the flow of the glacier as it calves icebergs into the sea.

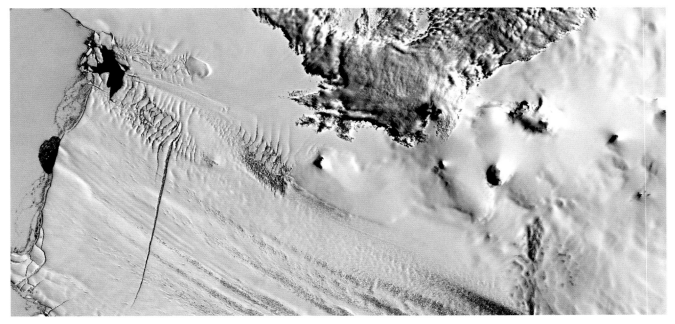

8

one in Antarctica) began 40 million years ago. It grew colder during the Pliocene and Pleistocene periods, starting around 3 million years ago, with the spread of ice sheets across the northern hemisphere. Since then, glaciers have advanced and retreated every 40,000 to 100,000 years. The most recent retreat of the glaciers ended about 10,000 years ago.

However, on their own Milanković cycles are not severe enough to alter global temperature by the amounts that are recorded in ocean sediments and ice cores, so there must be other processes, perhaps the greenhouse effect (**23**), that amplify the effects of fluctuations in the amount of solar radiation reaching the Earth. Terrestrial volcanism, continental drift and changes in the composition of the Earth's atmosphere all play a part. The recent increase in human-generated carbon dioxide and other greenhouse gases has started to cause changes that are quicker and more extreme than the Milanković cycles, increasing the Earth's temperature at an unprecedented rate.

20. Asteroid Impacts on the Earth

The true origin of craters on the Earth

A miner's doomed and all-consuming quest for riches under a strange hill in Arizona unearthed treasure of a different kind: proof of meteorite impacts on the Earth. From a monstrous fireball in Siberia to mysterious sparkling stones in the wall of a medieval church, the evidence of meteorite impacts has transformed our understanding of evolution and geologic change.

1 **Barringer Meteor Crater** The crater is 1,200 m in diameter and 170 m deep, and its rim rises 45 m above the surrounding plains. The scale of the crater can be appreciated in this evening picture under snow from the size of the Visitors' Centre on the rim of the far wall at the end of the road.

2 **The southern wall of the Barringer Meteor Crater** Barringer started by drilling at the centre of the crater, where the mining equipment and disturbed land can still be seen, but the meteorite in fact impacted at an angle under the far wall.

2

US Route 40 runs east from Flagstaff, Arizona, across the dry Colorado plateau and passes close to what from a distance looks like a low hill. This is Coon Butte. It is a complete surprise to reach the top of the hill and look out over a large, circular crater.

In 1891 the crater was mapped by Grove Karl Gilbert of the US Geological Survey. In 1895 Gilbert considered the possibility that the crater had been made by a meteorite impact, but rejected this explanation on two grounds. First, the amount of ejected material in the crater walls and on the surrounding plain was no larger than the hole in the ground. Overestimating how big a meteor would have to be in order to produce a crater of the size of Coon Butte,

Gilbert did not find the a large quantity of extra material he expected. Second, when he searched the crater floor with a magnetometer for a large mass of iron, he found nothing: there was no sign that an intact meteorite had buried itself in the ground. He therefore concluded that the crater was volcanic, like the nearby San Francisco mountains.

Nevertheless, in 1902, on a hotel verandah in Tucson, a Philadelphia mining engineer named Daniel Moreau Barringer heard local gossip from a government agent that the crater was meteoritic. His imagination was fired by the mention of meteors, and the businessman in him was attracted by the possibility that buried under the crater was a large iron-nickel mass,

1

3   **Location of the known terrestrial impact craters** Nearly 200 craters are plotted. It is hard to find meteor craters in the sea, in jungle or under ice, so there are empty areas.

4   **Gene Shoemaker** Planetary scientist, comet discoverer, astrogeologist.

which he could mine and sell – the nickel, in particular, was valuable. Samples of rocks found nearby on the surface of the plateau contained 5% nickel and traces of iron, mixed with the ejecta from the crater; the iron-nickel fragments were, therefore, evidently coeval with the crater's formation, and Barringer concluded that Coon Butte had been formed by a meteor.

Barringer and a partner, Benjamin Chew Tilghman, bought mining rights to the area containing the crater and began to search for the meteor mass below its centre. The crater was so nearly circular that it seemed logical to assume that the direction of impact had been straight down. By 1908 the pair had drilled 28 holes in the crater floor, but found no meteorite.

Tilghman then noticed that the ejecta littering the surrounding plain was asymmetrical, strewn to the south. Moreover, the southern rim of the crater was raised, as if the meteor had burrowed under it. Barringer experimented by firing

projectiles into earth. The craters he produced were always circular, even if the projectile impacted at an angle, but the ejecta from the crater continued the forward momentum of the projectile. On the basis of this experiment, the two men shifted the focus of their search, and drilled a mine shaft into the interior south wall of the crater. Yet they still did not find the mass of iron-nickel. In nearly every hole their drill encountered an isolated hard 'obstruction', which could well have been a fragment of the meteorite, but they were fixated on discovering the motherlode and paid no attention to these small pieces.

Undeterred by the repeated failures of their mining strategy, Barringer and Tilghman presented their theory that Coon Butte was a meteor crater to the Philadelphia Academy of Sciences in 1906 and to geologists at Princeton University in 1909. Barringer's style of argument was belligerent. He heaped scorn on Gilbert, who was a well-respected

5

6

5 **Shocked quartz** The lines or 'lamellae' are caused by the passage of a shock-wave that melts the crystal. This effect is characteristic of quartz that has been shocked by meteor impacts and nuclear explosions.

6 **Chicxulub crater** Imaged by NASA's Shuttle Radar Topography Mission (SRTM), a shallow trough (3–5 m deep), shown as a darker green semicircle terminating at the coastline, marks the crater's rim. The trough was formed by preferential erosion at the rim of limestone deposited after the impact.

geologist, and accused established scientists of blind prejudice. This did not endear him to academic audiences and on most occasions they could not stop from tittering.

Meanwhile, Barringer's investors were becoming increasingly alienated by his intemperate remarks, and demoralized by the continuing expense of the fruitless search. When the geologist George Merrill theorized that the Coon Butte meteorite would have shattered into small pieces on impact, leaving no single mass of iron-nickel to be found, they gradually withdrew from the project, abuse flung at their departing backs. Disillusioned by Barringer's demand to continue drilling, Tilghman also pulled out.

Barringer found new backers, but by 1928, nervous at the money that they had spent, they consulted the astronomer Forest Ray Moulton at the University of Chicago. Moulton estimated the mass of the Coon Butte meteorite

at 300,000 tons compared to Barringer's estimate of 10 million, which reduced the investors' potential return by 97%; the site would be even less profitable if the meteorite had fragmented into a large number of small pieces that would be impractical to collect. In an attempt to reassure them, Barringer sought a second opinion from Princeton astronomer Henry Norris Russell, who confirmed Moulton's calculations. Having spent a fortune of about US $1 million (more than ten times that amount at today's value) on his doomed hunt for the meteorite, Barringer died in 1929.

Barringer's discovery that the Coon Butte crater was meteoritic was potentially of greater value than the meteorite itself. Scientists began using Coon Butte as a model to explain similar craters in the Solar System, such as those on the Moon. Barringer's explanation received dramatic support when, in 1908, the region near the Tunguska river in Siberia was rocked

7

by explosions from a monstrous fireball. It was not until 1921 and 1922 that the Russian scientist Leonid Kulik was able to visit the region and found the impact point, an area of flattened trees, which had clearly been felled by a meteor. Kulik's work became known in the West by 1928. Nevertheless, the scientific establishment continued to assert that both Coon Butte and the craters on the Moon were volcanic.

It was only in 1957–63 that Barringer's theory about the origin of Coon Butte was decisively confirmed by astrogeologist Gene Shoemaker. He discovered and described the tell-tale geological clues that indicate a crater has been made by a meteorite impact rather than a volcanic explosion. The key sign is the presence of shocked quartz minerals such as coesite and stishovite, which are fused at much higher temperatures and pressures than those generated by volcanic action, and were first identified in nuclear test craters. Using these criteria, Shoemaker and his wife Caroline identified many of the 160 other impact craters known around the world. One of these was the Ries crater, which surrounds the village of Nördlingen in Bavaria. The walls of St George's church in Nördlingen are made of a rare and beautiful sparkling mineral called suevite, a strongly shocked impact material almost certainly generated by the meteorite impact.

Scientists now estimate that Barringer's Coon Butte crater was formed about 50,000 years ago when mammoths, sloths, bison and camels roamed the Colorado plateau. Animals as far away as 25 km would have been killed or injured by the blast. An impact of this size occurs on average only once in a thousand years on Earth. But once every million years or so the Earth is struck by a meteor large enough to devastate a continent. Some scientists attribute the extinction of the dinosaurs to the meteor impact that, 64 million years ago, created the Chicxulub crater in Mexico, which was discovered in 1978 by oil geophysicist Glen Penfield.

The realization that meteor impacts shape the landscape of the Earth resolved a 250-year dispute among geologists and evolutionary scientists. At the end of the 18th century, years before Darwin formulated his theory of evolution, the French palaeontologist George Cuvier proposed that individual catastrophes were the major engines of change in the Earth's natural history, causing phenomena such as continental drift and the extinction of species. This theory was widely accepted at the time because it was consistent with creationist biblical interpretations (for instance, those relating to Noah's Flood). The Russian-Jewish psychiatrist Immanuel Velikovsky revived similar views in the 1950s, knitting mythology, archaeology

8

7 **The Nördlinger Ries meteor crater** The flat crater floor stretches to the northern crater rim to the right. The viewpoint is on the southern crater rim near the village of Mönchsdeggingen. The meteor impact broke through and shattered the limestone rocks on the left.

8 **James Hutton** The father of modern geology, depicted in a cartoon by John Kay in 1787, hammering rocks that bear a not-so-subtle resemblance to the faces of his critics.

and pseudo-science into fantastic but very popular theories about floods and cosmic fires. Scientists feel instinctively uncomfortable with this world view, called 'catastrophism', because it suggests that everything on Earth – its ecosystems, weather, geology and geography – is subject to arbitrary events, a 'tale told by an idiot', and therefore cannot be analysed or modelled using scientific methods. Reacting against this approach, geologists James Hutton in the 18th century and Charles Lyell in the 19th century promoted the theories of 'uniformitarianism' and 'gradualism', insisting that geological change occurs slowly over long periods of time, more along the lines of Darwinian evolution.

Following Darwin, uniformitarianism became the dominant paradigm of geology for over a century, until Barringer's crater, Kulik's fireball and the Shoemakers' discoveries made their impact, forcing scientists to combine the two opposing views. Cosmic fireballs are not the stuff of legend, but are meteors, which cause dramatic effects on Earth. Scientists now believe that geological history is a process of slow and gradual evolution, punctuated by occasional catastrophic events.

W e have learned now that we cannot regard this planet as being fenced in and a secure abiding place for Man; we can never anticipate the unseen good or evil that may come upon us suddenly out of space.
H. G. Wells, *The War of the Worlds*, 1898

21.

The Origin of the Moon
Neither daughter nor sister of the Earth

The Moon is not just the Earth's nearest neighbour: it is the passive guardian of life on Earth, and its dead landscape, visible to the naked eye, holds the clues to our own planet's origins. When humans first walked on the Moon in 1969, the lunar dust and rocks began to reveal their secrets.

From time immemorial, people have speculated about the dark patches on the surface of the Moon: in folklore they are said to have the shape of a man, a rabbit or a crone carrying firewood. In the 17th century Galileo (**08, 09**) pointed his telescope at the Moon and discovered mountains and craters in a barren, dry desert. More recently, American astronauts and Soviet spacecraft explored the lunar landscape, and the geology of the rock samples they collected has shed light on the origins of the Moon. Despite its peaceful appearance in the night sky, the Moon has a violent past, having been born in a fiery collision between the Earth and another planet. Yet without this catastrophic event, the Earth would not be stable enough to support life.

The Apollo programme of human lunar exploration arose in response to a national challenge issued by President John F. Kennedy at the height of the Cold War in 1961, as the USA and USSR competed for primacy in the 'Space Race'. The exploration of the Moon was initially driven by nationalist aims and technological innovations, although a scientific research programme was eventually established to operate in tandem with the landings. The American lunar programme was initiated by a fleet of 22 robot spacecraft launched by NASA to the Moon: the Rangers crash-landed on the lunar surface; the Surveyors touched down softly; and the Lunar Orbiters, Explorers and early Apollos surveyed the Moon from orbit.

1　**Intrepid** The Apollo 12 landing module positions itself for its final descent.

2　**Full Moon** Dark flat plains (the 'maria' or 'seas') show as large, old craters (upper half), while a smaller impact, which created the crater named after Tycho (towards the bottom), must be recent because it has ejected rays of lighter coloured underground material that lie over the older terrain.

3　**Ptolemaeus** Half of the giant, old crater Ptolemaeus is in the foreground, with the smaller, newer crater Ammonius on its floor. This image was taken in November 1969 during the Apollo 12 mission.

3

Early Moon landings were manned by test pilots and military men with only basic training in geology, and the first landing sites were chosen for feasibility rather than scientific interest. Apollo 11 landed on a featureless plain in the Sea of Tranquillity, Apollo 12 in the Ocean of Storms. Later lunar missions targeted areas of geological interest in the hopes of gaining a clearer understanding of the origin and composition of the Moon. Apollo 13 and 14 were meant to land at Fra Mauro, a site made of debris thrown from the meteorite impact that created the massive crater of the Mare Imbrium basin, and would therefore include rocks ejected from deep inside the Moon. The Apollo 13 landing had to be aborted when the spacecraft's service module was damaged by the explosion of an oxygen tank. Apollo 14 was targeted at the same site, landing safely, but the astronauts got lost during the moonwalk, and did not collect the rock samples the geologists wanted. Apollo 15, 16 and 17 were more successful, landing near the Apennine mountains, the Descartes highlands and the Littrow valley in the Taurus mountains, sites chosen for their geological significance, with the Taurus-Littrow valley showing signs of recent volcanism. All the Apollo modules that landed on the Moon brought back rock samples for terrestrial analysis – over 300 kg altogether. Some small lunar samples from other sites were returned to Earth by three robotic explorers from the Soviet Luna series of spacecraft.

4

6

5

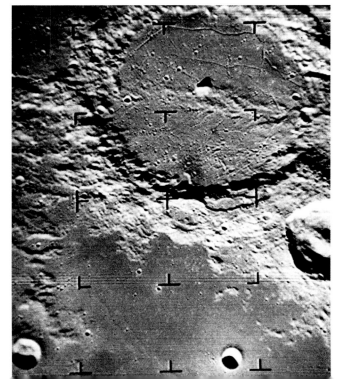

The ages of the lunar rocks recovered by American and Soviet spacecraft range from 4.6 billion years old for rocks from the highlands to 3.2 billion years old for rocks from the giant craters of the lowlands (or 'lunar maria'). One of the oldest is the 'Genesis Rock', which was collected by Apollo 15 astronauts James Irwin and David Scott and is representative of the rocks found in the lunar highlands. Scott was attracted by its lighter grey colour and, after cleaning off the dust, he saw the glints of embedded crystalline inclusions. The 'Genesis Rock' is made of anorthosite and contains a distinctive type of rock called KreeP, which is found only on the moon. KreeP rocks have high abundances of potassium (K), rare earth elements (ree), and phosphorus (P). The rock's makeup showed that the outer portion of the Moon was completely molten about 3.9 billion years ago (**22**).

Other rocks from the Moon were found to be made of materials that were melted more recently than the 'Genesis Rock', probably by the impact of a meteor that formed a lunar crater. At the Shorty Crater near the Apollo 17 landing site, astronaut Harrison Schmitt, a professional geologist, discovered 'orange soil' containing volcanic glass from an explosive volcanic fire fountain that erupted 3.6 billion years ago.

Before lunar rock samples were available for analysis, theories of how the Moon was made were very disparate, compromised by insoluble riddles, and – in the words of

7

4 **Harrison Schmitt on the Moon** The flag sags on the windless mast, under a black sky with the distant planet Earth half illuminated.

5 **Footprint** During the Apollo 11 mission in 1969, astronaut Buzz Aldrin's boot made an impression on the dusty surface of the Moon as if it was snow. If this footprint could be revisited, it would appear the same, undisturbed on the lifeless, weatherless Moon, although over aeons of time micrometeorites striking the surface nearby will erase it with splashed soil.

6 **Alphonsus** The space probe Ranger 9 crashed on to the Moon's surface in 1965, sending back progressively more detailed photographs of the lunar geology, including this one of the crater Alphonsus (108 km in diameter) taken at an altitude of 360 km.

7 **Genesis rock** Astronaut David Scott noticed the lighter colour of this rock while exploring the Apollo 15 landing site in 1971 and photographed it in situ, with a colour calibration scale in the field of view.

8 **David Scott** Both astronaut and rock safely back on Earth, Scott looks at the Genesis rock under study in the Lunar Receiving Laboratory in Houston.

8

9

10

chemist Harold Urey – 'tended to prove that the Moon did not exist'. George Darwin (son of Charles Darwin), Don Wise and John O'Keefe suggested that the Moon was the Earth's 'daughter': that it was formed from material that had spun off from the Earth's mantle. Other planetologists, such as Gerard Kuiper, thought that the Earth and Moon originated as two 'sister' planets during the early formation of the Solar System. Yet others held that the Earth captured the Moon after it had formed somewhere else, either drawing it into orbit intact (Urey), or as a kind of 'Saturn's ring' of pieces that later coalesced into a sphere (Ernst Öpik).

In 1984, at a conference in Kona, Hawaii, a dramatic and radically different theory was adopted to explain the origins of the Moon, incorporating the information from the analysis of lunar rocks. This was the 'collision model', first mentioned in 1946 by Harvard geologist Reginald Daly and revived in 1975 by planetologists William Hartmann and Donald R. Davis and astrophysicist A. G. W. Cameron. They held that the Moon and the Earth both formed from the glancing collision of two planets: the proto-Earth and a planet the size of Mars called Theia (named after the titan in Greek mythology who

9 Lunar breccia Collected by astronaut Alan Bean from a small crater during the Apollo 12 mission in 1969, the rocks are highly compacted and heavily coated with glass ('It's sort of unusual; it's got a lot of those little droplets on it', said Bean as he noticed it). It was probably formed in a distant impact and ejected as a fragment that fell back to make a small secondary crater.

10 Copernicus The first close-up photo of the crater Copernicus was taken by Lunar Orbiter 2 in 1966 as it scanned the Moon's surface for suitable sites on which to land the Apollo missions. In the foreground are the central peaks of the crater and in the middle distance the ramparts of its inside walls. The crater is 100 km across and the central peaks rise 400 m above the crater floor.

11 Shorty In 1972 during Apollo 17, the last manned lunar mission, the Lunar Rover is parked on the rim of the crater, Shorty. Astronaut Harrison Schmitt is working beside the Lunar Rover. A few metres beyond the Rover to the right is the spot where, only minutes after this photograph was taken, his geologist's eye was caught by the ancient 'orange soil'. Soon afterwards, the Apollo 17 crew returned to Earth, the last people to have walked on the Moon.

12 Birth of the Moon In this simulation, Theia approaches the Earth and collides with it. Its core (yellow) merges with the core of the Earth, while its mantle (grey) is mostly stripped off and ejected to form the Moon.

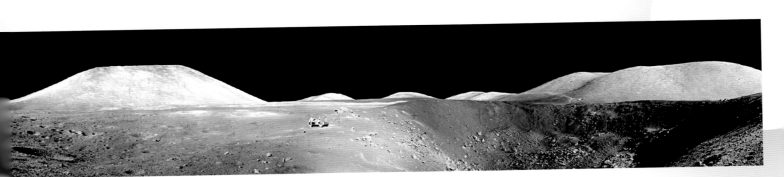

11

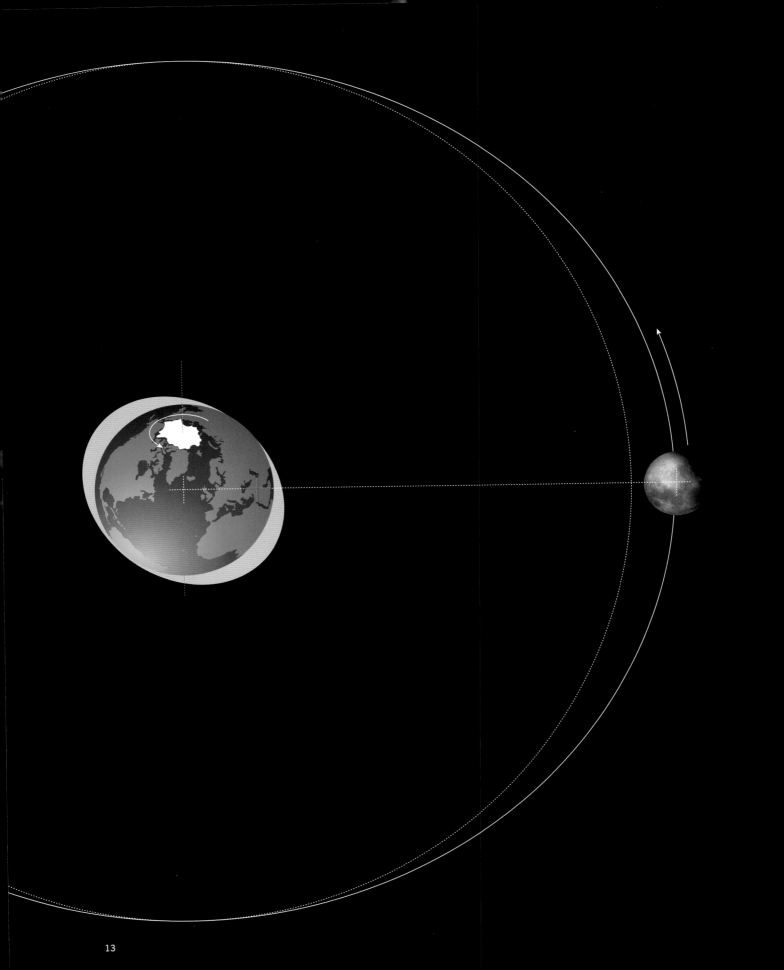

MOON TRAVEL

In all fairness to those who by training are not prepared to evaluate the fundamental difficulties of going from one planet to another, or even from the Earth to the Moon, it must be stated that there is not the slightest possibility of such journeys.

F. R. Moulton, 1935

gave birth to the moon-goddess Selene). During the collision, Theia and the proto-Earth broke up and reassembled as two new planets, the Earth and the Moon. Both Theia and the proto-Earth had molten iron cores, which coalesced to form the core of one of the new planets – the Earth – surrounded by a thin mantle. Fragments from the mantles of the two colliders coalesced into the other planet, now the Moon. This explains why some lunar rocks had a similar composition to rocks found on Earth, but others did not. The heat of the collision melted the mantle material that formed the Moon, generating KreeP rocks and encouraging volcanic activity. The high heat also vaporized all the water on the Moon, which is why it is so dry. The rapid rotation of the Earth and the tilt of the Moon's orbit were also consequences of the collision.

Without this violent freak event, life as we now know it would not exist on Earth. Compared to other satellites, the Moon is unusually large relative to the Earth and consequently its gravity is able to stabilize the Earth's rotation. As a result the Earth has seen less drastic changes in the tilt of its rotational axis than other planets and its climate has been unusually stable (**19, 23**). This has facilitated the evolution of life on Earth, especially the emergence of multicellular and mammalian life, which needs aeons of time to evolve. Moreover, the collision produced the large iron core of the Earth, which generates the strong and stable magnetic field that defends the Earth's atmosphere against erosion by the solar wind (**17**). It seems possible that without that collision 4 billion years ago there would be no human beings on Earth (**65**).

13 Lunar tides In this diagram, the Moon orbits the Earth (its axis under the Arctic ice cap tilted at 23.5º). The Moon's pull creates tides on the Earth and causes the axis to precess but also stabilizes the size of the tilt. It thus limits the changes to the global environment.

14 Astronaut David Scott Scott salutes his flag outside Falcon, the lunar lander of the Apollo 15 mission. His return journey will start in the small rocket that is the top half of the lander.

22. Mercury

The Late Heavy Bombardment

Even though Mercury is not far from the Earth, it is something of an enigma, as its close proximity to the Sun makes it difficult to observe. We do know that Mercury's surface, like the Moon's, was heavily bombarded by meteors at an early stage in the planet's development. But the cause of the bombardment – and what it might have done to the Earth – is still as much a mystery as Mercury itself.

2

1

1 **Caloris basin** The Caloris basin is one of the biggest impact craters in the Solar System. In this colour-coded image by the Messenger probe, the basin is the large yellow patch, pimpled with more recent impact craters, coloured blue. The southern (bottom) rim is surrounded by orange spots, which are lava fields from volcanoes triggered by the impact.

2 **Mercury** Mariner 10 images have been stitched together and mounted on a sphere to provide this global view.

3 **Ice on Mercury** Radar images by the Very Large Array radar telescope, made in 1991, show a bright region at the north pole (in red) like similar regions at the poles of Mars and some moons of Jupiter. It is a deposit of very cold water ice. Other red areas in this picture are regions that are particularly rough.

4 **Mariner 10** The exploratory craft that made the first reconnaissance of Mercury shelters under a flat sunshade to reflect solar heat.

Mercury is the smallest planet, not much bigger than the Moon. It is the closest planet to the Sun, which it orbits in 88 days. Greek astronomers at first had two names for Mercury, as they thought it was two separate planets, calling it 'Apollo' when it was to the east of the Sun and 'Hermes' when it was to the west. It was Pythagoras who realized that the two bodies were actually the same planet.

Even though Mercury is relatively near to the Earth, its close proximity to the Sun makes it very difficult to see using conventional telescopes. As Mercury is never more than 28° from the Sun, it is never high in the sky and can only be viewed from the Earth in the twilight. Nor is observation from space any easier. The Hubble Space Telescope is not permitted to view Mercury directly, as the Sun's heat and light could damage it. For much the same reasons, it is difficult to design a space probe capable of visiting Mercury. A spacecraft will quickly overheat as it approaches so close to the Sun, and will be peppered by storms of solar particles. Furthermore, as the spacecraft drops towards the Sun it will pick up speed as it is drawn down by the Sun's strong gravitational field, and can overshoot Mercury. This must be countered by burning large amounts of fuel or by approaching Venus in just the right direction, at just the right time in the orbits of both planets, so that Venus's gravity can be used to slow the spacecraft down.

Because of these constraints Mercury is one of the least-studied planets. Radar has proved a useful tool, but its capabilities are limited at such a distance. In 1965 it was used to show that Mercury turns exactly three times on its axis for every two orbits around the Sun. A solar day on Mercury (sunrise to sunrise) therefore lasts two Mercury years or 176 Earth days.

The first space probe successfully to visit Mercury was Mariner 10 in 1974–75, but it was more than 30 years before the second space visit was accomplished by the Messenger probe in 2008–9. Mariner 10 discovered that Mercury suffers extreme fluctuations of temperature, ranging from 90 °K (-183 °C) to 700 °K (427 °C), as the atmosphere is too thin to provide an insulating effect. The floors of some deep craters near the poles never see direct sunlight and never warm above 112 °K. There are patches in these craters that seem to reflect radar pulses in the same way that ice does. If there is indeed ice on Mercury, it may have been deposited by melting comets: a large comet impact on Mercury's surface could have generated a temporary steamy atmosphere, which may have condensed and frozen in the dark, cold craters.

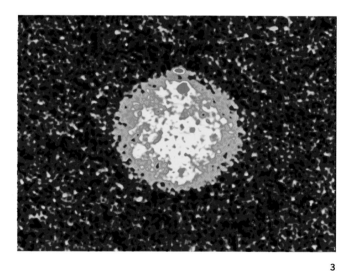

3

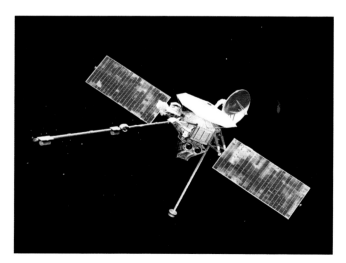

4

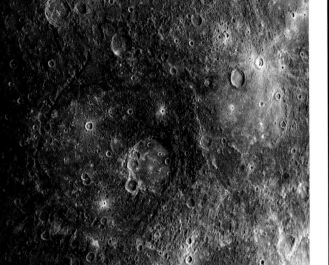

5

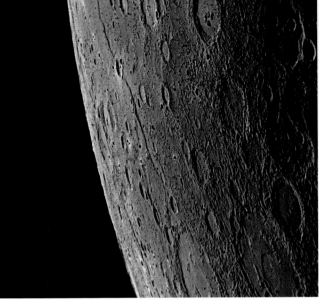

6

The thin 'atmosphere' is constantly being lost and replenished and is therefore properly termed an 'exosphere' rather than an atmosphere. Mercury's exosphere consists mainly of hydrogen and helium picked up from the Sun, but also contains less abundant atoms such as sodium and silicon that have been knocked off the surface of the planet by the solar wind, a process called 'sputtering'. A surprise discovery from Messenger is that the exosphere contains water, perhaps derived from the cometary ice at the poles.

Mercury's surface is heavily cratered, like the Moon's. In fact, the oldest surface areas of both bodies were cratered at the same time. To estimate the ages of the surfaces of different planets and satellites, astronomers count the number of meteor craters – the younger the surface, the fewer craters it has, especially large ones. These counts suggest that meteor impacts occurred more frequently throughout the Solar System during an early period in its history. This is confirmed by studies of rocks collected from the Moon by the Apollo astronauts and lunar meteorites collected on Earth (**21**).

Lunar meteorites are pieces of the Moon that were knocked off its surface by the impact of asteroids. None of them is older than 3.9 billion years. This posed a puzzle: given that the Moon was 4.6 billion years old, why were the oldest meteorites so much younger? Lunar rocks collected by Apollo astronauts showed that the crust of the Moon had been strongly heated 3.9 billion years ago. What had caused this catastrophic heating? In 1974–76 studies by

a number of planetologists, including Fouad Tera, Dimitri Papanastassiou, Gerald Wasserburg and Grenville Turner, suggested that 3.9 billion years ago asteroids and meteors had heavily bombarded the surface of the Moon for a discrete period of time (200 million years) and melted it. They called this event the 'lunar cataclysm'; it is also known as the Late Heavy Bombardment. The oldest parts of Mercury's surface were cratered during the same period.

No one knows why the bombardment occurred. There may have been a major collision between planets, or a disturbance in the outer Solar System that caused a rain of asteroids or Kuiper Belt Objects to fall inward towards the Sun. The focusing effect of the Sun's strong gravity meant that Mercury was pummelled especially heavily by infalling asteroids. A particularly large impact created the Caloris Basin, one of the largest craters in the Solar System, with a diameter of 1,550 km. As discovered by Messenger, the impact caused volcanoes to spring up around the rim of the crater. Mariner 10 had already discovered that shock waves from the impact travelled to the other side of the planet to create an unusual hilly region known as the 'Weird Terrain'.

Of course, if the Moon and Mercury suffered under the Late Heavy Bombardment, so did the Earth. The Bombardment produced about 1,700 craters on the Moon that were more than 20 km in diameter. Given the larger size of our planet, the bombardment would have generated ten times as many craters on Earth, some of which would have been as large

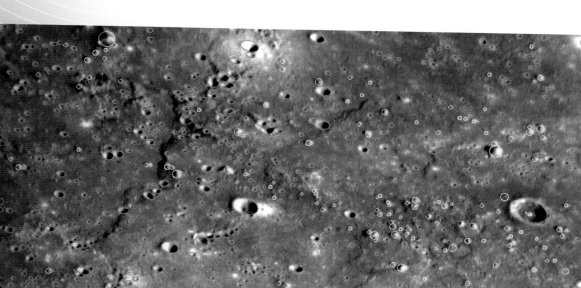

7

as 1,000 km in diameter. This scenario is supported by the discovery that extraterrestrial isotopes are especially abundant in deep sediments laid down in Greenland and Canada during the time of the Late Heavy Bombardment. It might also be significant that the fossil record of life on Earth seems to have started after 3.9 billion years ago. If life had begun to evolved before then, the bombardment would have interrupted the process and erased any earlier traces. Alternatively, the Late Heavy Bombardment may actually have triggered the evolution of life, bringing an abundance of organic molecules to the Earth on asteroids or comets (**18**).

5 **Mercury's crater history** The large smooth-floored crater near the centre of the image (from the Messenger probe) is Titian. The outline of a very large, eroded crater shows just above ground level, and smaller, sharper, more recent craters speckle the surface.

6 **Shrinkage crack on Mercury** Prominent towards the horizon in this view by the Messenger probe is a long cliff face, which extends for over 400 km. Mercury has a relatively large iron core and as it has cooled and shrunk over the aeons, the surface of the planet has cracked in numerous long faults.

7 **Crater counting** After Messenger had imaged previously unseen areas of Mercury's surface, astronomers draw up statistics of the craters on the surface in order to estimate the ages of the surfaces – the more craters, the older the surface. This image is 276 km wide and has 763 craters (green) along with 189 hills (yellow).

8 **Mercury** Mercury is a predominantly grey world, but the Messenger space probe used a camera with a more extended spectral sensitivity than the human eye to reveal subtle colour differences among its rocks. The Caloris basin is pinkish, visible on the upper right. Recent craters show as light colours, often light blue, with rays.

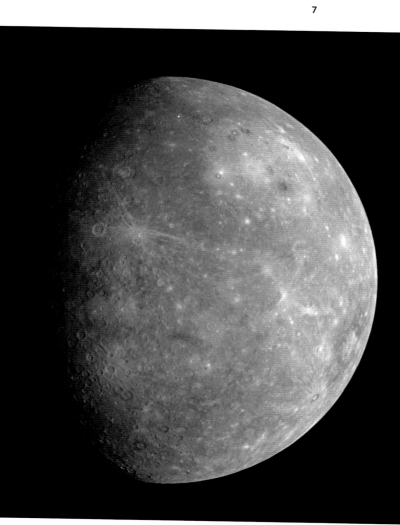

23

The Greenhouse Effect

Venus and the Earth

1

The greenhouse effect was discovered on the Earth and on Venus in parallel. It is a property of some of the gases in the the atmosphere that keeps the surface of the Earth at a comfortable temperature. The greenhouse effect is essential for life on Earth. However, man-made (anthropogenic) greenhouse gases threaten to upset its benign equilibrium. The surface of Venus is hot enough to melt lead. Could the same thing happen on Earth?

3

1   **Earth** Ever-changing cloud patterns over the brown soil of the continents (here Africa) would be an indication from space that our planet had an atmosphere that contained water vapour, evaporating from the blue seas.

2   **Venus** Dense cloud entirely obscures the surface of Venus, but its streakiness betrays the 'trade-wind' pattern of the circulation of its atmosphere.

3   **Carl Sagan** The planetary scientist and space visionary discovered the runaway greenhouse effect on Venus.

Venus is the second-nearest planet to the Sun, at about three quarters the distance of the Earth. It is Earth-like in its size and physical properties, such as its internal structure, solid surface, atmosphere and magnetic field. Seen through a telescope, Venus seems to have an almost uniform white surface. In 1761 the Russian scientist Mikhail Lomonosov observed a halo around Venus as it passed in front of the Sun and realized that the halo was scattered light radiating from Venus's upper atmosphere. The planet's white 'surface' was actually an opaque layer of thick white clouds.

Because it is highly reflective and close to both the Sun and the Earth, Venus is the brightest object in the sky after the Sun and the Moon. It was thus one of the first celestial objects whose spectrum (the bands of visible and invisible light it emits) was photographed. In 1932 Walter Adams and Theodore Dunham, Jr, imaged Venus's spectrum using the 100-inch telescope at the Mt Wilson Observatory in California and specially made emulsions supplied by C. Kenneth Mees of the Eastman–Kodak Company. They discovered that previously unknown lines in Venus's spectrum were generated by carbon dioxide, which proved that this gas was a major constituent of Venus's atmosphere.

The carbon-dioxide atmosphere of Venus had peculiar properties. Between 1923 and 1928 Edison Pettit and Seth

Nicholson measured infrared radiation that came from the tops of Venus's clouds and discovered that the temperature there was cold, ranging from -37 to -42 °C. In 1956 pioneer radio astronomer Cornell H. Mayer and his colleagues at the US Naval Research Laboratory measured the microwave radiation emitted from deep within Venus's atmosphere, much nearer to the ground. Mayer's team discovered that the surface of the planet had a very high temperature indeed, over 300 °C, hot enough to melt lead. This was a surprise: the dense and highly reflective atmosphere of Venus should have reduced the amount of solar radiation that was able to reach the planet's surface. In 1961 Cornell astronomer Carl Sagan put forward the currently accepted explanation for the high temperature: the atmosphere of Venus has a strong greenhouse effect, much stronger than the Earth's.

The French mathematician Joseph Fourier first noticed the greenhouse effect in the Earth's atmosphere in 1827. The Earth receives light from the Sun, of which about 70% is absorbed. This sunlight warms the land, atmosphere and oceans, which radiate energy back towards space as infrared light. But most of the infrared emitted from the surface does not escape into space; it is absorbed in the atmosphere by greenhouse gases and clouds. It heats the lower air, just as the glass roof of a greenhouse or the windows of a car allow sunlight in but trap the infrared light radiated by everything inside, causing the interior of the greenhouse or the car to become hot.

Throughout the 19th and 20th centuries, scientists gradually came to understand precisely how the greenhouse effect works on the Earth. In 1861 the Irish physicist John Tyndall discovered that water vapour and carbon dioxide are the most important greenhouse gases in the Earth's atmosphere. Astronomer Samuel Langley, director of Allegheny Observatory in Pittsburgh, Pennsylvania, used spectrum mapping in 1884 to show exactly how carbon dioxide and water vapour were opaque to infrared radiation. As Fourier and Tyndall had suspected, these two atmospheric gases allow incoming sunlight to pass through the atmosphere, but block much of the outgoing infrared radiation.

4

As they learned more about the properties of greenhouse gases, scientists began to wonder whether an increase in the amount of them in the atmosphere might raise the temperature of the Earth. In a series of publications from 1896 to 1908, the Swedish chemist Svante Arrhenius speculated that changes to the level of carbon dioxide in the Earth's atmosphere could alter its surface temperature through the greenhouse effect. Not only did geologically produced changes cause the Ice Ages, Arrhenius thought, but the burning of fossil fuels (such as coal) would produce anthropogenic global warming. Between 1928 and 1938, British meteorologists George Simpson and Guy Stewart Callendar finally succeeded in calculating the full extent of the greenhouse effect on Earth. By 1961 physicists Gilbert N. Plass and Lewis D. Kaplan were making increasingly realistic climate models for the Earth and the phrase 'greenhouse effect' began to be used in connection with global warming. In the mid-1960s the first conferences on global warming were held and the first official reports on global warming appeared.

Scientists were also applying the same ideas to other planets. Between 1908 and 1922, astronomers Frank Very, Charles G. Abbott and Edward A. Milne made various studies of the greenhouse effect in the atmospheres of Venus and Mars, on the basis existing knowledge regarding their compositions. Astronomer Rupert Wildt even speculated in 1937–40 that the surface temperature of Venus would be exceptionally high 'on account of the greenhouse effect of the carbon dioxide', estimating that large quantities of this gas could raise the surface temperature of the planet by up to 50 °C above what would otherwise be expected.

Compared to the Earth, the greenhouse effect on Venus is extreme, so much so that Carl Sagan suggested it had become unbalanced and run away with itself. Venus's atmosphere was originally carbon dioxide, but the greenhouse effect had warmed the planet so much that the high temperature changed the composition of its surface, generating more greenhouse gases in the atmosphere and raising the temperature still further. Accurate calculations were difficult in Sagan's day because too little was known about Venus in the 1960s. This gap in knowledge was a major impetus for the exploration of Venus by spacecraft.

The American Mariner fly-bys of Venus began in 1962 with Mariner 2. This mission confirmed the high temperature of Venus's surface and the high density of its atmosphere (93 times denser than the Earth's). In 1967 the Soviet Venera

5

4 **The greenhouse effect** Infrared and ultraviolet light are absorbed by the atmosphere but visible light itself reaches the ground and surface water and warms them. The surface radiates infrared, which is absorbed by the atmosphere. Greenhouse gases in the atmosphere include water vapour evaporating from seas, methane from decomposing vegetable matter and livestock, and carbon dioxide from forest fires, animals and human activities.

5 **Surface of Venus** This image from the Soviet Venera 13 lander in 1982 shows a hot, barren, rocky surface pressed down under a weighty and corrosive atmosphere.

6 **A volcano on Venus** Lava flows down the side of a Venusian volcano, imaged by radar techniques from NASA's Magellan spacecraft. Computer techniques used to generate this perspective view have exaggerated the heights of the surface terrain.

7 **Earthrise** Apollo astronauts photographed earthrise from their lunar orbiter, encapsulating the growing realization that our Earth is a planet with limited resources.

6

spacecraft series, initially designed by pioneering Soviet space engineer Sergei Korolyov, began their scientific explorations of Venus. The Venera missions were the first to venture onto the planet's surface, followed by American Pioneer probes. The spacecraft glided and parachuted through the atmosphere, measuring its properties, but they landed too fast to survive the impact. Later Venera missions between 1967 and 1982 successfully landed softly on the surface, as did two Russian Vega landers in 1985. These missions discovered that Venus's surface has a temperature of 740 °K (467 °C), compared to the Earth's 287 °K (14 °C). Its hot, dense atmosphere is indeed primarily made up of carbon dioxide. Clouds of sulphuric acid droplets, are thought to be responsible for the intensity of the present greenhouse effect on Venus. Other minor components include nitrogen and water vapour with hydrogen chloride and hydrogen fluoride. Like sulphuric acid, these last two substances are very powerful acids and the landers withstood their corrosive hot rain for only between a few minutes and two hours.

Owing to these severe technical difficulties, no landers have been sent to Venus since 1985. However, it has become possible to maintain orbiters around Venus for longer durations. The Magellan spacecraft lasted four years, from 1990 to 1994, and mapped the surface of Venus with radar that penetrated through the clouds. The space exploration of Venus continues with European Space Agency's Venus Express, which arrived at Venus in 2005.

Since 1827, when Fourier first identified the greenhouse effect on Earth, carbon dioxide in the Earth's atmosphere has increased by about a third due to man-made emissions, principally from industrial processes; an increase of the amount of methane generated by modern farming practices augments the change. A doubling of the amount of carbon dioxide by anthropogenic means has the potential to increase the temperature by 3 to 5 °C, causing serious global climate change. In the absence of man-made emissions, the moderate greenhouse effect on the Earth raises the planet's overall temperature by 33 °K. However, the runaway greenhouse effect on Venus raises its temperature by 500 °K, completely changing the planet's atmosphere and its climate. Venus is a horrific vision of what catastrophic climate change could be like on Earth.

7

24.

Mars
The dying planet

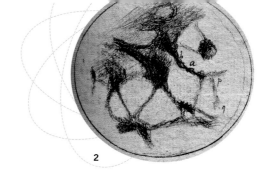

Why did Mars – the planet in the Solar System most similar to the Earth – fail to sustain life? Early science-fiction writers portrayed Mars as a dying planet inhabited by desperate warlike aliens. We now know that Mars suffered a global climatic catastrophe early in its development.

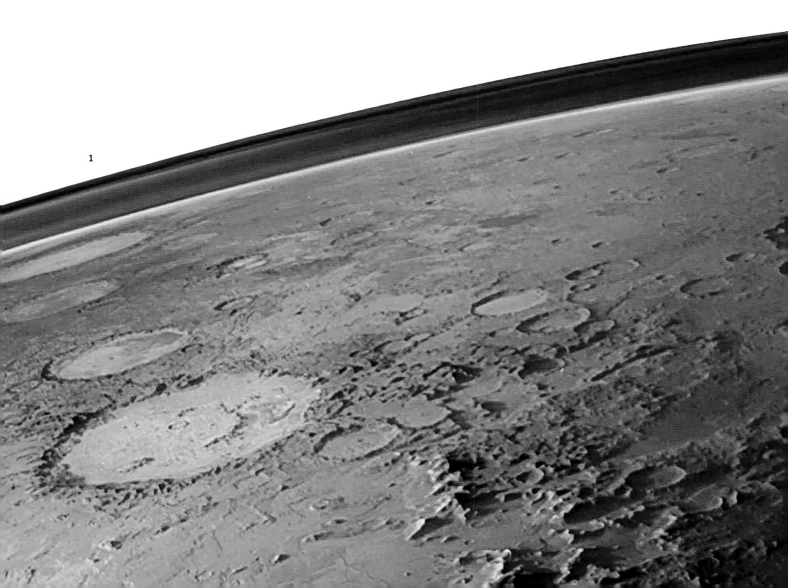

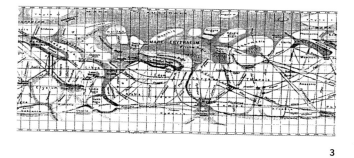

3

1 **Mars pictured by the Viking Orbiter in 1976** The hazy red lines over the horizon are due to dust in the Martian atmosphere.

2 **Mars as sketched by Giovanni Schiaparelli** In his observation notebook on 26 December 1879 he measured the positions of features, which he labelled with letters to transcribe onto a map (see Fig. 3).

3 **Giovanni Schiaparelli's map of Mars**, compiled between 1877 and 1886. Some of the names that he gave to surface markings are still in use.

4 **Percival Lowell's map of Mars** Lowell made Schiaparelli's 'canals' straighter and thinner, adding unwarranted credibility to the idea that they were artificial irrigation canals carrying water between darker 'cultivated' areas.

Mars is the red planet, the fourth from the Sun. It is smaller, colder and drier that the Earth and has a much thinner atmosphere. Galileo saw the disc of Mars with his telescope, but could not see its surface markings, which were discovered in 1659 by the Dutch astronomer Christiaan Huygens, who determined that Mars had a rotation period of about 24 hours – its days were almost the same length as the Earth's. In 1666 the Italian-French astronomer Gian Cassini discovered Mars's polar caps, which were assumed to be ice caps like the Earth's; 350 years later, space probes confirmed that the polar caps are deposits of ice and dry ice, 2 to 3 km thick.

In 1840 the German banker and amateur astronomer Wilhelm Beer and his colleague Johann H. von Madler made the first maps of Mars, showing dark areas that seemed variable in colour and intensity. Initially these dark areas were thought to be seas, but the French astronomer Emmanuel Liais suggested in 1860 that they could be large patches of vegetation, showing seasonal variations in colour. When Giovanni Schiaparelli mapped Mars in 1877, he labelled the dark patches as 'continents', 'islands' and 'bays', linked by numerous long, straight *canali* ('channels').

The *canali* led to speculation that Mars was inhabited by intelligent life. The American astronomer Percival Lowell (**13**) interpreted the *canali* as artificial canals, built as an irrigation system. Although other astronomers scorned Schiaparelli's overly detailed maps as works of fantasy, his vision of Mars took hold in science fiction and popular culture, where the planet was depicted as an old world, inhabited by warlike aliens looking to colonize the Earth because, depite their efforts at irrigation, their own world was dying. But Turkish-born French astronomer Eugenios Antoniadi proposed that that the canals were only psychological interpretations of faint, blotchy structures seen through the terrestrial atmosphere. In 1903, the Greenwich astronomer Edward Maunder used schoolboys as test subjects to demonstrate that a defective telescope causes an area with many point-like features (such as a group of craters) to appear as a network of lines.

During a fly-by mission in 1964, the American space probe Mariner 4 discovered that Mars's surface is indeed heavily cratered. Mars has no tectonic plates, so its surface is not regularly churned over in the same way as the Earth's, and weather erosion of the landscape is minimal because of the

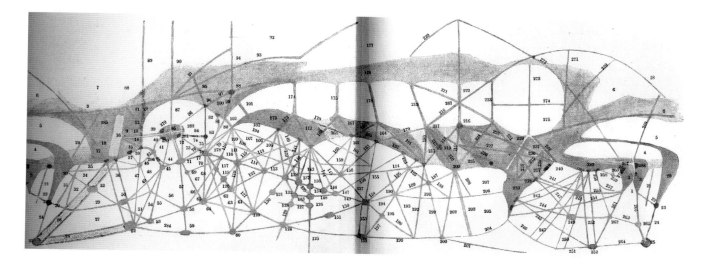

thin, dry atmosphere. Much of the crust of Mars is therefore very old; some terrains formed over 3.8 billion years ago and preserve traces of every subsequent meteoric bombardment.

There are a few newer terrains on Mars, chiefly ash-strewn volcanoes and lava flows. Mariner 9 discovered recent volcanic activity on Mars in 1971–72. The largest volcano is Olympus Mons, which at 24 km high is three times higher than Mt Everest. Other dramatic landscapes were shaped by water, including vast, now-dry flood plains and glacial features (**25**). Mariner 9 sent back pictures of the eponymous Valles Marineris, a huge canyon system 600 km wide and 7 km deep, which extends 4,000 km east–west along the Martian equator.

The Viking missions of 1976 were the first to land on Mars and to image its desert-like surface at close range. The landers looked for and failed to discover organic material in the soil; there did not seem to be life on the desert surface, although it is still possible that evidence may be found elsewhere on the planet. Mars Pathfinder and the Mars Rovers landed in 1997 and 2004, respectively, and Mars Global Surveyor (2001) and Mars Express (2003) have been mapping the surface at high resolution for several years, confirming that there were massive floods and glaciers on areas of Mars in the past and discovering that water and ice still produce changes on Mars (**25**).

5

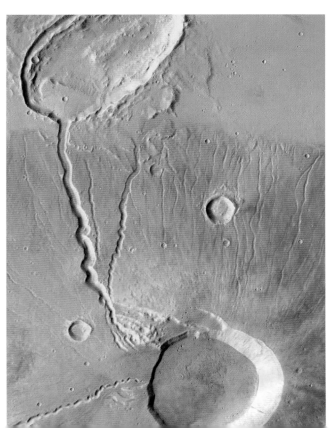

6, 7

The Martian atmosphere is made of carbon dioxide, nitrogen and argon. Because it is so thin, it is easy to see clouds, sand storms and seasonal exchanges of material between the polar caps. During the northern winter and southern summer, great dust storms sometimes cover virtually the whole planet. The orbit of Mars changes with time, causing the seasons and climatic cycles to vary dramatically. The Martian equivalents of Milanković cycles are therefore more extreme than those on Earth (20).

Mars has no magnetic field, but in 1999 Mars Global Surveyor discovered residual magnetism in old surface rocks in the southern hemisphere, laid down as they drifted over the convective iron core of the planet. At some time about 4 billion years ago the liquid core cooled enough to solidify, and the Martian dynamo died away.

Since Mars does not have a protective magnetosphere like the Earth's (17), its atmosphere is exposed directly to the solar wind and has gradually been stripped away, allowing ultraviolet light and solar particle radiation to reach the surface at levels that would be deadly to surface-dwelling life. The weak atmosphere means that the Martian air pressure is only 1% of the Earth's atmospheric pressure, too low for liquid water to remain liquid for long. Without a thick atmosphere to insulate it, the planet does not retain much of the heat it receives from the Sun, causing severe frosts at night, with temperatures plunging as low as -140 °C in the polar regions.

These discoveries suggest that Mars was once wet and warm (25). Its climate changed catastrophically when the planet lost its magnetic field and became the dry and sterile place that it is today. Yet it is still possible that life developed in the wet and warm era and survives in niche environments even now (65).

5   Sunset on Mars In 2005 the Mars Exploration Rover, 'Spirit', pictured sunset at the far rim of Gusev crater 80 km away. Because Mars is farther from the Sun than the Earth is, the Sun appears only about two-thirds the size that it appears from Earth. The Martian twilight lasts a couple of hours, a long time compared to Earth's, because of abundant high-altitude dust in the Martian atmosphere.

6   Martian volcanoes Ceraunius Tholus is a volcano in the Tharsis area of Mars, about the size of the Big Island of Hawaii. Several erosion valleys have been cut into its ashy flanks; the most prominent valley has drained into and partly filled an impact crater with ash and dust. The flanks of the volcano are peppered with meteor craters, showing that the volcano has not recently erupted and buried them.

7   Frost on Mars In 1976 Viking 2 landed at a northerly latitude (48° N) of Mars. An early-morning picture showed that a thin white deposit of frost had formed on the rock-strewn landscape overnight.

8   Storms on Mars Frosty water-ice clouds and orange dust storms show on this view of the planet in 2001, taken by the Hubble Space Telescope when Mars was at a close approach to Earth. A large storm system is in progress above the north polar cap (top), with a smaller dust storm nearby. Another large dust storm is spilling out of the Hellas impact basin in the southern hemisphere (lower right).

8

25

Water on Mars and Europa

Evidence for extraterrestrial life?

Water is the key ingredient for life on Earth. Wherever we find water in the Solar System, it is possible that we may also find evidence of life. 'Splosh' craters, traces of catastrophic floods, glaciers and channels carved by underground rivers suggest that life could at one time – or might even now – exist on Mars. But the most promising place for extraterrestrial life may be Jupiter's moon Europa, which contains more water than the Earth.

When Mariner 9 reached Mars in November 1971 it was the first spacecraft to enter into orbit around another planet. A Martian dust-storm was in progress at the time and all that could be seen in Mariner's first transmissions were the south pole and the tops of four high volcanoes. Controllers waited two anxious months before the atmosphere cleared and Mariner began photographing the surface of Mars. By the time it shut down in October 1972 the spacecraft had sent almost 7,000 images back to Earth.

Mariner's success paved the way for the two Viking missions that were launched in 1975. Upon reaching Mars, each spacecraft separated into a lander and an orbiter. The Viking Landers showed close-up images of a sandy, wind-blown desert, strewn with angular rocks that had fallen as debris from meteor craters. The Viking missions were followed by Mars Pathfinder in 1997, and in 2004 by the Spirit and Opportunity Rovers, which were mobile and able to range outside their immediate landing areas. Mars Phoenix Lander touched down in 2007 at the most northerly latitude of Mars yet explored. Meanwhile the Viking Orbiters mapped the surface in detail from a distance, a task later shared by Mars Global Surveyor (1997–2006), Mars Odyssey (2001–), Mars Express (2003–), and the Mars Reconnaissance Orbiter (2006–).

These spacecraft have discovered that the surface of Mars has wide-ranging networks of valleys. Some of these are dry riverbeds, which are the remains of an extinct drainage system. Unlike water courses on Earth, they seem to have no small streams and tributaries, only large rivers

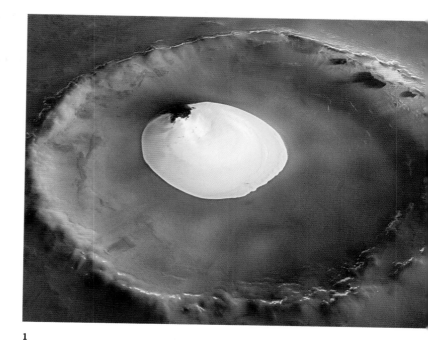

1

1 **Ice on Mars** In this photograph taken by the European Space Agency's Mars Express spacecraft, a field of ice lies on an elevated plateau inside an unnamed crater 35 km in diameter on the plain Vastitas Borealis, near the north pole of Mars. Frost coats the shadowed crater ramparts, which extend up to 2,000 m above the crater floor.

2 **The Martian landscape** A panoramic 360° view of Mars from the Mars Pathfinder Lander shows rocks of different colours ejected from different geological deposits. In the foreground are small, worn depressions 'gardened' from the soil by meteor strikes, while the walls of a crater called Twin Peaks stand above the horizon. In July 1997, the Sojourner Rover descended down a ramp to study the largest rock, called 'Yogi'.

2

that emerge full-size at their source. There are also numerous glacial features on Mars. It seems likely that the planet's riverbeds were not made by the runoff of rain, but were carved by ground-water flow: rivers flowing at first underground, then emerging from beneath icy glaciers that channelled the water. Other valleys were created when the permafrost above ground was melted by geothermal springs, causing the roof structures of subterranean rivers to collapse.

In addition to its rivers and springs, Mars also had massive floods that lasted for a matter of weeks, producing lakes and even seas. Mars Global Surveyor saw crusty, dried-up lake beds, and stepped cliffs on the interior walls of craters: platforms cut by waves, showing that the craters had once been filled with water. A startling discovery by Viking Orbiter was of crater-crowned 'islands' standing above a dry plain at Ares Vallis in the Chryse Planitia region. The lozenge shapes of these islands and the fact that their steep cliffs are between 400 and 600 m high suggests the islands were carved from impact craters by a flood of catastrophic proportions. Mars Pathfinder also saw rounded rocks and boulders of many different compositions that must have been deposited in their present location by floodwater.

Some of these floods were truly awesome in scale. In one case where a natural ice dam had collapsed, something like 100,000 km^3 of water was released in only a few days. In comparison, a typical flood on Earth is only a few km^3, and the largest known flood in the geological record of our planet released somewhere between 100 and 1,000 km^3 of water.

Where is this Martian water now? Some of it is frozen in the polar ice caps, but much of it could be hidden under the surface of the planet. At its cold, high-latitude landing site, the Mars Phoenix Lander exposed small quantities of water-ice, which lay millimetres under the surface, by scratching at the ground with a scoop. It is possible that more water-ice is present in a permafrost layer many metres deep.

Occasionally the ice melts. Some recent craters on the surface of Mars, like the one known as Yuty, are surrounded by outward-flowing lobes, which look like the petals of a flower – features not found on the Moon or Mercury. These craters are called 'splosh' craters, and seem to have been formed by projectiles that crashed into mud. This is further evidence that the subsurface soil of Mars may contain ice, which was melted into mud by the meteor's impact, flowed outwards and then re-solidified. Geothermal activity – that is, heat generated beneath the planet's crust – may also be responsible for some of the melting ice, which

3

4

5

6

3 **Drainage channels** The Viking spacecraft provided the first evidence that there had been water on Mars. In this image from the Southern Highlands, showing an area 160 km across, the Mars Reconnaissance Orbiter saw dry drainage channels, cut by water flowing from high ground and merging into a stronger stream, which cut deeper chasms.

4 **Spirit on Husband Hill** The Spirit Rover on the flank of Husband Hill in April 2005. Husband Hill lies in the Gusev crater on Mars and is a barren, windswept landscape of rocks, outcrops and sand dunes.

5 **Disappearing ice** A trench dug by the scoop of the Mars Phoenix Lander mission in June 2008 exposed water ice from below Mars's surface (white deposit). Within four Martian days ('sols'), some of the ice had disappeared. It had 'sublimated' (changed from solid to gas) when it was brought from below the surface of Mars and exposed to the thin air.

6 **Mars Phoenix Lander** The scoop of the Mars Phoenix Lander has dug into the soil near the north pole of Mars, and is poised to drop a sample into the spacecraft for analysis. On 31 July 2008, onboard instruments confirmed the presence of water in a sample of Martian soil. It is possible that microbes might live in the soil, below the surface and out of the drying and sterilizing effects of the Sun.

7

7 **Melas Chasma** Melas Chasma is a basin 1,200 m above the floor of Valles Marineris. It once was filled by a lake and now contains deposits left when the lake dried. This image from the Mars Odyssey Orbiter is colour-coded to identify the nature of the terrain. Swirling rusty streaks indicate hardened sediments that have been exposed by waves lapping at the edge of the lake.

8 **Recent water activity on Mars** In 2004 the operators of Mars Global Surveyor were startled to see changes to a crater in the Centauri Montes region, compared to images taken in 1999. The floor of a gully on the inside wall had been coated with a lighter deposit as meltwater from a spring flowed down the slope, fanning out as it reached the crater floor.

Mars Global Surveyor has seen sporadically staining the Martian ground as water drains from springs in cliffs.

The presence of water opens the possibility that life might have developed on Mars at some point in the past. In limited areas of the planet conditions exist that would favour its survival even now (65). Since 2004, astronomers using ground-based telescopes in Hawaii and Chile, and the European Space Agency's Mars Express satellite have seen methane in Mars atmosphere, which is released from several vents during the summer, when the ice melts. On Earth, methane is puffed into the atmosphere by volcanoes and deep ocean vents – but it is also produced by bacteria and animals. It is too early to say which process (if either) is responsible for the methane on Mars.

Mars may not be the only place in the Universe where life might once have existed, nor perhaps even the most likely place for astronomers to search for traces of life. There is water elsewhere in the Solar System – lots of it.

Jupiter's satellite Europa, which is about the size of Earth's Moon, was photographed by both the Voyager and Galileo spacecraft. Their images revealed that Europa has a completely spherical, nearly smooth surface with a grooved pattern that looks like crazy paving. Its surface is predominantly water ice, covered with icy plains. The grooves are cracks in the ice, where floes the size of cities have broken off from the main sheet. The floes are mobile because they float on water. Frozen 'puddles' of ice smooth over older cracks and warmer material bubbles up from below the surface. Evaporative salts tint the white ice a reddish-brown in some areas.

The ice layer of Europa could be a kilometre or more thick. The pressure of the heavy layer of floating ice combines with the radioactive and tidal heating of Europa's interior to liquidize the water beneath the ice. Europa has more water than the total amount found in the oceans on Earth.

A salty ocean warmed under an icy surface: if there is indeed life anywhere else in the Solar System, it could well be here on Europa (65).

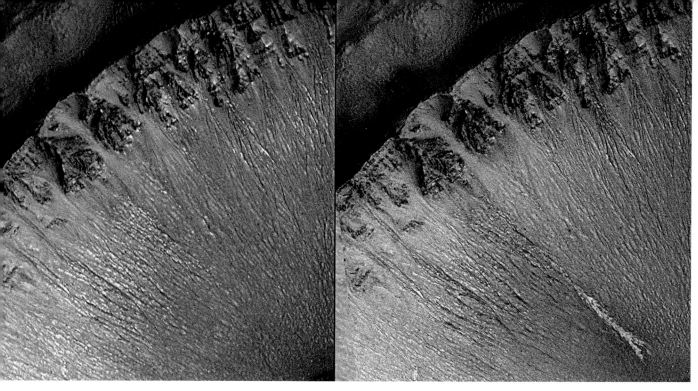

9 **Europa** Jupiter's moon is covered with cracked ice floes. The ice becomes tinted (dark brown in this colour-enhanced image) by meteors and by evaporative salts from water that wells up through the cracks from an ocean underneath the ice. Large meteor strikes throw out fresh snow, but the snow soon loses its whiteness and only a few craters can be seen, like the one called Pwyll (the dark spot in a white 'splash' at the 5 o'clock position).

10 **Ice floes on Europa** Europa is covered with an ice sheet that cracks as it shifts under tidal forces from Jupiter. New sharp cracks appear at every tide, while old ice floes are shoved aside, the cracks becoming fuzzy as they are iced over again.

11 **Mars splosh crater** Yuty is a 'splosh' crater on Mars. The crater and the sunflower-like lobes that surround it were made in the icy soil when a meteor impacted the surface. Briefly melted mud surged outwards, and froze again, causing the 'splosh' effect.

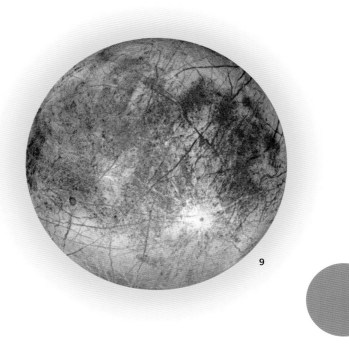

9

10

11

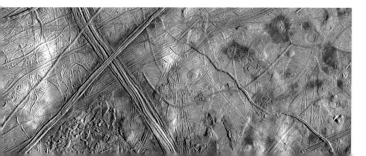

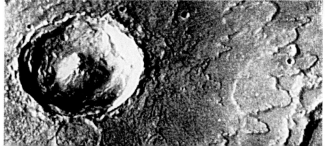

26. Volcanoes on Io

Linda Morabito makes a chance discovery

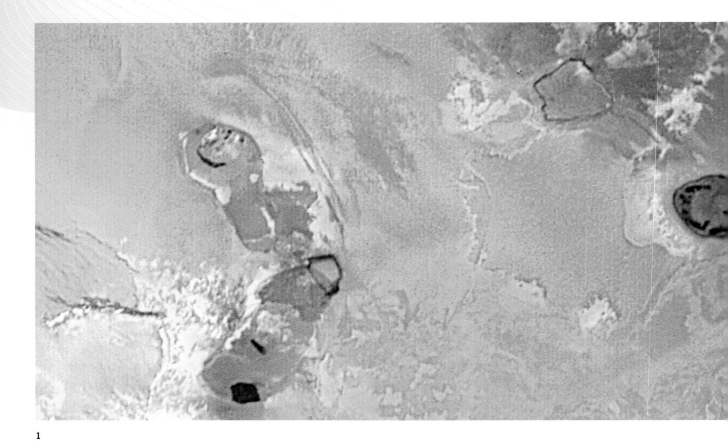

1

 'It was a moment that every astronomer, every planetary scientist lives for,' recalled Linda Morabito of the instant when she guessed what was causing a mysterious crescent-shaped blob to appear on photographs of one of Jupiter's moons. 'I had the sense that I was seeing something that no one else had seen before.' That evening at dinner, Morabito told her father what had happened. 'He looked at me and said: "Do you realize you may have discovered the first volcanic activity outside the Earth?"'

1 **Volcanic calderas on Io** This desolate volcanic landscape was imaged by Voyager 1 and shows a chain of calderas (left) in land covered by sulphurous deposits, with blue clouds of venting gas in the topmost caldera.

2 **Culann Patera** The Galileo spacecraft pictured this volcano on Io in 2002. The volcano (top half of the mosaic) has produced red and black lava flows, and yellow and green sulphur patches. White areas are sulphur dioxide snow. Tohil Patera (lower half) is an older volcano and rises 6 km high.

3 **Discovery of Io's active volcanoes** Morabito's discovery image shows the volcano Pele ejecting white ash clouds to a height of 250 km above the horizon. Loki's eruption is the intense white spot on the terminator.

4 **Linda Morabito** posing in front of a model of the Voyager spacecraft.

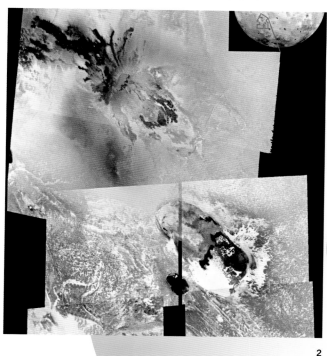

2

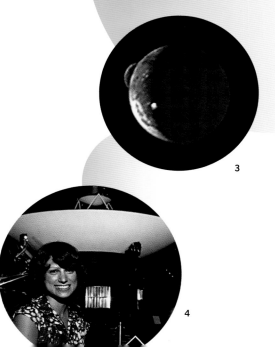

3

4

Jupiter's innermost satellite was discovered in 1610 by Galileo Galilei and named 'Io' by Simon Marius, after one of the Roman god's mythological lovers (**08**). Until the advent of large telescopes in the last years of the 19th century, Io remained a featureless point, its status as a world or moon 3,600 km in diameter (over a quarter the diameter of Earth) a matter of theoretical inference. In the 1890s, using the Lick Observatory telescopes at the University of California, Edward E. Barnard saw changes in Io's brightness that suggested the colour and reflectivity of its surface varied from area to area (equator and pole). In the 20th century, R. B. Minton was able to show that Io had reddish-brown poles and a yellow-white equator – its surface colours were very different from the other four ice-covered Galilean satellites. The prevailing theory was that Io's surface was covered with sodium and sulphur salts.

When the Pioneer space probes passed Jupiter in 1973–74, the mass of Io was calculated from the amount by which it deflected the spacecraft off course. It was clear that Io was denser than Jupiter's other satellites, with less ice and more rock. Pioneer 11 took a fuzzy picture as it flew over Io's pole, showing Io's yellow colour and some mysterious dark patches.

The next spacecraft to view Io were the Voyager 1 and 2 probes, which flew past Jupiter in 1979. On 5 March 1979, Voyager 1 was only 20,000 km above Io's surface and transmitted magnificent close-up images back to Earth. The surface was brightly coloured with reds, oranges and yellows. There were no meteor craters, showing that the moon's surface was young, and that some form of geological process had erased any craters that had formed earlier. However, Io had pits that resembled calderas, as well as mountains taller than any on Earth and what seemed to be lava flows. It looked like a volcanic landscape. At the press conference immediately after the encounter, the scientists made much of a multicoloured heart-shaped feature that they thought was indeed a volcano. Proof of this would be discovered a few days later, not by an experienced planetary geologist, but by a young navigation engineer processing routine data.

Linda Morabito had begun working at the Jet Propulsion Laboratories when she was a student at the University of Southern California, and by 1977 had been appointed a navigation engineer for Voyager 1. During the encounter with Jupiter, Morabito was working fourteen hours a day in the Voyager navigation area at the control centre where 'data was falling down on us like rainfall and the images were coming in at all hours of the day and night'. Morabito's immediate task during the encounter was to identify background stars in the images from the spacecraft and use them to determine its position so that its trajectory could be corrected in real time. Afterwards the images would be analysed to reconstruct the trajectory as accurately as possible. On the morning of 9 March, after the encounter with Io, Morabito set about this routine analysis. She began processing several images taken by the Voyager 1 spacecraft as it was looking back 'over its shoulder' for one last view of the Jovian system. One image, taken from a distance of 4.5 million km, had been put up on the monitors for everyone to see. Morabito 'stretched' the image – increased its contrast to look for a particular dim star – and noticed something that no one had been able to see on the ordinary picture: an 'anomaly' to the left of Io, just off the rim of the moon. The anomaly was crescent-shaped and extremely large relative to the overall size of Io.

Careful scientist that she was, Morabito first considered whether the anomaly might be caused by a piece of debris or a blemish on the camera. When these had been eliminated, only one possible explanation remained – the anomaly had something to do with Io. Io itself was overexposed on the image and it took some work to discover the exact area of Io's surface where the mysterious object had appeared. The area proved to be the large heart-shaped feature, now called Pele, which had been shown at the press conference. The anomaly was a plume from a volcano, an ash cloud rising more than 260 km above the satellite's surface; the heart-shaped feature was the volcano itself, with its bright orange and green slopes, yellow ejecta and black lava flows, surrounded by a ring of red

material ejected over the years and extending up to 600 km from the volcano. Nine such plumes were discovered later in footage from the same fly-by, and a second plume was actually visible in Morabito's discovery image, above a dark patch called Loki, where the volcanic cloud was high above the night side of Io, catching the rays of the rising Sun. The Loki volcano has been continuously active since it was first observed in 1979.

Current estimates suggest that there are at least fifteen active volcanoes among the hundred or so mountains on Io. The low gravity of Io allows volcanic plumes to reach as high as 500 km above the surface. Although the Galileo spacecraft was prevented from passing too close to Io for fear that the moon's hostile plasma environment would damage it, Galileo nevertheless saw ten plumes in 1998.

Why – uniquely among Jupiter's moons – does Io have volcanoes, and why does it have so many? According to calculations made in 1979 by Stan Peale, Patrick Cassen and R. T. Reynolds the interior rocks of Io are compressed and expanded by the gravity of Jupiter and its other moons. The resulting friction and pressure heat the rocks into liquid magma, triggering volcanic eruptions. The eruptions create pits and mountains on Io's surface, coating it with sulphurous deposits and covering meteor craters with lava flows hundreds of kilometres long. The scale of these flows is enormous: they contain hundreds of times the volumes of lava produced by volcanoes on Earth. Gases released in the eruptions give Io its thin atmosphere, and leak along Jupiter's magnetic field lines, producing striking aurorae.

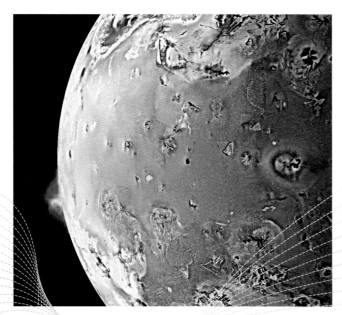

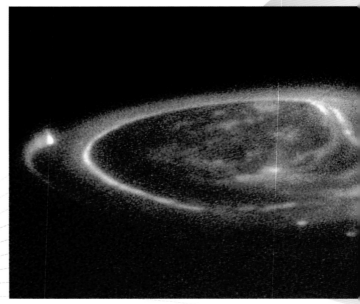

4

5

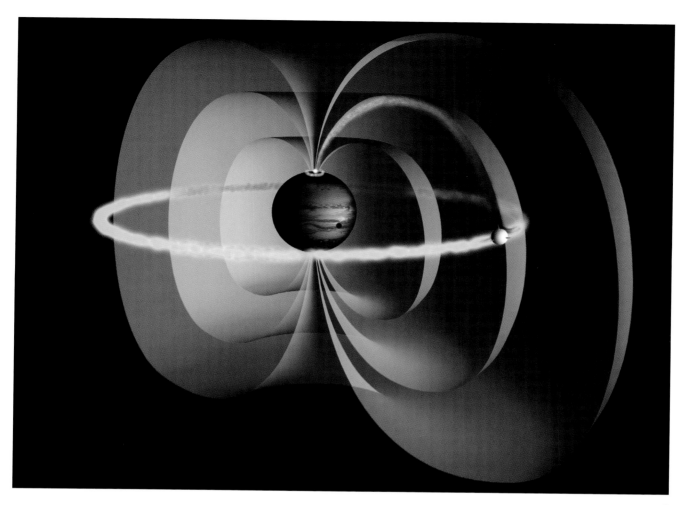

6

4 **Pillan Patera** A volcanic plume rises over a caldera on Io called Pillan Patera. The plume seen by the Galileo spacecraft in 1997 is 140 km high.

5 **Jovian aurora** The Hubble Space Telescope imaged aurorae that sparkled around Jupiter's north pole in 1998. The aurorae are mainly confined to a kinked oval on the cloud tops. Bright spots ('auroral footprints') mark where the magnetic field leads from Jupiter's satellites to the top of Jupiter's atmosphere; here they stretch from Io (along the left edge), Ganymede (the bright spot well inside the oval), and Europa (just below and to the right of Ganymede's auroral footprint). The aurorae are produced by electrical particles from the satellites (especially Io's volcanic emissions).

6 **Jupiter's magnetosphere** Io and the other satellites orbit inside Jupiter's huge magnetosphere. Volcanic and other material trailing behind them gets caught by the magnetic field and drawn down to Jupiter's atmosphere, causing the magnificent aurorae.

7 **Tvashtar Catena** The chain of giant volcanic calderas was active in 1999 when it was pictured by the Galileo spacecraft. Hot black lava (extreme left) has flowed in two rivulets, and is being fed into bright, hot areas, which are 60 km long. New, dark volcanic ash has been strewn over the older orange and yellow deposits on the surface of Io in the left half of the picture.

27. Saturn and the Gas Giant Planets

Lords of the rings

The four outermost planets in the Solar System – Jupiter, Saturn, Uranus and Neptune – are strikingly different from the terrestrial planets that are closer to the Sun. They are exceptionally large and made mostly of gas and ice rather than rock and dust. When Galileo first saw Saturn through his telescope, he was astounded to find that the planet had 'handles'. The 'handles' were, of course, Saturn's rings. Recent discoveries reveal that all the gas giant planets have these peculiar systems of rings. But where did the rings come from?

1

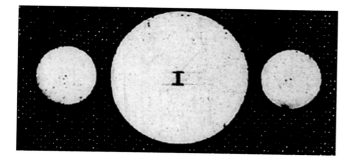

2

1 **The Rings of Saturn** The 'ring' around Saturn is in fact many rings of particles. Saturn's satellites control where the particles can orbit, and shepherd errant particles out of the gaps. Somehow the particles have been sorted into rings by size. In this image, green rings have smaller particles (1 cm or less) than purple (from 5 cm to several metres across).

2 **Galileo's interpretation** In 1610 Galileo described the rings as 'handles' or large moons.

3 **Christiaan Huygens** The Dutch astronomer and physicist also invented the pendulum clock.

4 **Saturn's rings, unresolved** Christiaan Huygens collected representations of Saturn's rings prior to his own discovery of their nature. This image is from Huygens's 1659 book *Systema Saturnium* ('Systems of Saturn').

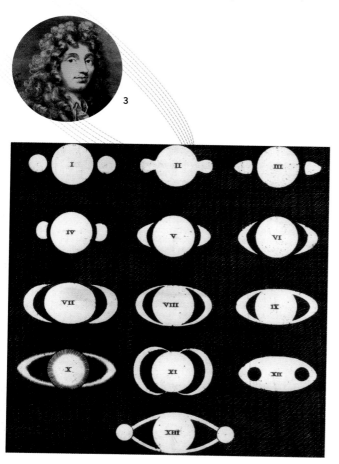

3

4

Jupiter, Saturn, Uranus and Neptune have always been much larger than the rest of the planets in the Solar System. They originally formed from parts of the solar nebula that were too distant from the Sun for ices to be vaporized by its heat. The gas giants consequently retained their icy material; this is why they are much larger than the inner terrestrial planets, whose ices were 'cooked' away by the newly formed Sun, leaving only small quantities of rock and dust to form into planets.

Each of the gas giants, most visibly Jupiter, acted as a centre for the formation of a miniature planetary system of small planet-like moons. In addition to its satellites, each also has a system of planetary rings, discs of particles that follow nearly circular orbits near its equatorial plane. Saturn's distinctive rings are well known and can be seen with a good pair of binoculars, but the rings of the rest of the gas giant planets were only discovered in the last decades of the 20th century.

When Galileo turned his telescope on Saturn, his equipment was not powerful enough to reveal the true nature of the rings. In 1610 he described what he saw as *ansae* (Latin for 'handles'), which he interpreted as large moons. 'I have observed [Saturn] to be triple-bodied. To my very great amazement I saw that Saturn is not a single star, but three together, which almost touch each other.' Two years later the 'moons' had disappeared. 'I do not know what to say about so surprising a case, so unexpected and so novel,' he exclaimed. Saturn's rings were edge-on to the Earth at the time, and could not easily be seen. In 1616 he saw a complex shape, because his telescope had improved and the aspect of the rings had again changed. 'The two companions are no longer two small perfectly round globes...but are much larger and no longer round...there are two half ellipses with two little dark triangles in the middle, each contiguous to the middle globe of Saturn, which is always perfectly round.' Galileo was unable to offer a solution to the puzzle.

In 1655 Dutch astronomer Christiaan Huygens took an interest in Saturn and discovered its largest moon, Titan, which lined up with Saturn and its handles. Huygens and his brother produced a more powerful telescope, and in February 1656 Huygens saw that the handles were in fact a ring around the planet. To establish precedence for his discovery, while buying himself time for further study of the ring, Huygens inserted a cryptic remark in his book about Titan, *De Saturni luna observatio nova* ('New Observation of a Moon of Saturn'), that he had found an explanation for the *ansae*, inviting anyone else in the know to step forward. He then described his discovery using an anagram (into which he

5

5   **Neptune's rings** Voyager 2 imaged Neptune's three rings in 1989.

6   **Uranus's rings** Voyager 2 looked towards the Sun to see the rings of Uranus, and beyond them stars trailed by the exposure of the picture. In this light, the rings merge into a continuous disc of fine dust.

7   **Saturn's rings** The Cassini spacecraft pictured the rings in natural colour in 2007. The rings are labelled on the bottom edge, the named gaps on the top, and the distance from the F ring to the D ring is 70,000 km.

6

had put minimal creative effort): *aaaaaaa ccccc d eeeee g h iiiiiii llll mm nnnnnnnnn oooo pp q rr s ttttt uuuuu*, which he later revealed meant: 'Annulo cingitur, tenui, plano, nusquam cohaerente, ad eclipticam inclinato' ('it is surrounded by a thin flat ring, nowhere touching, and inclined to the ecliptic').

In 1787 the French mathematician Pierre-Simon Laplace suggested that Saturn's ring was a set of thin solid ringlets, because a single solid ring could not orbit around the planet. However, in 1849, the French scientist Édouard Roche calculated that if any solid satellite was in orbit too close to its planet, it would break up under the tidal forces that the planet exerts. The 'Roche limit', within which a solid satellite cannot survive, is 2.44 times the radius of the planet, 'a little farther than the external radius of Saturn's rings'. This made it unlikely that Saturn's ring could be a solid structure. In 1857 the Scottish physicist James Clerk Maxwell showed that 'the only system of rings which can exist is one composed of an

indefinite number of unconnected particles, revolving around the planet with different velocities according to their respective distances.' This was confirmed in 1895 when Allegheny Observatory director James Keeler found that the inner part of the ring was orbiting Saturn faster than the outer part.

As telescopes improved in the 17th and 18th centuries, astronomers saw that Saturn's ring was in fact a series of rings. In 1675 Cassini saw two main rings, which were later named 'A' and 'B', separated by a gap now called the Cassini Division. In 1850 the Harvard Observatory father-and-son team William and George Bond discovered a third ring inside A and B, called the 'Crepe' or C ring. Keeler discovered in 1888 a 325-km gap in the A ring, which he named the Encke Gap after the German astronomer Johann Encke, who had studied the rings. The gap is actually the orbital path of a small moon, Pan, which sweeps aside or accretes the dust of the rings as it orbits Saturn. The Pioneer 11, Voyager and Cassini

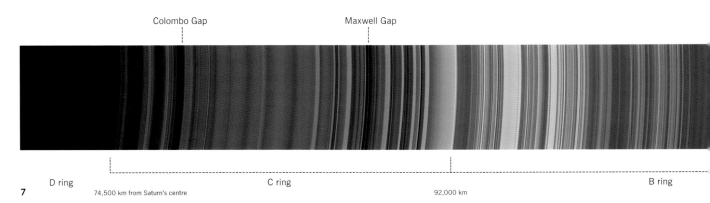

Colombo Gap   Maxwell Gap

D ring   C ring   B ring

7   74,500 km from Saturn's centre   92,000 km

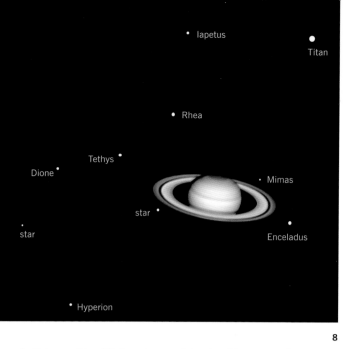

Iapetus
Titan
Rhea
Tethys
Dione
star · Mimas
star
Enceladus
Hyperion

8

8 Saturn and ten of its largest moons Saturn has 60 moons, or, if you counted the individual particles in the rings, a near-infinite number.

spacecraft discovered further, thinner rings around Saturn, and a marvellously detailed structure within the rings themselves. The total mass of Saturn's rings is approximately equivalent to the mass of its satellite Mimas, which fits with the theory that the rings are pulverized remains of a former moon. They consist of fragments 1 cm to 5 m in size, made primarily of water ice, or of micron-sized particles. The beautiful structure of the rings is the result of the complex pulls of Saturn's larger satellites which, like Pan, act as 'shepherds', ushering particles out of some zones to leave gaps.

The planet Uranus has more than ten rings. The first of these rings was discovered when Uranus passed across a star by chance in 1977. The telescopes of the Gerard P. Kuiper Airborne Observatory (a modified C-141A jet transport aircraft flying at stratospheric altitudes) were preparing to observe the way that the star's light faded as it passed behind Uranus's atmosphere. But before the planet reached its predicted

position in front of the star, the starlight unexpectedly dimmed. It had been temporarily blocked by Uranus's planetary rings. After the star exited from behind the planet, it was blocked again by the other side of the rings, in reverse order. The Voyager 2 spacecraft directly imaged Uranus's rings in 1986.

After the discovery of Uranus's rings, it seemed likely that Neptune might also have rings. Using the same technique, astronomers carefully observed stars for signs of premature dimming as they were occulted by Neptune, and in the 1980s found the evidence they were looking for. To confirm this discovery, NASA engineers reprogrammed the orbit of Voyager 2 as it approached Neptune in 1989, both to learn more about the ring system and also to avoid the risk that dust from the rings would endanger the spacecraft. The Voyager images showed that the planet has four, perhaps five, dusty narrow rings, which have since been named Adams, Le Verrier, Lassell, Arago and Galle for the astronomers involved in the discovery of Neptune (**11**). The Adams ring is incomplete and has three main arcs named Liberté, Egalité and Fraternité.

Jupiter's rings were discovered by the Voyager spacecraft in 1979, and further studied by the Galileo mission. They are very thin and faint but have been imaged by the Hubble Space Telescope.

Rings of particles and debris do not orbit a planet forever, but fade away, relatively quickly compared to the age of the Solar System. Saturn's rings may have been caused by the break-up of a 200-km satellite less than 500 million years ago; other planets' rings formed from fragments of small icy satellites or captured comets, and some are periodically replenished by sprays of dust from meteoroid impacts on rocky satellites. Our own planet almost certainly had a ring system at least once in its history, after the Moon was formed (**22**). The rings would have presented a beautiful spectacle in the sky, whether they were seen only by the uncomprehending eyes of dinosaurs, or no creatures had yet evolved that were able to see them at all.

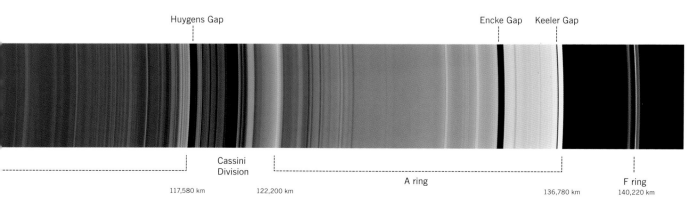

Huygens Gap
Encke Gap   Keeler Gap

Cassini Division
A ring
F ring

117,580 km   122,200 km   136,780 km   140,220 km

Discoveries of the Dynamic Universe

28. Helium

The cosmic element

1 H																	2 He
3 Li	4 Be											5 B	6 C	7 N	8 O	9 F	10 Ne
11 Na	12 Mg											13 Al	14 Si	15 P	16 S	17 Cl	18 Ar
19 K	20 Ca	21 Sc	22 Ti	23 V	24 Cr	25 Mn	26 Fe	27 Co	28 Ni	29 Cu	30 Zn	31 Ga	32 Ge	33 As	34 Se	35 Br	36 Kr
37 Rb	38 Sr	39 Y	40 Zr	41 Nb	42 Mo	43 Tc	44 Ru	45 Rh	46 Pd	47 Ag	48 Cd	49 In	50 Sn	51 Sb	52 Te	53 I	54 Xe
55 Cs	56 Ba	71 Lu	72 Hf	73 Ta	74 W	75 Re	76 Os	77 Ir	78 Pt	79 Au	80 Hg	81 Ti	82 Pb	83 Bi	84 Po	85 At	86 Rn
87 Fr	88 Ra	103 Lr	104 Rf	105 Db	106 Sg	107 Bh	108 Hs	109 Mt	110	111	112	113	114	115	116	117	118

57 La	58 Ce	59 Pr	60 Nd	61 Pm	62 Sm	63 Eu	64 Gd	65 Tb	66 Dy	67 Ho	68 Er	69 Tm	70 Yb
89 Ac	90 Th	91 Pa	92 U	93 Np	94 Pu	95 Am	96 Cm	97 Bk	98 Cf	99 Es	100 Fm	101 Md	102 No

1

The discovery of helium in 1868 was a transformative moment for chemists and astronomers. The revelation came as a faint yellow line of light, observed during an eclipse of the Sun. It was emitted by a major ingredient in the makeup of stars, which was later found also to be an important building block of substances on Earth. Copernicus was right: the Earth and the heavens were made of the same basic materials.

2

1   **Periodic Table in its modern form**

2   **Original Periodic Table** A draft of the earliest version by Dmitri Mendeleev
    (dated 17 February 1869).

3   **Joseph von Fraunhofer** Optician, physicist, astronomer.

3

In the Middle Ages, astrology – the arrangement of the planets in the zodiac – was part of the study of alchemy, a primitive form of chemistry whose original purpose was to turn base elements into precious metals. The relationships that alchemists perceived between planets and chemicals were reflected in the old names for certain metals. Quicksilver is still called mercury but other cosmic names have fallen into disuse: copper was once known as Venus, iron as Mars, tin as Jupiter, lead as Saturn, gold as the Sun, and silver as the Moon. As recently as a century ago, household pipes were still stamped with the astrological symbol for the planet Saturn to indicate that they were made of lead.

Eventually alchemy evolved into the modern science of chemistry. The French chemist Antoine Lavoisier, in a series of experiments conducted between 1782 and 1789, discovered by carefully weighing his chemical compounds before and after they participated in reactions that some chemicals were never broken down into lighter ones. He called these chemicals 'elements', publishing in 1789 a list of 33 elements (although not a list modern chemists would entirely agree with). Through the work of 19th-century Russian chemist Dmitri Mendeleev, this list evolved into a precursor of the modern Periodic Table, with the known elements grouped into columns and rows according to their observed chemical properties (which we now know are dictated by their atomic structures).

The empirical arrangement of the elements in the early Periodic Table left a number of empty holes, sparking a search for the missing elements, many of which were subsequently named for celestial objects, reflecting the traditional association of chemicals with planets. Uranium was named for the planet Uranus in the 18th century; a few years later, palladium and cerium were named after the recently discovered asteroids Pallas and Ceres. Neptunium and plutonium were named for the planets Neptune and Pluto, tellurium from the Greek word for the Earth, and selenium for the Moon. Some of the celestial names that were given to the new elements are no longer used, including aldebarium and cassiopeium, names derived from the star Aldebaran and the constellation Cassiopeia, for the elements now called ytterbium and lutecium. Denebium (from the star Deneb) was a name given to a rare earth element whose existence was later disproved.

Such associations between newly discovered elements and the cosmos were only products of a poetic and fanciful naming convention. The first clues that the terrestrial elements were actually to be found elsewhere in the cosmos came from the spectrum of sunlight.

4

5

6

7

In 1802 William Wollaston discovered that the spectrum of sunlight had seven gaps, which he regarded as boundaries between the natural colours of the spectrum. But in 1814 the German optician Joseph Fraunhofer invented a spectroscope with superior resolution and discovered not seven but hundreds of gaps in the solar spectrum. The gaps are now known as the 'Fraunhofer lines'. Fraunhofer accurately measured their wavelengths (as an aid to making accurate optical instruments) and labelled the more prominent gaps with letters. The German chemists Robert Bunsen and Gustav Kirchhoff discovered that many of the Fraunhofer lines represented light that had the same wavelength as the light that materials emitted when they were heated and vaporized. For example, the Fraunhofer D-lines in the solar spectrum were identical with the yellow sodium emission from salt. This suggested that sodium was a component of the material heated in the atmosphere of the Sun.

By the end of the 1880s, spectral emissions from 50 of the then known elements had been discovered in the solar spectrum. This proved that the Sun was made of similar elements to the Earth. Applying spectroscopy to the brighter stars, Henry Draper and William Huggins showed that stars also had dark lines in their spectra. The spectroscopists Father Angelo Secchi, H. C. Vogel and Edward Charles Pickering developed schemes for classifying the spectra of stars and listed the Fraunhofer and other spectral lines that were found in each. The same elements that had been found on the Earth were present not only in the Sun, but also in the stars.

4   **Norman Lockyer** Solar astronomer, founder of the scientific journal *Nature* and co-discoverer of helium.

5   **William Huggins** Merchant and astronomer, photographed in his observatory with his refracting telescope and spectroscope.

6   **Huggins's observatory at his house in Tulse Hill, London** William Huggins's wife, Margaret Lindsay Huggins, is standing in the dappled shadow of the trees, at the base of the observatory building, near the double doors.

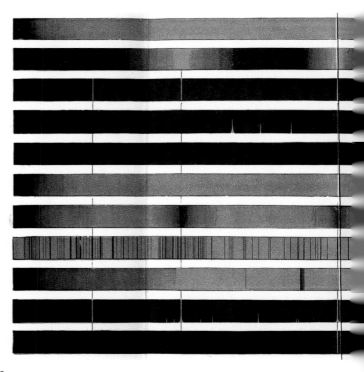

A FAMOUSLY INACCURATE FORECAST
ABOUT THE STARS

On the subject of stars, all investigations which are not ultimately reducible to simple visual observations are…necessarily denied to us. While we can conceive of the possibility of determining their shapes, their sizes, and their motions, we shall never be able by any means to study their chemical composition or their mineralogical structure.

Auguste Comte, *Cours de la philosophie positive*, 1835

7   **Jules Janssen** Astronomer, co-discoverer of helium.

8   **Astronomical spectra** Norman Lockyer compiled this montage to show the appearance of the spectra of a hot solid body (top), laboratory gases such as hydrogen and sodium, stars (Vega, Alpha (α) Cygni, the Sun, Alpha (α) Orionis), and nebulae (the Cat's Eye and Orion nebulae). He traced how some spectral lines, like the pair of sodium lines (central in the yellow area of the spectrum) appeared in the spectra of many stars. By 1870 he had proved that there were three lines at almost the same position in every stellar spectrum: two from sodium and one from the new element, helium. This is the frontispiece from Lockyer's textbook, *Elementary Lessons in Astronomy*, first published in 1864.

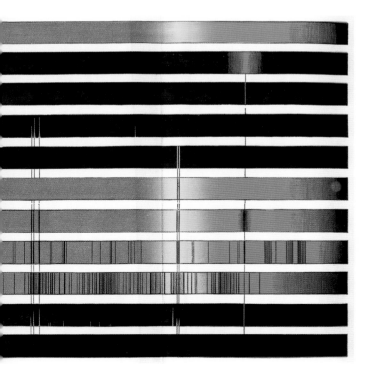

In 1868–69 there was another dramatic development. Astronomers Norman Lockyer and Jules Janssen observed the solar chromosphere (the Sun's denser, lower atmosphere) during the total solar eclipse of 1868 and discovered a strong spectral emission at a wavelength near to the sodium D-lines. To make it practical to measure the wavelength of this light, Lockyer and Janssen developed a technique for viewing the spectrum of the chromosphere in the absence of a solar eclipse. Measured at leisure in the solar observatory, the wavelength of the light emitted by the Sun's lower atmosphere was proved to be different from the sodium D-lines. In fact, the light did not correspond to the emissions from any of the then known elements.

Janssen and Lockyer realized that they had discovered a previously unknown element. Lockyer named it 'helium', after the Greek word for the Sun, *helios*. This element was isolated on Earth in 1895 by the Scottish chemist William Ramsay as he studied radioactive minerals that give off helium as they decay. Helium is thus unique among the elements in the periodic table, having been discovered in a cosmic object before it was identified on Earth. Most terrestrial helium is made by the radioactive decay of heavier elements on Earth – the same process Ramsay observed in his experiments – but most cosmic helium originated during the Big Bang (**59, 60**), or is generated inside stars (**46**).

In the wake of the discovery of helium came many similar claims for new elements, most of which turned out to be cases of mistaken identity. As conditions on the Sun are so different from conditions in a laboratory, its spectrum is easy to misread. New spectral lines were discovered in the chromospheric spectrum during the total solar eclipse of 1869, but the new element, 'coronium', which was invented to account for them, was shown in 1941 by the astronomer Walter Grottrian to be nothing more than iron heated to high temperatures and low densities. Similarly, some spectral lines in nebulae were once attributed to the element 'nebulium', but turned out to have been produced by oxygen and other common elements (**49**).

In both the discovery of helium and the rectification of spurious claims for other new elements, scientists used new astronomical discoveries to explain terrestrial phenomena, and vice versa. This was an extension of Copernicus's realization that the Earth was a planet like others: the material that the Earth is made of is the same as the material of the rest of the Universe, and the scientific laws that apply here are the same as those that apply everywhere.

29. Gravitation

Determinism and chaos

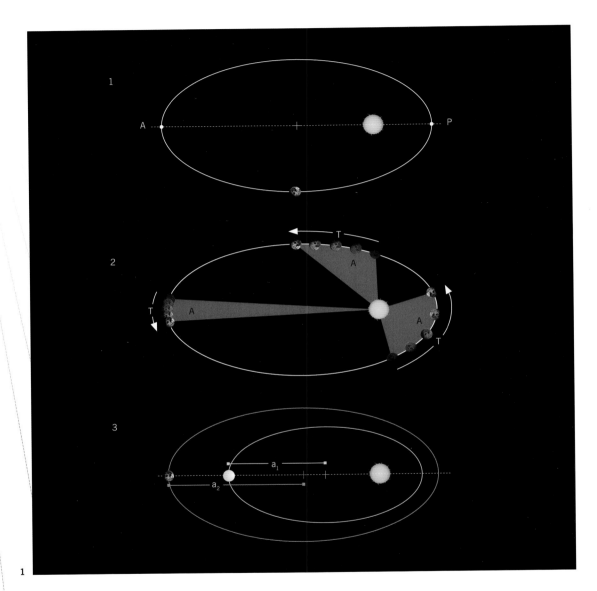

1

When the theory of gravitation emerged in the 17th century it seemed that mathematics had infinite power to see the future. But Newton could only predict the behaviour of one small planet orbiting a lone, absolutely spherical star. The real Universe has many planets, stars and irregular shapes, and by the 20th century 'chaos theory' reigned supreme.

Gravitation

Between 1609 and 1621, Johannes Kepler discovered three laws of planetary motion, which he derived from the accurate measurements of the orbit of Mars made by his teacher, Tycho Brahe (**6**). Kepler found that the path of each planet around the Sun is not perfectly circular, but elliptical, with the Sun at one of the foci of the ellipse. He formulated a further two mathematical laws, describing the rate at which each planet moves along its orbit as dependent on the size of the orbit.

In his 1687 treatise *Principia*, Isaac Newton showed that Kepler's laws were underpinned by the more fundamental theories of dynamics and gravitation, and added an all-important principle about gravity to existing theories of dynamics: all bodies in the Universe attract one another across space, and the force of this attraction between any two bodies varies according to the inverse square of the distance between them (that is, 1 divided by the distance squared). The popular story of this discovery is that in 1666 Newton was musing in the garden of his mother's house in Lincolnshire when he saw an apple fall. He began to think that the power of gravity was not limited to a certain height above the ground but extended to the Moon and beyond. The anecdote dates from the year before Newton died and may be the poetic reminiscence of an old man; the echoes

of the Genesis account of the Tree of Knowledge add to its repeatability. Whatever the circumstances, Newton had a flash of insight about gravity that sparked his investigation into how the Moon orbited the Earth and the Earth the Sun. His findings eventually would transform every branch of science and alter the fundamental human view of the Universe.

Newton's assertion that objects attracted each other across space was controversial. The concept was hard for everyone to grasp and was ridiculed by philosophers, particularly Cartesians, who thought that space was filled with a substance called the *plenum*, swirling in vortices that transmitted force from one body to another. But Newton's theory of gravitation worked. It explained the motions of the planets and even the shape of the Earth (**04**). It enabled Edmond Halley to predict the return of his eponymous comet (**18**), and in due course led Urbain Le Verrier and John Couch Adams to the discovery of Neptune in 1846 (**11**).

Modern astronomers use Newton's theory not only to account for the motions of planets in the Solar System but also to calculate how satellites move in orbit. These calculations are essential in planning complicated orbital tours that loop a spacecraft via nearby planets to more distant ones using the 'gravity-assist' technique to make the spacecraft

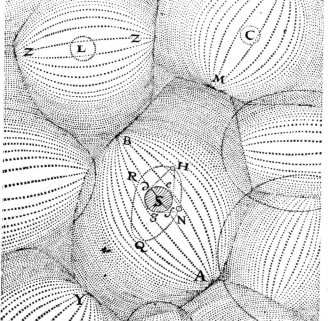

1 **Kepler's Laws** The orbit of a planet is an ellipse with the Sun at one focus (1). The line joining the Sun to the planet sweeps through equal areas of the orbit in equal periods of time (2). The ratio of the squares of the orbital periods for two planets is equal to the ratio of the cubes of the semimajor axes (half the longest diameter of each orbital ellipse) (3).

2 **Johannes Kepler** Astronomer and astrologer, pupil of Tycho Brahe and imperial mathematician to Emperor Rudolph II.

3 **Isaac Newton** Mathematician and physicist, Britain's greatest scientist.

4 **Woolthorpe Manor** Newton was born in his grandfather's small manor house near Grantham in Lincolnshire and returned in 1665 when plague closed Cambridge University. It was here that he allegedly saw an apple drop, related it to the fall of the Moon in its orbit, and hit on the idea of universal gravitation.

5 **Cosmic vortices** In his theory of gravity, René Descartes envisaged that the Universe was filled with a *plenum* of material moving in vortices that pushed or pulled from one object to another.

pick up (or lose) speed – for instance, when approaching the planet Mercury (22). Although computers enable more complicated calculations to be made accurately, the essential theory has remained the same for over 300 years.

Newton's theory of gravitation was considered superior to Kepler's laws of planetary motion, which only apply to cases where two bodies are interacting (usually the Sun and a planet, but also two stars or two galaxies). A comet may orbit through the Solar System controlled mainly by the Sun and obey Kepler's laws, but it may pass close to a planet and be pulled from its elliptical orbit, at which point Kepler's laws fail to predict its movements because three bodies are involved in the interaction. Moreover, the orbits of planets are actually more complex than the simple ellipses assumed by Kepler – for instance, there are perturbations in the Earth's orbit that cause the Milanković climate cycles (19). In principle, Newton's theory could be applied to any type of orbit and any number of planets, stars and galaxies.

In practice, however, the extension of Newton's theory from two bodies to even just three proved difficult, indeed, intractable. Entering an 1887 competition to solve what by then had become known as the 'Three-Body Problem', the French mathematician Henri Poincaré found that he could not give an exact prediction for the orbits of three stars or planets mutually attracted by gravity. He was able to calculate the orbits numerically – we would nowadays do this by computer, he did it by hand – but the paths were 'so tangled that I cannot even begin to draw them'. Moreover, he found that when the three bodies were started from slightly different initial positions, the orbits would be entirely different. 'It may happen that small differences in the initial positions may lead to enormous differences in the final phenomena. Prediction becomes impossible.' Poincaré had discovered a concept that we now term 'chaos theory'.

Poincaré's work has been confirmed by modern computer techniques. The planetary orbits, especially those of the inner planets, are 'chaotic'. If you displace one of the planets by just a single centimetre from its initial position, you might logically expect to find the same single centimetre difference in the planet's final position in 10 million, or even 100 million years. But in practice the planet's final position could be anywhere in its orbital range.

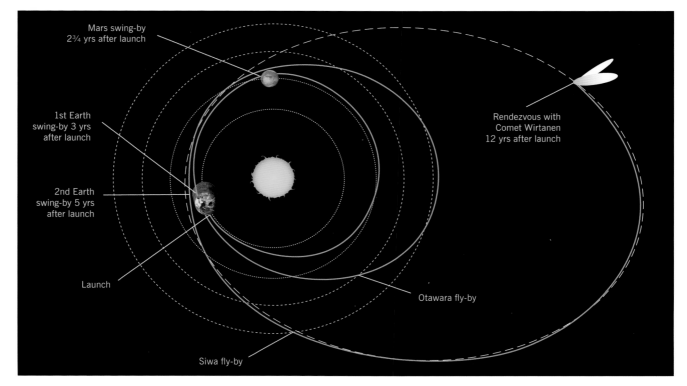

Mars swing-by
2¾ yrs after launch

1st Earth
swing-by 3 yrs
after launch

2nd Earth
swing-by 5 yrs
after launch

Launch

Siwa fly-by

Rendezvous with
Comet Wirtanen
12 yrs after launch

Otawara fly-by

In modern physics, 'chaos' is the word used to describe behaviour that is predictable in the short term but that, in the long term, depends so much on the starting conditions that minuscule changes can have enormous long-term effects that are impossible to calculate. Weather can be predicted, more or less accurately, one day or one week ahead. However, since something as negligible as a butterfly flapping its wings can set off long-term changes in the air currents, and since there are millions of butterflies all over the world constantly flapping their wings, meteorologists cannot predict at any given moment whether a hurricane will strike Texas a year later. This phenomenon was discovered in 1963 by Edward Lorenz, an MIT meteorologist. He was trying out a new computer with a weather model and found that, when he re-entered the initial data including fewer decimal places (it was just a test-run) and ran the simulation again, the weather patterns were completely different. It was not a problem with his model, nor with his new computer; it is an innate practical limitation of mathematics. Lorenz called the problem 'the Butterfly Effect'; James Yorke coined the name 'chaos'.

In principle, Newton's theory of mutual attraction could be used to calculate the future state of the entire Universe. The French mathematician Pierre-Simon Laplace therefore imagined a demon who would be able to predict everything that would happen, down to the movement of the tiniest atom. But because the Universe is so large and made up of so many bodies and particles, Newton's theory – for all the astounding discoveries it has made possible – cannot actually predict the future. The ultimate secret of the Universe is still a secret.

TRUTH IN SIMPLICITY, FALSEHOOD IN CONFUSION

Truth is ever to be found in the simplicity, and not in the multiplicity and confusion of things.
Isaac Newton

CHAOS

Prediction is difficult, especially the future.
Niels Bohr (attr.)

6   Gravity-assist manoeuvre Newton's theory of gravitation is used to calculate the orbits of spacecraft through the Solar System, especially for complex gravity-assist orbits, where a spacecraft picks up speed from a close fly-by of a planet. The figure shows an early plan for the Rosetta comet-chaser spacecraft to rendezvous with Comet Wirtanen (the plan had to be altered because of a rocket launch failure and Rosetta is headed for Comet Churyumov–Gerasimenko with which it will rendezvous in 2014).

7   Henri Poincaré French mathematician, statistician, physicist and engineer.

8   Edward Lorenz American mathematician and meteorologist, creator of chaos theory.

30. Relativity

The nature of space and time

According to Galileo and Isaac Newton, who followed the classical Greek philosophers, space and time are separate from each other, and together form a framework within which events occur. But Albert Einstein thought space and time were more closely connected as a single entity: space-time. This space-time is not simply the framework or stage on which events unfold – it affects *how* they unfold. Einstein's theory would inform all major 20th-century discoveries about the nature of the Universe.

1

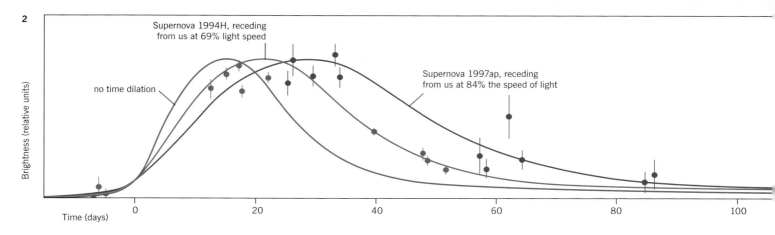

2 Supernova 1994H, receding from us at 69% light speed

no time dilation

Supernova 1997ap, receding from us at 84% the speed of light

Brightness (relative units)

Time (days)   0   20   40   60   80   100

1   **Albert Einstein** Mathematician, physicist, philosopher, pacifist.

2   **Time dilation in supernovae** The explosion of light from nearby supernovae (of a particular type) effectively dies away after 40–50 days (green curve). The explosion of light from more distant supernovae (red and blue curves) persists for longer – up to twice as long in these examples – because the supernovae are in galaxies that are receding considerably slower than the speed of light. This increases the time dilation effect of Special Relativity making these stellar explosions appear (to us) to happen more slowly.

3   **The eclipse of 1919** observed by Arthur Stanley Eddington and Edward Cottingham from Principe off the coast of West Africa. The positions of stars behind the Sun in this picture had been altered by the gravitational lensing effect predicted by Einstein's theory of General Relativity.

4   **Einstein Ring** If a quasar lines up behind the centre of a galaxy, the matter in the galaxy will act like a lens and create a distorted image of the quasar. If the alignment is almost exact, the image of the quasar will form a so-called Einstein Ring around the galaxy.

5   **Einstein Cross** If the alignment is just slightly off, the image of the quasar usually splits into five images, one fainter than the others: an Einstein Cross.

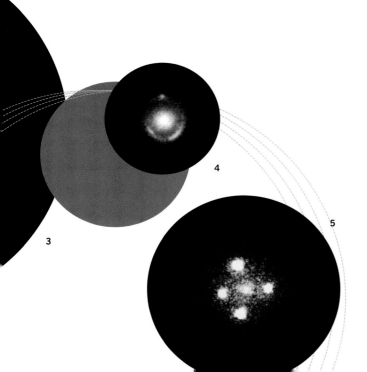

3

4

5

Einstein's theory of Special Relativity is based on two principles. The first principle is that every law of nature has the same mathematical form. The second principle is that the speed of light (represented by the letter c in Einstein's equations), is the same to all observers who move at constant velocity relative to one another.

Relativity is a more important concept in astronomy than in everyday life. For example, because the distances between places in space are enormous, light takes years, perhaps millions of years, to move between them (**37**). Two events that are simultaneous for one observer may therefore occur at different times according to other observers. A double-star eclipse in the Milky Way and the explosion of a nova in another galaxy, which on Earth we see happening in the sky on the same night, may appear at completely different times for observers in a distant galaxy.

Time dilation is another effect of relativity that has immediate astronomical consequences. The faster an object moves, the slower it seems to experience time – at least to the observer. A clock that is moving (for instance, aboard a spacecraft in orbit) runs slow when compared with an identical clock that is stationary on the ground. Terrestrial clocks are used to time the ticks of pulsars (**35**); however, terrestrial clocks are not actually stationary, but run fast and slow depending on the speed of the Earth as it revolves around the Sun in its eccentric annual orbit. This time dilation – a difference in the measurable passage of time due to an object's physical motion – must be taken into account when terrestrial clocks are used to measure astronomical processes. The effects of time dilation are not limited to man-made terrestrial clocks, but apply to any physical process of fixed duration. Supernovae have a 'clock', namely the time that it takes for their light to fade from maximum brightness. Because of time dilation, distant supernovae

5

travelling at high speeds inside their parent galaxies, which are receding further from the Earth as the Universe expands, fade more slowly than nearby supernovae in our own Galaxy, which are travelling at the same speed as the Earth (**41**).

In the case of cause-and-effect events, however, the signal that triggers the effect cannot travel faster than the speed of light. As a result, the cause always precedes the effect, no matter how fast you are travelling when you see the two events. This rule is used in astronomy to estimate the maximum size of a source of variable light, such as a pulsar. If a star varies in intensity during a clearly defined period of time (T) – say, 1 minute – its size cannot be greater than cT (that is, 1 minute multiplied by c, the speed of light, or 1 light-minute). This is because different parts of the star must be able to communicate with each other so that their brightnesses vary together. This argument was used to discover the small size of quasars, because some vary over only a day and therefore must be less than a light-day (about the size of the Solar System) at their maximum dimension; pulsars vary in less than a second and therefore must be less than a light-second in diameter – not much bigger than a planet (**35**, **55**).

The relationship between mass and energy in the theory of Special Relativity is fundamental to understanding the nature of stars. When a body's mass reduces by a given amount (represented in equations by m), its energy (represented by E) reduces by an equivalent amount. This is the iconic $E = mc^2$, equation, which was discovered by Einstein in 1905.

The equation expresses the direct relationship between an object's mass and its energy that underlies the whole of nuclear physics. c, the speed of light, is a big number and c^2 is even bigger, so a little mass makes a lot of energy. The equation explains why huge amounts of energy are released when hydrogen fuses to helium, and powers the stars (**45**).

The theory of General Relativity was discovered by Einstein between 1907 and 1915. It is effectively Special Relativity with gravitation taken into account. General Relativity says that the gravity generated by an object's mass bends the space-time 'framework' around it. This bending determines the paths that particles and rays of light follow around the object, a phenomenon called gravitational lensing (**62**), which was first seen by the British astronomer Arthur Stanley Eddington during a total eclipse of the Sun in May 1919, when he noticed that light from the stars in the background had been bent by the pull of the Sun. The same phenomenon has been observed more recently when a distant quasar happens to align with a nearby galaxy, which bends the quasar's light into a ring (an 'Einstein Ring'), a quadruple pattern (an 'Einstein Cross'), or irregular arcs and loops.

Einstein's theory of General Relativity represents a fine-tuning of the work of Galileo and Isaac Newton. It incorporates the Principle of Equivalence discovered by Galileo, which Newton had incorporated into his theory of gravitation. Galileo knew that the gravitational force on a body is directly proportional to the mass, and that the body's resistance to

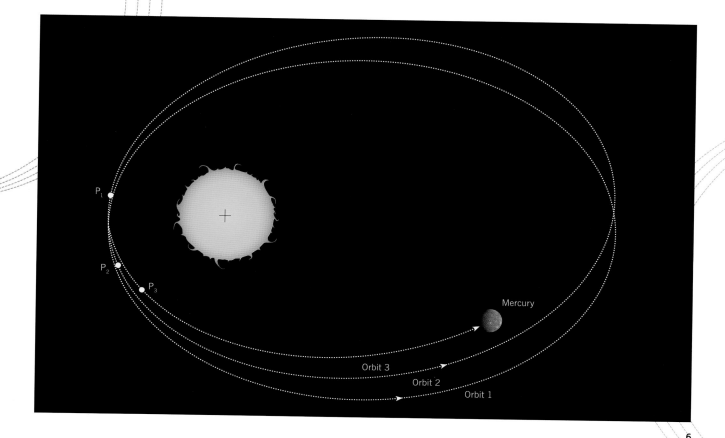

P₁

P₂

P₃

Mercury

Orbit 3

Orbit 2

Orbit 1

6

being moved by gravity (physicists call this 'inertia') is also proportional to the mass. The Principle of Equivalence claims that these two factors precisely cancel each other out. This is why all bodies falling in gravity fall together, no matter what their mass. Galileo is said to have dropped two weights of different sizes from the Leaning Tower of Pisa to demonstrate this: in the Earth's gravity both weights fell at the same rate and hit the ground at exactly the same time. The Apollo astronaut David Scott repeated the experiment on the airless Moon in 1971 with a hammer and a feather, with the same result.

Newton's theory of gravity works well enough for most calculations of the orbits of the planets (**29**) but on rare occasions – for instance, in the case of the planet Mercury – its accuracy subtly and mysteriously fails. Einstein discovered the reason for this in 1915. He realized that General Relativity alters the orbit of Mercury by an extra 43 arc seconds (43/3600 of a degree) per century because of its close proximity to the Sun, which exerts a strong gravitational force. This discrepancy had confounded astronomers since the 19th century. The French astronomer Urbain Le Verrier had thought it might be caused by an undiscovered planet, which he called Vulcan and unsuccessfully tried to locate inside Mercury's orbit.

For Einstein the realization that General Relativity could explain the longstanding mystery of Mercury's orbit gave him confidence to publish his theory, which has since facilitated countless astronomical discoveries. General Relativity is used to explain how black holes work (**36, 57**), and to calculate the orbits of the binary pulsars, which test the theory to the limit (**64**). In the 21st century General Relativity will be essential in solving the problems of Dark Energy (**63**) and gravitational waves (**64**).

5 **Principle of Equivalence on the Moon** Astronaut David Scott repeated Galileo's experiment at the Leaning Tower of Pisa in an airless environment, showing in these stills from a video recording that even a feather would fall just as fast as something heavy.

6 **Advancement of the perihelion of Mercury** In successive orbits Mercury's orbit rotates and its nearest approach to the Sun (the perihelion, P) shifts by an angle of 43 arc seconds per century (one complete revolution in 12 million orbits). When Einstein discovered that the discrepancy could be explained by his theory of General Relativity, he reported that 'for a few days I was beside myself with joyous excitement'.

31. Radio Waves

A new window on the Universe

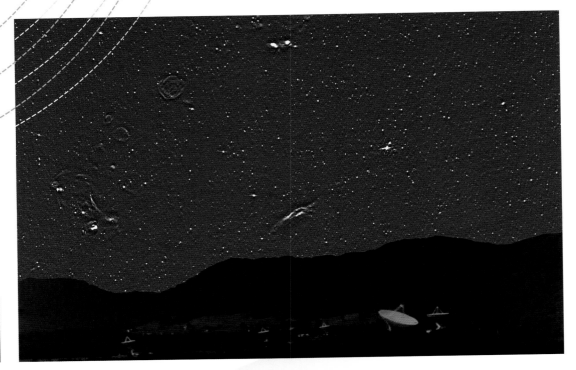

1 3

Radio waves are a type of invisible radiation at the extreme end of the electromagnetic spectrum. Stars and galaxies emit radio waves of various lengths, along with the full spectrum of visible and invisible light. When a radio engineer in New Jersey heard a faint, mysterious static on his home-built antenna, a window began to open on this unseen universe.

1 **The 'Merry-Go-Round'** Karl Jansky stands in front of his radio telescope, which he built at Holmdel, New Jersey, in the 1930s. There is now a full-size replica at the National Radio Astronomical Observatory in Green Bank, West Virginia.

2 **Karl Jansky** Radio engineer, pioneer of radio astronomy.

2

4

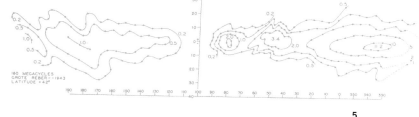

5

3 **Radio night sky** Above the National Radio Astronomy Observatory, the sky, as if seen by an extraterrestrial being with eyes that are sensitive to radio waves, is indeed an alien view. The Milky Way would stretch from lower left to upper right, and would seem to be made not of massed stars, but of vast nebulae, which are clouds of hydrogen energized by luminous stars and circular shells of exploding supernovae. The numerous individual radio 'stars' scattered over the sky are not stars at all; they are radio galaxies or quasars, at distances as far away as billions of light years.

4 **Grote Reber and his reconstructed telescope** at the entrance to NRAO's Green Bank site. Reber first built the telescope in Wheaton IL in 1937.

5 **Discovery of the radio Milky Way** Reber mapped his data as contour maps in 1947. The pair of maps is aligned on the Milky Way (horizontal across the middle of each). The brightest part is towards the centre of the Milky Way in Sagittarius (right panel), and there are other bright radio sources in Cygnus and Cassiopeia. They were not located accurately enough to be identied by Reber's early work but later proved to be an exploding galaxy and a supernova remnant.

In 1928 radio engineer Karl Jansky started work at Bell Telephone Laboratories in north-east New Jersey. His job was to investigate the sources of interference that might affect transatlantic telephony, such as electrical equipment and automobile ignition systems, but also natural sources, such as thunderstorms. To this end he built an antenna that was sensitive to emissions at a wavelength of 15 m (the sort of radio waves that you can pick up on a shortwave radio) and had a certain amount of directional discrimination. The antenna consisted of an open rectangular wooden frame with aerial wires strung over it, which rotated around a track on wheels, earning it the name 'the Merry-Go-Round'.

By 1932 Jansky had found three natural sources of 'static', or radio noise. The first type of interference was clearly associated with local thunderstorms. A second type had similar characteristics but was weaker and steadier, with occasional peaks. He realized that this static corresponded to distant thunderstorms in the tropics, and that it reached New Jersey via radio waves that bounced off the ionosphere.

But the third type of natural static was a mystery. Jansky described it as 'a very steady hiss-type static'. It was so weak that it had little practical effect on radio telephony, but his curiosity was aroused and he decided to investigate it further.

As he rotated the Merry-Go-Round, Jansky noticed that the mysterious static at first seemed to peak in intensity when the antenna was directed towards the Sun but as the year progressed this correspondence broke down. He began to study astronomy textbooks and concluded that the source of the hiss was not the Sun, but an unknown object fixed in space. In 1933 he presented a paper to fellow radio engineers called 'Electrical Disturbances Apparently of Extraterrestrial Origin'. A press release on the subject by Bell Laboratories led to immense publicity, but astronomers who picked up the discovery puzzled fruitlessly over the precise origin of the static, and it was Jansky himself who discovered that the steady hiss came from a band along the Milky Way. However, the economic stress of the Great Depression put a stop to his research and forced him to turn to work of more practical benefit to his employer.

Grote Reber, a radio engineer who pursued astronomy as a hobby, was one of the few of Jansky's colleagues who continued to investigate the static. In a suburban lot in Wheaton, Illinois, Reber built a parabolic reflector 9 m in diameter, which was able to measure the strength of the mysterious radio emissions from the Milky Way at metre and centimetre wavelengths. Reber's reflector attracted much curiosity: when a light plane circling the radio telescope suffered an engine failure and had to make an emergency landing, a rumour circulated that he was transmitting a death ray. Reber was the first to map the Milky Way using radio waves. Later he pioneered investigations into very long-wavelength radio astronomy, and emigrated to Tasmania, where, because of the island's position in the magnetic field of the Earth, it is easier to study this type of radiation.

Radio astronomy grew rapidly following the Second World War, aided by the significant wartime investment in radar technology. The Sun was shown to be a strong radio source. The Milky Way's radio emission proved to be of two kinds: some is diffuse emission from interstellar space – electrons releasing radio waves as they gyrate around the Galaxy's magnetic field; other radio emissions come from isolated sources in the Galaxy. The strongest radio source in the constellation of Taurus, called Taurus A, was the first to be identified: it was actually the Crab Nebula, a supernova remnant (**48**). Another radio source, Cygnus A, proved to be a galaxy far beyond the Milky Way – we now know that it has an active nucleus (a black hole) (**55**). Such galaxies are called 'radio galaxies' because of their strong emissions; they provided the first proof that the Universe was truly evolving and had an origin in a Big Bang (**59**).

Jansky's discovery made it possible to view the Universe outside the spectrum of visible light. It opened up the study of astronomical objects that were previously invisible and drew attention to spectacular objects that, in visible light, had looked unremarkable. Astronomers realized that it could be equally profitable to use other radiations – such as infrared, X-rays or ultraviolet light – to explore the Universe, and began developing new kinds of telescopes, detectors and space vehicles. It was like opening a window inside a house and seeing, for the first time, expansive views of an entire new world beyond the four walls.

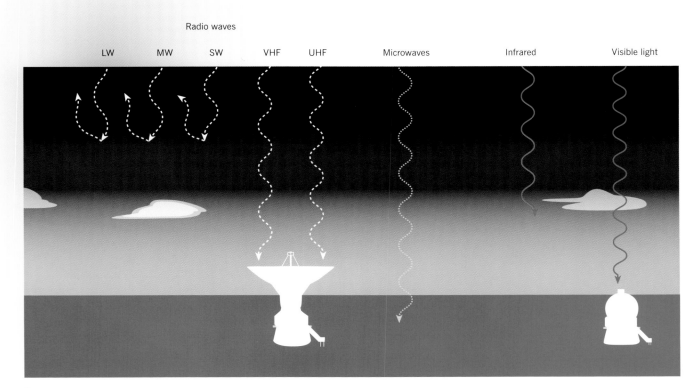

Radio waves

LW   MW   SW   VHF   UHF   Microwaves   Infrared   Visible light

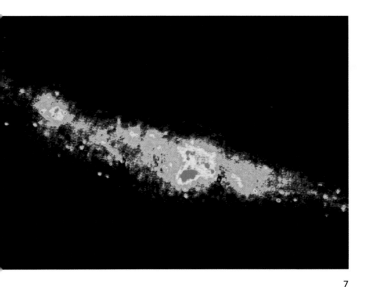

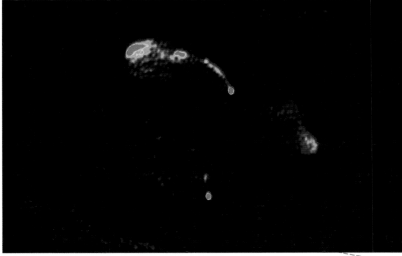

7

8

6 **Windows on the Universe** The atmosphere is opaque to most electromagnetic radiation, except visible light and some radio waves.

7 **The galaxy M82** A false-colour radio image, with red patches showing areas of intense radio emission, and yellow, green and blue identifying progressively less strong emissions. The small bright patches in the galaxy are the remnants of numerous supernovae that have occurred in the last thousand years, and are not visible in optical telescopes due to the large amounts of opaque dust in the galaxy.

8 **The double quasar 0957+561** The first gravitational lens, the so-called 'double quasar', was discovered as a radio star in 1979 by the Lovell Telescope. This false-colour picture shows two images of the quasar (small red dots) and the jet that is seen to extend from one (at the top).

9 **The Lovell Telescope** When built by Bernard Lovell near Manchester in 1957 this was the largest single-dish radio telescope. Britain was a pioneering nation in radio astronomy, its cloudy skies offering no disadvantage.

9

X-rays from Space

The energetic Universe

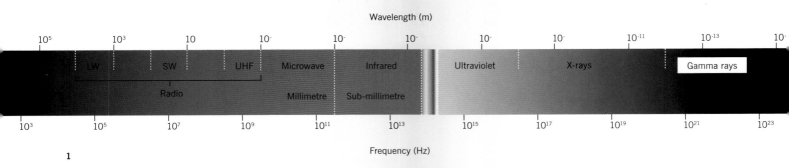

Wavelength (m)

| 10⁵ | 10³ | 10 | 10⁻ | 10⁻ | 10⁻ | 10⁻ | 10⁻ | 10⁻¹¹ | 10⁻¹³ | 10⁻ |

LW   SW   UHF   Microwave   Infrared   Ultraviolet   X-rays   Gamma rays

Radio   Millimetre   Sub-millimetre

| 10³ | 10⁵ | 10⁷ | 10⁹ | 10¹¹ | 10¹³ | 10¹⁵ | 10¹⁷ | 10¹⁹ | 10²¹ | 10²³ |

Frequency (Hz)

1

X-rays make it possible to see stars and galaxies as well as broken bones and suitcase contents. Like radio waves, X-rays are a type of invisible light given off by stars and other space objects, but unlike radio waves they can only be studied at exceptionally high altitudes, using detectors mounted on rockets or satellites. X-ray astronomy is therefore one of the great achievements of the space age – another new window opening onto the Universe.

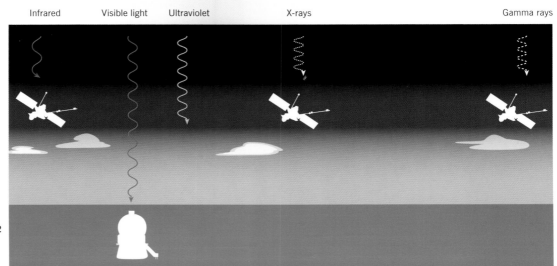

Infrared   Visible light   Ultraviolet   X-rays   Gamma rays

2

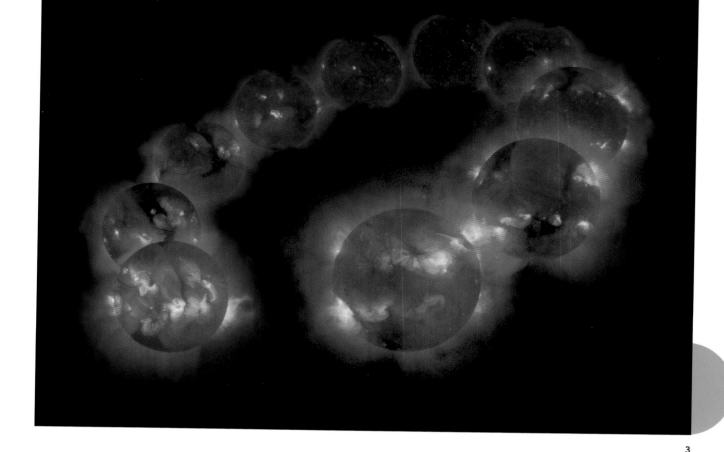

1   **Electromagnetic spectrum** Radio waves, ultraviolet light and X-rays have very different effects but are all different forms of electromagnetic waves, differing only in their wavelength. The boundaries between the different types are a bit arbitrary but the generally accepted divisions are given here.

2   **X-rays** Like gamma rays, X-rays are absorbed in the upper atmosphere.

3   **The Sun emits X-rays from its hottest active regions** The amount of activity on the Sun varies over an eleven-year cycle, as shown in this sequence of X-ray images obtained by the Japanese Yohkoh satellite between its launch in 1991 and its destructive re-entry into the Earth's atmosphere in 2005.

X-rays lie between ultraviolet light and gamma rays on the electromagnetic spectrum. The Earth's atmosphere completely absorbs X-rays before they reach the ground, so X-ray astronomy is exclusively a space activity.

X-rays from the Sun were first measured by the American scientist Herb Friedman in the 1940s, using a Geiger counter mounted on a V-2 rocket that had been captured from Germany at the end of the Second World War. The experiment was organized by the Naval Research Laboratory (NRL) in Washington, DC, as part of a programme to discover how the ionosphere (the layer of electrically charged plasma in the Earth's upper atmosphere) affected the propagation of radio waves. There is a very hot outer layer in the Sun called the corona, and the director of the research programme at the NRL, Edward O. Hulbert, suggested that X-rays from the Sun's corona produced the Earth's ionosphere. The first attempted observation of solar X-rays on 28 June 1946 may actually have been successful, but the data could not be retrieved because of an unforeseen hitch: the rocket carrying the equipment

re-entered the atmosphere at supersonic speeds and buried itself 10 m into the ground, smashing the detectors to unidentifiable pieces. In later experiments, the instruments were moved to the tail of the rocket where they could be safely jettisoned before impact. In September 1949 Friedman was finally able to prove that the Sun was indeed a source of X-rays.

The launch of Sputnik 1 by the USSR provoked the USA to expand its space programme, and several organisations, including NASA and American Science and Engineering (AS&E), were founded to carry out space research. One early AS&E recruit was an Italian particle physicist, Riccardo Giacconi. Following a suggestion made in 1959 by Bruno Rossi, an influential scientist in the American space programme, Giacconi turned his attention to X-ray astronomy – a field that was then completely empty of objects to study except the Sun. Mindful of the exciting astronomical discoveries that had recently been revealed by radio waves (**31**), Giacconi and Rossi suspected that X-rays had similar potential.

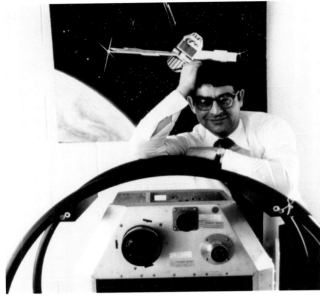

4

5

Teaming up with fellow scientists at AS&E, including Rossi Herb Gursky and Frank Paolini, Giacconi developed X-ray detectors and telescopes, and persuaded the US Air Force that it was worth investigating whether X-rays came from the Moon. The team already knew that the Moon is cold and emits no X-rays of its own accord, but they suspected that streams of solar particles might hit its surface and produce X-rays. This could provide a good way to monitor the flow of particles from the Sun. The US Air Force made large Aerobee rockets available at its White Sands launch site in New Mexico, and in June 1962 Giacconi and Gursky successfully launched their rocket-mounted detector, which spun to allow the detector to scan the sky in all directions.

As they monitored the progress of the flight from the launch-site blockhouse, the crew in the control room were able to see readings from the detector via the rocket's radio telemetry. Almost immediately after the doors in the rocket opened, they saw a large peak in the X-ray count rate as the rocket spun past a point in the southern sky. Some of the crew were jubilant: they had succeeded in detecting the Moon! But Gursky wasn't so sure. The source was too bright. 'I knew what the rate should have been and I knew we would have to add all the data together before we had a chance to determine the signal accurately. So I felt we were in trouble,' he later recalled. As they frantically processed the data, the team eliminated instrumental effects and the Moon as possible sources of the signal, which was actually coming from a position 30° off from the Moon, in the constellation Scorpius. About 60° from the main peak was another strong source of X-rays, located in the constellation Cygnus.

By late August, the AS&E group were confident enough to announce their discovery of these sources of cosmic X-rays, and, with Friedman's group, quickly confirmed their results in three rocket flights in 1962 and 1963. The X-ray sources were called Scorpius X-1 and Cygnus X-1, following the established convention of naming radio sources after the constellations in the sky in which they were situated. Because the early detectors and telescopes had very poor angular discrimination, few of the first X-ray sources that were discovered could be matched to specific known space objects within their general constellation areas. However, Friedman's group was able to identify Taurus X-1 as the Crab Nebula, because it was such an extraordinary object, standing out so prominently in the area in which the X-ray source was located (**48**). Scorpius X-1 eventually turned out to be a blue neutron star.

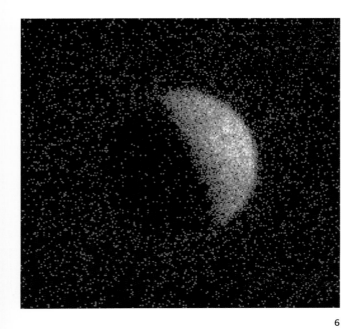

6

7

With the advent of satellite observatories, which enabled astronomers to make observations that lasted much longer than the few minutes of a rocket flight, X-ray astronomy took a great leap forward. The first satellite entirely devoted to X-ray astronomy was Uhuru, a project led by Giacconi and named after the Swahili word for 'freedom', as it was launched from Kenya in 1970 on the twelfth anniversary of the nation's independence. Uhuru was a spinning spacecraft, able to survey the whole sky; it discovered 339 new sources of X-rays. Giacconi was eventually awarded the Nobel Prize in 2002 for his 'pioneering contributions to astrophysics, which have led to the discovery of cosmic X-ray sources'. Satellites like Uhuru and its successors enabled astronomers to identify many X-ray sources as binary stars. The X-rays are generated by the infall of matter from one star into the strong gravitational field of its companion, usually a neutron star or black hole. Scorpius X-1 and Cygnus X-1 are in this class (**36**). Other sources of X-rays are supernova remnants (**41**), Seyfert galaxies and quasars (**55**, **56**), clusters of galaxies, and gamma-ray bursters (**58**). Surely this gallery of extraordinary discoveries justified Rossi's declaration of faith: nature is infinitely more imaginative than man.

REALITY BEYOND IMAGINATION

I [have] a deep seated faith in the boundless resourcefulness of nature, which so often leaves the most daring imagination of man far behind.
Bruno Rossi

4   **Riccardo Giacconi** with the Uhuru X-ray astronomy satellite, *c.* 1970.

5   **Rocket launch** at sunset from White Sands, New Mexico.

6   **X-ray moon** Distant quasars are distributed over the sky beyond the Solar System (corners of this picture) and provide a background to this X-ray image made in 1990 by the German satellite ROSAT. The Moon is opaque and blocks X-rays from behind it, which is why the unlit sector of the Moon in the picture is black – though not entirely black, as particles in the solar wind excite the material of the lunar surface. The sunlit side of the Moon's surface reflects solar X-rays and causes it to shine brightly in X-ray images.

7   **Centre of our Galaxy in X-rays** This image of the centre of the Milky Way was produced by NASA's Chandra X-ray astronomy observatory. Point-like X-ray sources are neutron stars, black holes, white dwarfs, foreground stars and background galaxies. The diffuse X-ray glow extending from the upper left to the lower right is gas heated by supernova explosions.

33. Variable Stars

Discovery of star systems

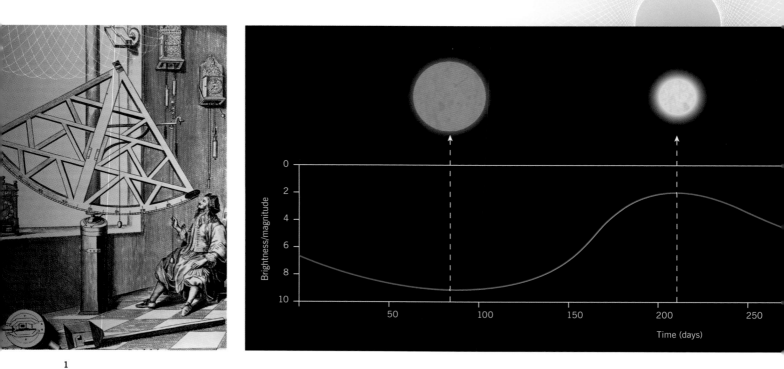

1

For many centuries stars were thought to be constant and unchanging. Astronomers were puzzled when they first noticed that some stars varied in brightness or even faded away entirely only to appear again later. Early Arab astronomers called one of these stars 'the Demon Star'. Although there is nothing demonic about the stars' periodic disappearance, the modern explanation for their 'ghostly' behaviour is every bit as astonishing.

1 **Quadrant** As Fabricius was doing when he discovered Mira, an astronomer measures the position of a planet with a quadrant, sighting between it and a reference star and measuring the angle between them.

2 **Mira's light curve** The brightness of Mira cycles with a period of nearly a year as it pulsates, throbbing in size like a beating heart (but also changing in its temperature, as indicated at the top of the picture). For half the time it cannot be seen by the naked eye (magnitude 6 on the left hand scale).

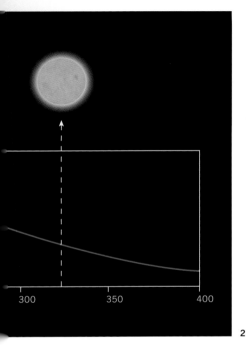

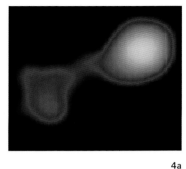

4a

4b

3 **Geminiano Montanari** Italian polymath, astronomer, physicist, engineer and lawyer; he is known for the discovery of the variability of the star Algol and for writing an almanac full of random astrological predictions to demonstrate that they had no foundation in reality.

4a **Mira** Mira is a double star, with Mira A (right) being the variable red giant star. Its companion, Mira B (left), is a white dwarf. The Chandra X-ray Observatory caught this image showing X-rays from the bridge of gas that is flowing from Mira A to Mira B.

4b **Mira system, artist's impression** Mira A is losing gas from its upper atmosphere. Mira B tugs the gas into a bridge between the two stars. Gas pours into an accretion disk around Mira B.

In 1596, while observing the planet Mercury, David Fabricius of Friesland in the Netherlands, a German pastor and a disciple of Tycho Brahe, noticed that a star, which he had earlier used as a positional reference, had inexplicably brightened and then faded away. At first he believed it to be a nova, like the one that had been observed by Brahe in 1572 (**41**), but the star then reappeared. Jan Fokkens Holwarda (sometimes called Johann Phocylides), also of Friesland, discovered in 1638 that Fabricius's star faded and came back every 11 months. In 1642, it was named Mira (Latin for 'wonderful') by Johannes Hevelius of Danzig; it is also called Omicron (o) Ceti. Fabricius did not live to enjoy the recognition for his discovery, since he was murdered in 1617 by a peasant whom he had accused of stealing a goose.

In 1667 the Italian polymath and astronomer Geminiano Montanari noticed that the star Beta (β) Persei also varied in brightness. Beta (β) Persei is traditionally called 'Algol' (Arabic for 'The Ghoul', or 'The Demon Star'), which suggests that early Arab astronomers had observed these mysterious

changes. Its variability was rediscovered in 1744 by a farmer and amateur astronomer, Johann Georg Palitzsch, who lived near Dresden, and was recorded again in 1782 by the English amateur astronomer John Goodricke.

Goodricke was born in 1764 in Gröningen, the son of a British diplomat and a Dutch woman. At five years of age he caught scarlet fever and became deaf. His parents had him educated at a special school for the deaf in Edinburgh; he learned to lip-read well enough to study at the Warrington Academy near York, where he became interested in astronomy. On 12 November 1782, when he was only eighteen years old, Goodricke recorded in his journal his discovery of the variability of Algol, which he observed was very regular. The star's brightness usually stayed near a magnitude of 2.1 but every 2.867 days it suddenly dropped to a minimum brightness of magnitude 3.3.

When Goodricke reported his observations in 1783 to the Royal Society in London, he offered two alternative explanations for the star's peculiar behaviour: either that Algol

5

6

was periodically occulted by another, dark body; or that Algol rotated and had a big spot on one side that made the star appear darker when the spot was facing Earth. We now know that the second theory does not apply to Algol, but it is a good explanation for the behaviour of other types of variable stars.

Goodricke's first theory was correct. Algol is in fact not one but two stars, one bright and one dim, each orbiting around the other. The orbit is almost edge-on, and when the larger, dimmer star completely covers the smaller, brighter star, it causes the most acute drops in brightness. When the smaller brighter star passes in front of part of the larger dimmer one, it produces a smaller dip in brightness exactly halfway between the main minima. This second dip was too slight for Goodricke to see and was detected much later with a photometer, in 1910.

The most puzzling feature of Algol's double-star system is that the less massive (dimmer, orange) star is more advanced in its evolution than the more massive (brighter, white) star. Usually it is the other way around: the more massive stars typically go through their life cycles more quickly than their less massive sister stars. The inexplicable reversal of these circumstances in Goodricke's star is known as the 'Algol paradox'.

The mystery was explained by American astronomer John Crawford in 1956. Crawford proposed that the more massive star had indeed evolved faster, in the usual way. But when this star expanded and became an orange giant, some of its material leaked onto the less massive, less evolved star close by, increasing its mass. Many stars are in binary systems and many are close enough for this exchange of material to happen, with outcomes that can be surprising. The discovery of variable stars has helped astronomers to account for a variety of exotic specimens in the astronomical zoo.

5 **John Goodricke** points to a diagram of his discovery as reported in a heavy volume. Letters from other astronomers lie on the table behind him, which, together with his powdered periwig and dandy's uniform, give him a certain gravitas beyond his years in the portrait.

6 **Perseus** The Greek mythological hero holds the head of the Gorgon in this constellation figure from Hevelius's star atlas. Algol is the main star in the Gorgon's head.

7 **Algol** The light curve of the star shows two dips as one star passes in front of the other, as diagrammed in the top row.

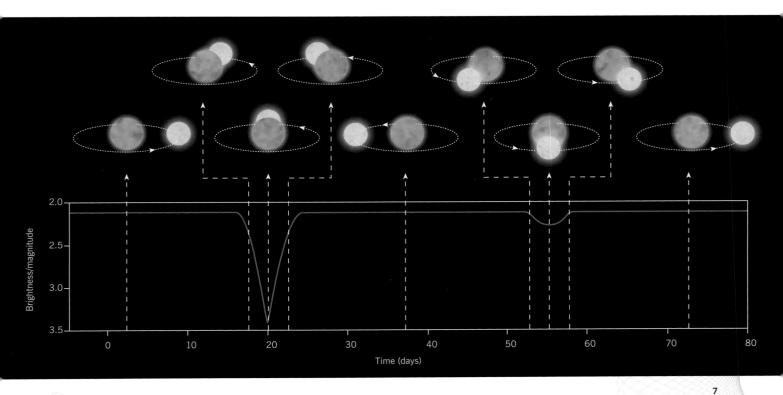

This night looked at Beta-Persei (Algol) and was much amazed to find its brightness altered. It now appears to be fourth magnitude... I had never heard of any star varying so quick in its brightness. I thought it might be perhaps owing to an optical illusion, a defect in my eyes or bad air, but the sequel will show that its change is true and that it was not mistaken.

John Goodricke's journal, 12 November 1789

34. Sirius B and White Dwarfs

Discovery of stellar cinders

An astronomer's offhand remark to a colleague and a student's mathematical puzzle to pass the time during a long sea-journey led to the discovery of white dwarfs: dying stars so small and dense that they can throw other stars out of orbit, or implode into black holes.

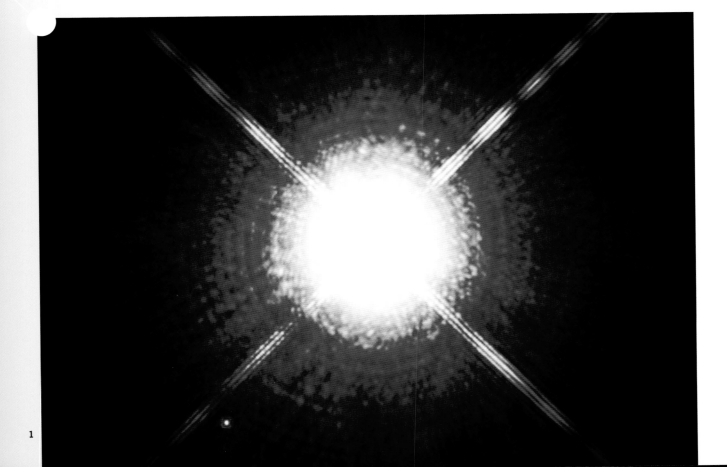

As a young man, Friedrich Bessel worked as an accountant for a shipping company, where he developed interests in navigation and then astronomy. At the age of 26 he became the director of the Königsberg Observatory in Prussia and for the rest of his life he measured star positions with the observatory's telescopes. In 1844 Bessel noticed that the brightest visible star, Sirius, was progressing across the sky in a wavy motion. He realized that Sirius had an unseen companion star that was pulling it back and forth in its orbit. This was the first star to be discovered by means of its gravitational effect on another, although Sirius's companion star was not actually seen until eighteen years later.

In 1862 the American telescope maker Alvan Clark was testing a new refracting telescope he had made for Dearborn Observatory in Evanston, Illinois, by inspecting the image that it formed of Sirius. His son Alvan Graham Clark, Jr, saw the faint satellite star, which was almost lost in the white glare of Sirius. By chance, the faint companion star, Sirius B, happened to be at the point in its orbit when it was furthest from its much brighter parent, and therefore easiest to see. The orbital period of Sirius B around Sirius A is 50 years. In 1914, at the time of Sirius A's next large separation from B, the first spectrum of Sirius B was obtained by the Mt Wilson astronomer Walter Adams. Adams's spectroscopy showed that Sirius B was a little hotter than Sirius A although 10,000 times less bright. Since Sirius A and B are both at the same distance from Earth, Sirius B has to be much smaller than Sirius A – less than 1% of its size, which is a shade smaller than the Earth and very tiny as stars go.

Sirius B is actually a white dwarf: the burnt cinder of a dying star. The first white dwarf ever identified was a star called 40 Eridani B, which was of similar brightness and temperature to Sirius B. William Herschel discovered 40 Eridani to be a double star (it is in fact a triple).

1 **Sirius** Sirius A is the bright star, in the centre of this Hubble Space Telescope image, so overexposed that it shows faint artefacts from the telescope and camera – the cross-shaped diffraction spikes and concentric rings. But the overexposure also shows up Sirius B, the dim star at lower left, a white dwarf star which orbits its much brighter companion.

2 **Hertzsprung–Russell diagram** In 1913 Henry Norris Russell plotted the brightness of nearby stars (left axis, brighter stars at the top) relative to their temperatures (bottom axis, hotter stars towards the left). Ejnar Hertzsprung had published a different version of the diagram a few years earlier, hence its name. Most stars lie on the diagonal, called the Main Sequence, including 40 Eridani A. Russell noticed the stars plotted in the lower left corner, one of them 40 Eridani B. If it was cool it could well be as dim as the stars in the lower right of the diagram, and not unusual.

But if it was hot and of normal size it would be as bright as the stars upper left. Because it is faint, if it was hot it must be very small. This analysis had led Russell in 1910 to consult Pickering and Fleming about 40 Eridani B; they provided the exact measurement plotted here on the horizontal scale.

3 **Williamina Fleming** Of Scottish birth, she emigrated aged 21 to the USA with her husband, who abandoned her with a baby. She gained a job as a maid for Edward Pickering, the director of Harvard College Observatory. Discovering her true abilities, Pickering hired her as a scientific assistant and she worked as an astronomer for the remainder of her life.

4 **Edward Charles Pickering** For 40 years he was the director of Harvard College Observatory, and had a number of female assistants who carried out the laborious tasks of measuring and classifying vast numbers of stars.

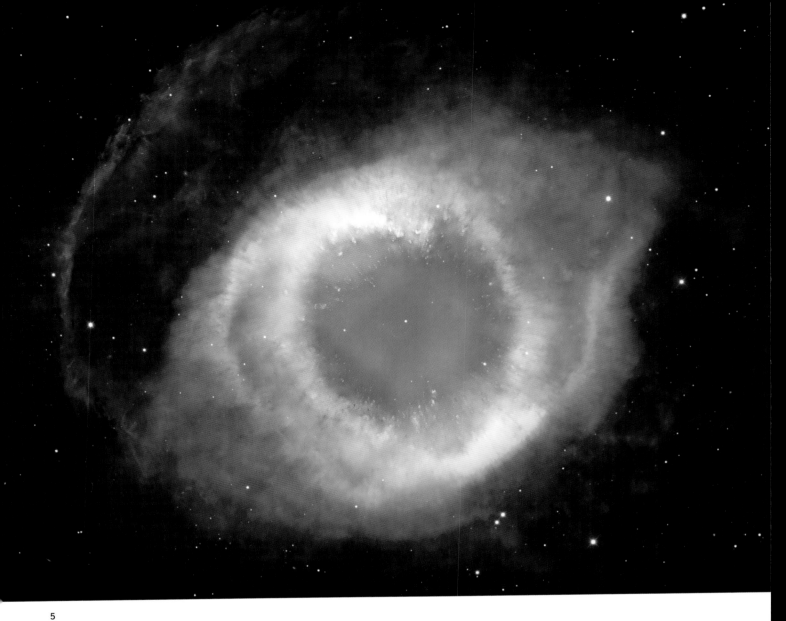

In 1910 on a routine visit to Harvard, Princeton astronomer Henry Norris Russell pointed out to Harvard Observatory director Edward Pickering how faint 40 Eridani B was, and that it must be rather small, mentioning rather wistfully how useful it would be to know the star's temperature so that its size could be determined. Pickering happened to be directing a mass-photography project to find the temperatures of large numbers of stars, so Russell's 'Eureka!' moment came as a mundane telephone call to Pickering's assistant Williamina Fleming, as Russell later recalled. 'In half an hour she came up and said "I've got it here…" I knew enough, even then, to know what it meant…At that moment, Pickering, Mrs Fleming and I were the only people in the world who knew of white dwarfs.' The temperature of 40 Eridani B was very high: it was 'white' hot and very dim, which meant that it was very small, a 'dwarf',

confirming Russell's suspicion that the star was a similar size to the Earth, a mass not unusual for a star in a binary system.

95% of stars end their lives as white dwarfs. (Our Sun will.) A typical star becomes a red giant, then a planetary nebula (49) and then a white dwarf, which passively fades away to a dark, dense stellar cinder. A white dwarf's mass is typically similar to the Sun's but its size is much smaller, which makes it exceptionally dense – a matchbox filled with white dwarf material would weigh a tonne – and the force of gravity at its surface is very strong. The material inside a white dwarf star is extraordinarily strong as it has to withstand the tendency of the star to collapse under its own weight. In 1925 a young British physicist, Ralph Fowler, discovered that the material in white dwarf stars is 'degenerate': all the electrons in the material are packed together as closely

6

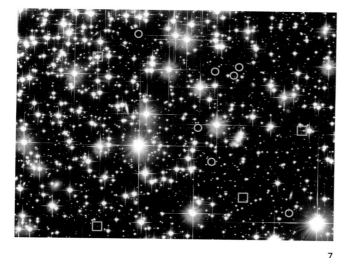

7

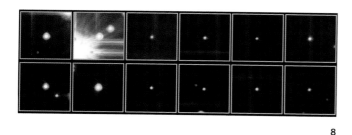

8

5 **The Helix Nebula** This dramatic nebula is centred on the star that generated it. It used to be a red giant (when it ejected the material of the nebula) and is now on the way to being a white dwarf, emitting the ultraviolet light that excites the surrounding nebula, creating the beautiful display.

6 **Subrahmanyan Chandrasekhar** The Nobel-prize-winning Indian-born astrophysicist worked for most of his life on white dwarf stars, neutron stars and black holes at the University of Chicago. The Chandra X-ray Observatory, which studies these objects, is named for him.

7 **White dwarfs** NGC 6397 is an ancient globular star cluster, which, at 8,500 light-years away, is one of the closest globular clusters to Earth. Most of its brighter stars are red giants, but some have turned into white dwarfs, with blue squares indicating the younger white dwarfs, which are still very hot, and red circles showing the older ones, which have cooled.

8 **NGC 6397** A montage showing 12 of the 84 white dwarfs discovered in the cluster.

as is physically possible. The pressure generated by the degenerate material stops the star from collapsing.

Fowler's principles were applied to the structure of white dwarfs by a nineteen-year-old Indian mathematician, Subrahmanyan Chandrasekhar, who in 1930 was on his way from India in an ocean liner to study at Trinity College, Cambridge. Through the calculations he made to pass the time on the journey, Chandrasekhar discovered that there is a counter-intuitive relationship between the mass and the radius of a white dwarf: the more massive the star, the smaller its size. This means that there is a maximum mass above which a white dwarf cannot exist. This limit is known as the Chandrasekhar mass, and it is about 1.4 times the mass of the Sun. If a white dwarf gets more massive than this it shrinks to a point and becomes a black hole (**36**). As

proposed in 1973 by the young British astronomer John Whelan and American theoretician Icko Iben, this is the scenario that creates some types of supernovae. Extra material dribbles on to a white dwarf from a nearby star, increasing its mass above the maximum, which causes it to collapse, release huge amounts of energy and finally explode.

However, when Chandrasekhar presented his results to his colleagues in 1935, he was publicly humiliated by the most distinguished astronomer in Britain at the time, Arthur Stanley Eddington, who called the result 'stellar buffoonery'. In reaction to this incident, Chandrasekhar abandoned his intention to work in Britain and emigrated to the USA where he worked at the University of Chicago for the rest of his life. Chandrasekhar was awarded the Nobel Prize in 1983 'for his theoretical studies of the physical processes of importance to the structure and evolution of the stars', in particular for his work on white dwarfs and black holes.

Walter Adams, director of the Mt Wilson Observatory, carried out work on other aspects of white dwarfs, of which Eddington was much more supportive. In 1925 Adams had discovered that light from the surface of Sirius B was redshifted. That is, light that set out from the surface of the white dwarf lost energy as it climbed out of the gravitational field of the star. As the light lost energy, it became redder. This was a predicted effect of General Relativity, and verified the high mass and small size of Sirius B. Elated, Eddington reported that 'Adams...has confirmed that matter 2000 times denser than platinum [degenerate matter] is not only possible but is actually present in the Universe.'

35. Neutron Stars and Pulsars
Lighthouse stars

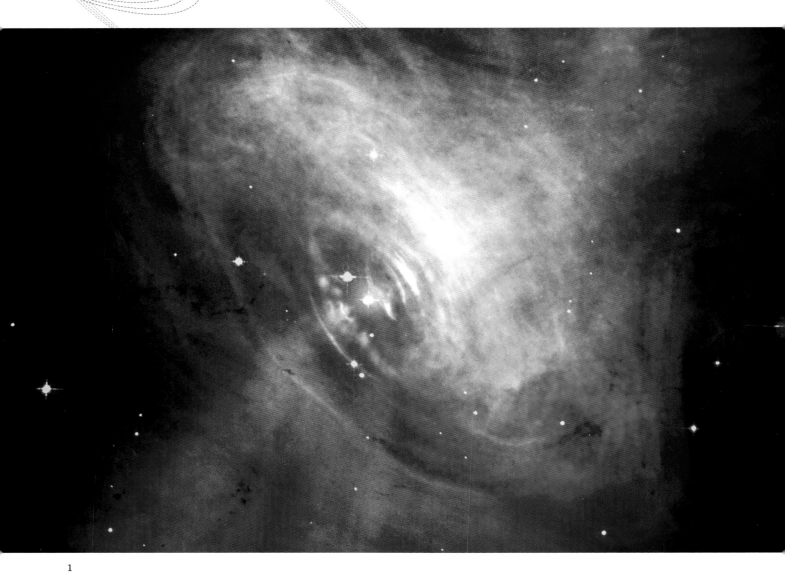

1

The discovery of pulsars was a serendipitous by-product of an investigation with completely different aims, intended to study the twinkling of radio stars. Then a young PhD student noticed an odd 'bit of scruff' on a data chart. Among the twinkling stars was a strange astronomical object that had never before been seen.

2

3

In the 1960s radio astronomer Antony Hewish and his colleagues in Cambridge built a radio telescope known (in splendidly archaic units) as the '4½ Acre Telescope'. The telescope was intended to look at the scintillation, or twinkling, of radio 'stars'. The twinkling of ordinary stars is caused by irregularities in the Earth's atmosphere; the scintillation of radio stars is caused by irregularities in the plasma from the Sun that pervades the Solar System. Radio sources scintillate if they appear point-like, which is often the case if they are a long way away. Hewish's telescope was meant to identify twinkling radio sources that were quasars: exploding black holes, embedded in galaxies at vast distances. To see the twinkles, the telescope had to respond to changes in a radio source's intensity on a very short time scale, which required a large collecting area – hence the 4½-acre surface of wire netting that covered the stationary radio telescope. The enormous 'mirror' looked straight up and surveyed a strip of the sky as it rotated above the telescope.

By 1967 the telescope was ready and Hewish assigned a PhD student, Jocelyn Bell, the job of analysing the data. She surveyed a 120 m strip of chart paper for signs of the scintillating radio sources, rejecting terrestrial radio interference from aircraft or TV stations. In October 1967 she noticed what she called 'a bit of scruff'. It was passing through the beam of the radio telescope in the middle of the night when scintillation caused by the Sun was at a minimum, which pointed to it being terrestrial interference. Bell was, however, not convinced: 'Sometimes within the record there were signals that I could not quite classify. They weren't either twinkling or manmade interference. I began to remember that I had seen this particular bit of scruff before …'

Bell and Hewish decided to use a faster recorder to get a clearer view. By November she had a satisfactory recording, which showed clearly that the 'scruff' was a burst of pulses almost exactly 1.3 seconds apart, similar to many kinds of terrestrial interference. When Bell told Hewish, he said, 'Oh that settles it. It must be man-made.' Nothing of the sort had been seen in astrophysics before that varied so quickly and with such regularity.

However, as Bell studied the 'bit of scruff' further, it became obvious that it wasn't man-made. It stayed exactly in the same position in the sky, so it was celestial. It exhibited no signs of motion, so it was not in orbit around the Sun – it lay beyond the Solar System, among the stars. Bell had soon discovered three more sources of 'scruff', which she confirmed by backtracking through 5 km of paper recordings.

1 **Crab Nebula and its pulsar** The Hubble Space Telescope and the Chandra X-ray Observatory cooperated in 2001 to make an optical and X-ray image of the centre of the Crab Nebula, showing how the pulsar (central brighter star) generates concentric bright rings around a central cavity. A turbulent jet lies perpendicular to the rings (lower-left to upper-right corners of the picture).

2 **Jocelyn Bell** and the 4½ Acre Telescope in Cambridge with which she observed twinkling quasars and discovered pulsars.

3 **Antony Hewish (left) and Martin Ryle** The pair shared the Nobel Prize in 1974 for the discovery of pulsars and the development of radio astronomy techniques.

For a brief period the Cambridge radio astronomers even wondered whether the sources of scruff were interplanetary craft, possibly navigation beacons, and jokingly numbered them LGM 1, 2 etc (for Little Green Men). The lack of any orbital motion seemed to rule this out, since the sources were nowhere near any other star or sun. Finally the true identity of the mysterious objects was announced, both in a sensational paper for the magazine *Nature* and as a dry appendix to Bell's thesis on the interplanetary scintillation of radio sources.

Bell had discovered the first examples of new astronomical objects called 'pulsars', a contraction of 'pulsing radio stars'. A pulsar is a small rotating neutron star. The pulsations arise because there is a kind of rotating lighthouse beam of radio waves on the neutron star that sweeps in the direction of the Earth once per rotation. The period of the pulsations is the rotation period – the star rotates about once per second. It can only do this because it is so tiny compared to other stars – a neutron star is roughly of radius 10 km, so it would fit over a typical large city.

Bell's discovery actually confirmed an older theory that by the 1960s had been all but forgotten. In 1933 California astronomers Walter Baade and Fritz Zwicky suggested that the release of gravitational potential energy as an ordinary star collapsed to a neutron star was the source of the energy of supernovae. In 1939 physicists Robert Oppenheimer and George Volkoff calculated the structure of a star made of neutrons, realizing that it was so compact that its gravity was governed by General Relativity. The internal pressure that held the star up was generated by the repulsion that one neutron has for another, as if the star was a gigantic atomic nucleus. At the time Oppenheimer and Volkoff thought that nature had no way of actually making neutron stars and that their calculation was entirely theoretical. Their suggestions were all but forgotten until revived to explain pulsars.

In 1974 the discovery of pulsars was recognized by the award of a share of the Nobel Prize for Physics to Antony Hewish for 'pioneering research in radio astrophysics' for his 'decisive role in the discovery of pulsars'. The fact that Jocelyn Bell was not awarded the prize jointly was a matter of controversy, with the provocative astronomer Fred Hoyle criticising the circumstances. Hoyle wrote: 'Miss Bell's achievement … came from a willingness to contemplate as a serious possibility a phenomenon that all past experience suggested was impossible. I have to go back in my mind to the discovery of radioactivity by Henri Becquerel for a comparable example of a scientific bolt from the blue.' (In 1896 Becquerel

4

was experimenting with a uranium-bearing crystal and left it in a drawer with photographic paper. When he opened the drawer some time later and developed the paper, the crystal had made its own photograph from radiation given off by the radioactive decay of the uranium.) That Jocelyn Bell had been disregarded and the Nobel Prize for her discovery awarded only to her male supervisor remains to this day a feminist issue, but the affair changed the Nobel Prize Committee's rules and nowadays prizes for discoveries made during PhD studies are awarded to both the supervisor and the student.

4 **Fritz Zwicky** In the 1930s Zwicky theorized how neutron stars were made in supernovae.

5 **'Be Scientific with Ol' Doc Dabble'** Harold Detje drew a popular-science cartoon series for the *Los Angeles Times*, which on 19 January 1934 carried the first published reference to supernovae and neutron stars, after Detje had paid a visit to Cal Tech and spoken to Fritz Zwicky.

6 **Pulsar model** A spinning magnetic neutron star creates a lighthouse beam of light and radio waves, which flashes or pulses if the beam passes across the Earth. The star's magnetosphere generates electric currents that clear the surrounding area of space and control the gases around it.

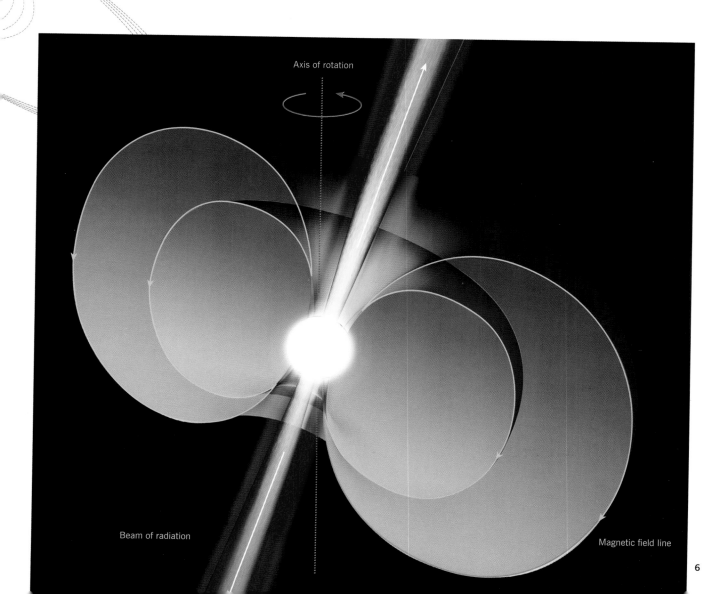

Axis of rotation

Beam of radiation

Magnetic field line

6

Black Holes

A solution looking for a problem

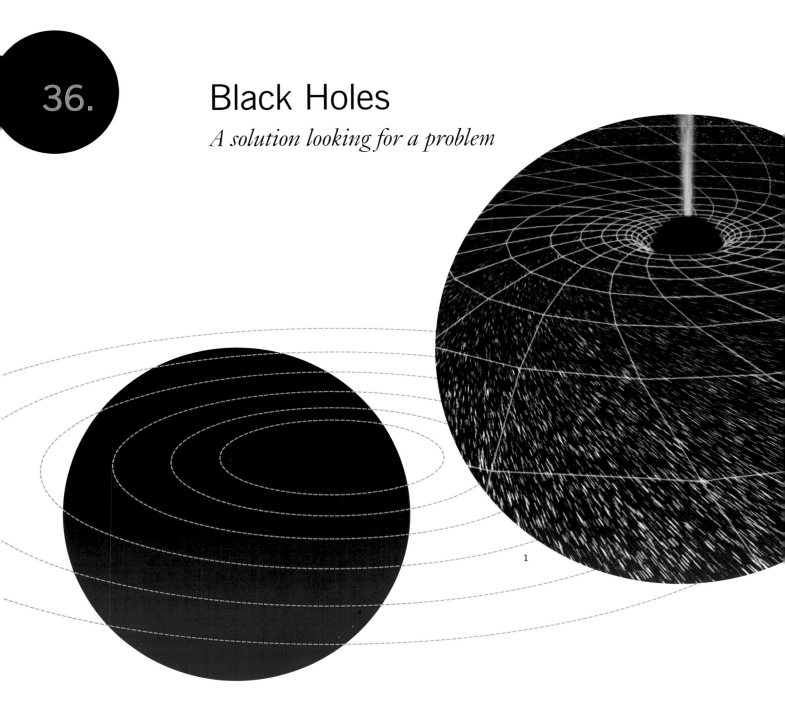

1

The existence of black holes was predicted as early as the 18th century, but it was not until the 1970s that it became possible for astronomers to observe them.

1 **Black hole** A black hole warps space-time in its vicinity and may draw down nearby material, which can disappear forever if it does not get rejected in a jet as it is squashed and heated by the strong gravity.

Imagine a star in space. A projectile is flung from its surface, like a ball thrown up from Earth. If the projectile is thrown at low speed, it rises from the surface and falls back. There is a speed called the escape velocity, which is just fast enough to allow the projectile to escape from the star's gravity. The escape velocity of the star depends on its mass and its size – the more massive and smaller it is, the faster its escape velocity.

A star with a combination of small size and large mass might have an escape velocity faster than the speed of light. Nothing can travel faster than the speed of light (according to the theory of Special Relativity), so it would be impossible to throw a projectile into space from such a star. The projectile would always fall back. This is the basic idea underlying the concept of a black hole. It is a body in space – a planet, a star, or something similar – whose mass and size combine to give it an impossibly large escape velocity.

Cambridge cleric and professor of geology John Michell discovered the concept of black holes in 1783. He speculated about the effect of gravity on light from the Sun. If the principle of escape velocity applied to light in the same way that it did to solid projectiles, the Sun's gravity would slow the flow of light out into space. For our modestly-sized Sun the effect would be small, but Michell calculated that if the Sun was 500 times its actual size, so that its mass was 100 million times heavier, its gravity would be so strong that light would slow to a halt – it would not make it as far as the Earth and we would not be able to see the Sun. Alternatively, if the mass of the Sun shrunk into a sphere with a radius of only 3 km, it would generate the same effect. Pierre-Simon Laplace, a director of the Paris Observatory, put forward the same concept in 1795. However, since Relativity had not been discovered, Michell and Laplace did not know that the speed of light is constant, so these theoretical explanations for black holes were not entirely satisfactory.

Around 1910 the modern theory of black holes was expressed more correctly in terms of Albert Einstein's General Relativity by the German mathematical physicist Karl Schwarzschild. Schwarzschild put forward the following scenario: space curves around a massive body due to the gravitational distortion of space-time, which causes light to follow curved paths (geodesics) (**30**). If a body is sufficiently massive and sufficiently small, then light from the surface of the body could curve so tightly that it might reach no more than a small distance from the body. The body would then be black, because light would never leave it.

Although their theoretical basis had become quite

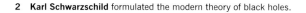

2

2 **Karl Schwarzschild** formulated the modern theory of black holes.

3 **MCG-6-30-15** This galaxy contains a massive spinning black hole surrounded by an accretion disc – material from the galaxy that is falling onto the black hole. A strong magnetic field links the black hole to the accretion disc and transfers rotational energy from the black hole to the surrounding gas (artist's impression).

3

4

advanced, black holes had never actually been observed in nature, so for a long time they were a solution looking for a problem. We now know that nature makes black holes in at least two ways: by supernova explosions in stars, and in the nuclei of active galaxies (**56, 57**). Isolated black holes are dark and difficult to see. However, if matter (gas) falls into a black hole, it releases gravitational energy, which heats the gas. This can happen if the black hole has a companion star that leaks gas onto it, or if other stars get drawn near the black hole, break up and then fall into it. These two scenarios make some black holes visible as, on the one hand, X-ray binary stars (**32**) or, on the other, active galactic nuclei (**56, 57**).

X-ray binary stars are close binary stars that orbit around each other in short periods ranging from minutes to days. One component of the binary system is more-or-less normal star and the other is a much smaller star like a white dwarf, neutron star or a black hole. The normal star transfers matter (gas) onto its compact companion via an accretion disc: matter spirals inward and falls onto the surface of the compact companion. The impact of the matter on the surface of the companion star makes the gas hot enough (a temperature of 10 million $^{\circ}$K) to radiate X-rays.

The orbit of the normal star depends on the mass of its companion, which in turn depends on what type of star it is. White dwarfs have a maximum mass of 1.5 times the mass of the Sun and the theoretical maximum mass of a neutron star is 3.2 times the mass of the Sun (although, no known neutron star has more than twice the Sun's mass). If the mass of the compact object exceeds 3.2 times the solar mass, it can only be a black hole. To apply this method in practice, astronomers must have a clear view of the normal companion in the binary system, and it must be a common type of star so its mass can be estimated accurately.

In 1971 I was working with Louise Webster at the Royal Greenwich Observatory. The Observatory had developed a new spectrograph for the 2.5-m Isaac Newton Telescope, and

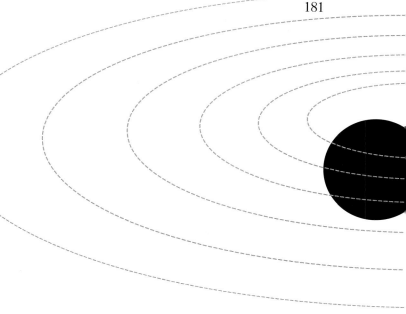

5

4 **Cygnus X-1** This binary system contains a star-sized black hole that is fed material by a companion star, HDE 226868, around which it orbits. A stream of gas falls from HDE 226868 to an accretion disc that orbits the black hole. The gas spirals down into the black hole and on its way in is squashed and heated, generating the X-rays.

5 **Paul Murdin** The author stands by the 2.5 m Isaac Newton Telescope at the Royal Greenwich Observatory in Herstmonceux in the 1970s, before it was moved to La Palma in the Canary Islands. Using this telescope, an RGO team discovered the black hole that was the companion of HDE 226868.

we were testing the instrument by measuring the motion of various stars, including the star called HDE 226868, which appeared in the same direction of the sky as Cygnus X-1. I suspected there might be a connection between them. HDE 226868 is a normal blue supergiant star and we did not think there was anything about it that would cause it to emit X-rays. However, we thought that its motion would change if it was circling around an X-ray emitting companion.

The first two or three spectra that we took were disappointing – there was no change of motion – and we considered giving up. Then we found a spectrum that showed a large change. We had unlimited access to the telescope because we were testing the new instrument, so we continued, and with the next few spectra the cyclic change of motion became clear. HDE 226868 was moving around a companion. Later we realized that by unlucky coincidence we had taken the first few spectra at times when the star was at the same position in its orbit, which made it

appear to be stationary. We also realized that the initial large change that had re-motivated us was actually a false reading – we were using a new instrument and were not practised with it. Luckily, the wrong 'un had inspired us to go on.

Because HDE 226868 is a supergiant, it has an unusual evolutionary history, so there is considerable uncertainty about its mass (estimated at twelve to twenty times the mass of the Sun). Nevertheless we had discovered that the other X-ray emitting star in Cygnus X-1 is actually a black hole, with a mass greater than four solar masses (no less eminent a physicist than Stephen Hawking has said that he is 95% certain that it is a black hole). Louise and I did not know that Tom Bolton, a Canadian astronomer working independently in Toronto, was following exactly the same train of thought and was coming to the same conclusion; he is also credited as a co-discoverer of the black hole in Cygnus X-1. Together we had found the problem for which Michell, Laplace and Schwarzschild had already provided the solution.

Discoveries of our Galaxy and its Stars

Distances of the Stars

The radiance of a star that was shining years ago

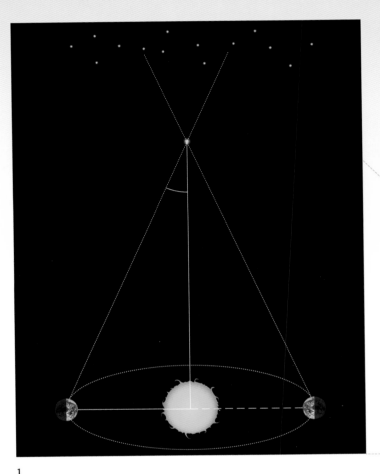

1

When we look at distant stars in the night sky, we are actually looking into the past. From antiquity astronomers knew that if a star was observed regularly from the same position, it ought to appear to move very slightly each year. They sought to discover these small movements to measure the distances of the stars from Earth. Most stars are so far away that their light takes years to reach the Earth.

Efforts to calculate the distances of the stars from Earth date back more than 2,000 years. Aristarchus (4th century BCE) and Copernicus (1543) both realized that if the Earth moves around the Sun, the fixed stars should move in a reflection of the Earth's motion. This is manifestly not the case – the stars move very slightly, but nowhere near as much as the Sun's rising and setting. Both men came to the same conclusion: the stars are much further away than the Sun, and consequently the radius of the Earth's orbit is negligible compared with the distance of the stars.

The apparent movement of a star due to the Earth's motion around the Sun is called the star's annual parallax. 'Parallax' means the apparent shift of something due to the motion of the observer. Hold your finger up at arm's length, and keep it still, but move your head from side to side. The finger moves against the background. The angle by which it moves is its parallax.

In 1580 the Danish astronomer Tycho Brahe built a vast pre-telescopic sighting instrument at his observatory in Uraniborg on the island of Hven to measure star positions. It was called the Great Mural Quadrant and Brahe attempted to use it to determine the annual parallax of the stars. He could not detect the parallax of any star and concluded that the stars were more than 700 times more distant than the Sun.

Dutch scientist Christiaan Huygens took a different approach. If the Sun is a star and all stars are the same brightness, then the reason why the stars are so much fainter than the Sun is that they are further away and their light is diminished by distance – in fact by the amount of their distance squared. Huygens thought that if you could measure the relative brightness of the Sun and a star such as Sirius, then by finding the square root of the difference you could calculate the relative distances of the Sun and the star. Huygens tried to measure the brightness of Sirius relative to the Sun by covering the Sun's face with a card pricked with different-sized holes. He matched the appearance of sunlight through the smaller holes with Sirius. His estimate, published posthumously in 1698, was that Sirius is 27,664 times the distance of the Sun.

This is a difficult measurement to make because the contrast in brightness between the Sun and Sirius is so great. The Scottish mathematician James Gregory tried a variation on Huygens's technique in 1668 by comparing the brightness of Sirius to that of a planet. He chose a time when a given planet was at its greatest distance from the Earth and roughly the same brightness as the star. He then waited until the planet was much nearer the Earth, and bright enough to compare with the Sun. He used his knowledge of the planet's distance to link

2

1 **Parallax** A nearby star appears to shift in angle when viewed against a background of more distant stars as the Earth orbits the Sun through a year. The bigger the angle, the closer the star.

2 **Uraniborg** Tycho Brahe directs operations at his multi-storey observatory, 1580–97. The largest instrument, the Great Mural Quadrant, was mounted on a wall built precisely north–south to measure the altitude of stars as they passed due south. It had a brass measuring scale with a 2 m radius, and was at the time the most accurate instrument ever built to measure star positions, but even Brahe could not determine the parallaxes of stars, as they were so far away.

186

3 **Mural quadrant** John Flamsteed's 3 m mural quadrant, built in 1676 to measure star positions, proved dangerous to use. Flamsteed complained that with its complicated sighting mechanism (protruding from the centre of the arc at lower left) he had torn his own hands 'and had like to have deprived Cuthbert [his assistant] of his fingers.'

4 **Light travel times** Light from a distant galaxy may have left it when dinosaurs were still stalking the Earth. Light from the stars of the constellation Orion left when the Roman Empire was at its height in the 1st century CE. Only light that originates within the Solar System shows us something as it is today, and even then what we see is microseconds to hours old.

3

together all his measurements. Using this approach, Gregory estimated the distance of Sirius at 83,190 times the distance to the Sun. Isaac Newton likewise calculated the distance to a typical bright star as 1 million times the distance of the Sun, but regarded this number (which was in fact much closer to modern estimates) as controversial and did not press it.

Unfortunately, this method for measuring the distances of the stars was fatally flawed: it incorrectly assumed that all stars are of the same brightness. With the rise of instrumental technology in the 18th and 19th centuries – telescopes that produced sharp images over large areas of the sky, and had finely calibrated scales for measuring angles – the geometrical method became the most promising technique. In 1669 Robert Hooke built a telescope on the side of a wall of his London house, intending to measure the parallax of Gamma (γ) Draconis, a star that passes directly overhead in London, but gave up after a few inconclusive measurements, illness and an accident to the telescope lens. Samuel Molyneux and James Gregory tried again in 1725 and established that the parallax of Gamma Draconis was less than 1 arc

second (1/3600 of a degree). On this basis they concluded only that the star was at a distance from the Earth further than 200,000 times the Sun's distance – they were right, Gamma Draconis is in fact 5 million times more distant.

Molyneux and Gregory's experiments made it clear that a nearby star needed to be selected as the target if the geometric method was to have any chance of success. But without knowing the relative distances of the stars beforehand, how would astronomers select a nearby star on which to try the measurement technique? Using a much larger telescope than was available to Molyneux and Gregory, but using a similar technique, Wilhelm Struve, a German astronomer in Dorpat (now Tartu, Estonia) chose to measure the parallax of the bright star Vega in 1837 on the assumption that bright means close. Struve first measured a parallax of 1/8 of an arc second, but got 1/4 of an arc second when he repeated the experiment. This discredited his methods. Thomas Henderson, a Scottish astronomer working in South Africa in 1832–33, measured the parallax of the star Alpha (α) Centauri and would have got quite an accurate result, but did not analyse his data

4

until he returned to Britain and even then could not quite believe what he had found and kept reworking his analysis.

The German astronomer Friedreich Bessel had a better result when he chose the star 61 Cygni because it tracked quickly across the sky. This indicated that it was close to Earth, much as a close-passing bee will move more quickly across your field of vision than a high-flying aircraft. He also used a new technique. Essentially he measured the angle between 61 Cygni and a star that was very nearby, and watched how this angle varied throughout the year. In 1838, at the observatory in Königsberg, Prussia (now Kaliningrad, Russia), Bessel's measurement of 61 Cygni with a parallax of 1/3 of an arc second was believed. Bessel's work boosted Henderson's confidence in his own result for Alpha Centauri and he published his measurement in 1839. Historians rightly credit Bessel as the astronomer who discovered the true distances of the stars.

The most accurate stellar parallaxes have been measured from space by the Hipparcos satellite, led by its project scientist Mike Perryman of the European Space Agency. From 1989 to 1993 the satellite precisely measured the distances of 120,000 stars and of many more to a lower degree of accuracy.

As we measure the distances to the stars, we actually study the past. Expressed in terms of the time that light takes to travel, the distance of the Sun is 8 light minutes, but the distance of 61 Cygni is 9 light years. That is to say, the light that glimmers down at us from 61 Cygni is actually the light that was emitted by the star 9 years ago. The lights that we see when we look up at the night sky originated at many different times long ago, and some show the memory of stars that have since reached the end of their lives and no longer exist. Our view of the sky is thus a view that is peculiar to us, a mosaic of various distances and epochs that meld into a single view for us here and now. We see the stars in Orion (2000 light years distant) as they were when Christ walked in Jerusalem. To us they are new-born stars; but to anyone in the Andromeda galaxy (2.5 million light years away), these stars haven't formed yet. On the other hand, there are stars in Andromeda that have already exploded and died in their own galaxy, but still shine on us in ours.

38

The Discovery of our Galaxy

Stars in an island universe

The discovery of nebulae in the 17th century opened the doors onto the rest of the Universe. Viewed through the lenses of the newly invented telescopes, the Milky Way resolved into individual stars and astronomers realized that we were not alone. Beyond our Sun there were other suns, and beyond our Galaxy, other galaxies.

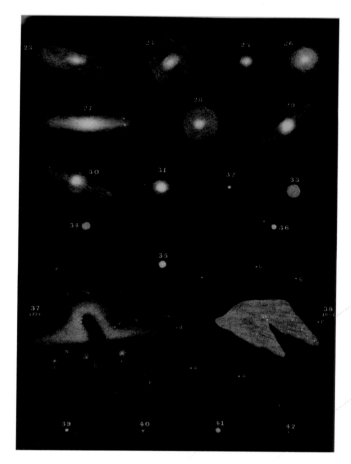

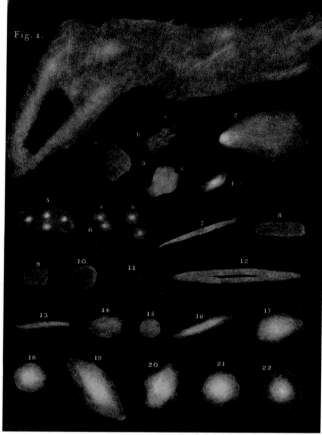

1 **Andromeda galaxy** The Andromeda constellation in a medieval Islamic copy of drawings by the Persian astronomer Abd al-Rahman al-Sufi; in pre-Islamic cultures the constellation was depicted as a fish. At the fish's mouth is the first known representation of the Andromeda Nebula, which al-Sufi called 'the Little Cloud' in his *Book of Fixed Stars* in 964 CE.

2 **Galileo's sketch of Orion** With the aid of his telescope, Galileo was able to see many more stars in the Galaxy. In his 1610 *Sidereus nuncius* ('Starry Messenger'), he sketched the Belt of Orion (the three brighter stars in an almost horizontal row at the top) and Sword (the area vertically below the central star of the Belt), and many other previously unseen stars.

3 **Nebulae by Herschel** These are engravings by James Basire of some of Herschel's observations of nebulae, some of theme recognizably galaxies, some of them a planetary shape, some of them amorphous. The variety of shapes represents the difficulty in the 18th and 19th centuries of figuring out what these fuzzy blobs, seen through a telescope eyepiece, really were.

In 1609 and 1610 Galileo turned his telescope on the night sky, and was amazed when he looked at the Milky Way. The milky luminescence, visible to the naked eye and thought at the time to be a seam, a road or a cloud (03), split into thousands of stars. Galileo announced his discovery in his 1610 treatise *Sidereus nuncius* ('Starry Messenger'): 'We are at last free from earthly debates about the nature of the Milky Way. It is, in fact, nothing but a collection of innumerable stars grouped together in clusters. Upon whatever part of it the telescope was directed, a vast cloud of stars is immediately presented to view. Many of the clouds are rather large and quite bright, while the number of smaller ones is quite beyond calculation.'

Excited by Galileo's report, other astronomers followed him in inspecting the sky, chancing on interesting clouds or 'nebulae'. The Bavarian mathematician Simon Marius discovered the Andromeda Nebula in 1612, describing it as a looking like a 'flame seen through horn' (lanterns in Marius's day had windows made from thin sheets of horn) as the nebula had a characteristic elliptical shape. The Andromeda Nebula is just visible to the naked eye and had actually been discovered by the Persian astronomer

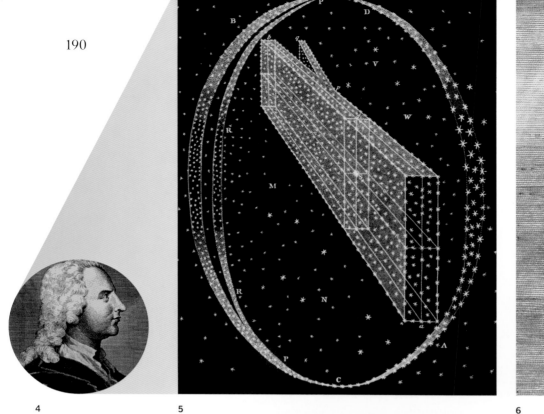

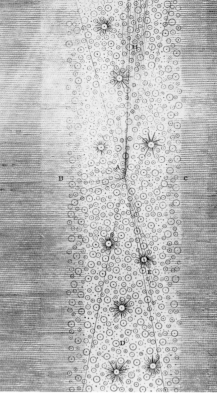

4   5   6

Abd al-Rahman al-Sufi in 964. It was rediscovered in 1654 by the Sicilian priest Giovanni Batista Hodierna, who compiled a list of up to 40 similar nebulae.

The list of 'nebulous stars' grew longer as astronomers started observing with telescopes systematically, measuring the positions of all the stars in the sky and noting what they looked like. The astronomer Johan Hevel from Danzig, known as 'Hevelius', measured 1564 stars, listing sixteen as nebulous in his posthumous catalogue of 1690. John Flamsteed, the first British Astronomer Royal, catalogued 2,935 stars and mentioned several that were nebulae; his successor Edmond Halley added six more, including the 'star' Omega (ω) Centauri that he observed from St Helena, where it was possible to see more of the southern part of the sky. In an expedition to the Cape of Good Hope to survey the whole of the southern sky, Abbé Nicholas Louis de Lacaille discovered dozens of nebulae while he was measuring the positions of uncharted stars and inventing new southern constellations.

Between 1764 and 1781 the comet-hunter Charles Messier compiled all these discoveries, along with some of his own, into a catalogue of nebulae that eventually totalled over 100 entries. The catalogue became an essential tool for astronomers, a source list of objects on which to test new theories and instruments (7).

Messier's catalogue was sent to William Herschel for review on the same day that he was elected to the Royal Society for his discovery of Uranus, which inspired him to examine all the entries with his new telescopes. He then began to to search for and classify other fainter nebulae. His method was to sweep the telescope over the sky in a raster pattern (a scanning pattern of parallel lines looking like a rake), noting double stars and nebulae from his perch high on the telescope, shouting down the details to his sister Caroline, who took notes at a table below. Herschel described this process (in 18th-century spelling) as 'star gaging'. He had expected to add only a few nebulae to Messier's list, but in the end he found more than 2,000. With his telescope, Herschel was able to resolve many of the nebulae he found into individual stars, and, like Galileo, thought that all would eventually succumb to his increasingly powerful instruments.

Meanwhile, in 1750, the English mathematics teacher Thomas Wright had developed his theory of the Milky Way. He modelled it as a band of light that arises from a flattened slab of stars. Look from within the slab, in the plane, and you see many stars and much starlight; look across the slab and you see fewer stars, and thus less starlight. An account of Wright's theory was published in a newspaper in Königsberg where it was read by the philosopher Immanuel Kant, who was inspired to work on the problem. He proposed in 1755 that the Milky Way star system was flattened because it was rotating, and was held together by its own gravity. Kant also proposed that other nebulous patches like

4 Thomas Wright The astronomer, mathematician and architect proposed in 1750 that the appearance of the Milky Way is 'an optical effect due to our immersion in what locally approximates to a flat layer of stars'.

5 Wright's theories of the Milky Way Wright explained the Milky Way as a slab of stars but had to elaborate the slab into various circular or spherical structures in order to 'reconcile it to our Ideas of a circular Creation, and the known Laws of orbicular Motion…a Number of Stars, so disposed in a circular Manner round any given Centre, may solve the Phænomena before us'.

6 Plane of the Milky Way Another of Wright's drawings, showing the Galaxy as 'a vast Gulph, or Medium, every way extended like a Plane, and enclosed between two surfaces'.

7 Milky Way Galaxy If seen from outside, the Milky Way galaxy might look like this, with a bar across the middle of its tight two-armed spiral pattern. It is difficult to be sure what the Milky Way looks like from outside because we see the Galaxy from within its dusty plane, but most astronomers seem convinced it has a bar; recent infrared observations suggest that the bar is very large, like this, making an angle of 45° to a line joining the Sun and the galactic centre.

7

the Andromeda Nebula were similarly rotating masses of stars like the Milky Way, held together in the same way.

The implications of Kant's second suggestion were astounding. Although by the 18th century astronomers were comfortable with Galileo's revelation that the Earth was not the centre of the Solar System and some even had speculated that there might be more than one planetary system, the idea that there could be more than one galaxy required another powerful transformation of perspective. Before they could convince the public, however, astronomers needed to prove that the theory was correct.

Herschel gave quantitative expression to Kant's suggestions as he investigated the structure of the Milky Way by counting the density of its stars. He thought that the number of visible stars in a given area of the sky would indicate the extent of the Milky Way in that direction. Using this technique, Herschel discovered that the stars filled a flattened circular structure that he compared to the shape of a grindstone. He suggested that the Andromeda Nebula is another 'Milky Way' – a galaxy like our own, seen at an angle; a closely compressed cluster of stars that would eventually be individually resolved from the nebulosity. As time went on he became less sure of this, and made contradictory statements; when he died in 1822, no one was really sure what Herschel thought he had discovered.

Herschel became confused because in his day the classification of 'nebulae' was complicated. Originally the word meant something that looked amorphous, like a cloud. Some nebulae, such as the Orion Nebula, are indeed true clouds of gas that can never be resolved into stars, although stars may be embedded within the clouds (**39, 49**). Modern astronomers still call these gaseous objects 'nebulae', but other nebulae are in fact clusters of stars, which may be densely packed together or distant. Only the power of the instrumentation at the astronomer's disposal or the enormous distance of the cluster prevents this second type of 'nebula' from being resolved into individual stars (**40**).

At the beginning of the 20th century, the development of larger telescopes, like the 24-inch telescope at the Boyden Observatory in Arequipa, Peru, and the 60- and 100-inch telescopes at the Mt Wilson Observatory in California, allowed astronomers to see individual stars in some of the closer 'nebulae' when the night air was still and clear. Some were variable stars, Cepheids, whose distances could be determined (**43**), paving the way for American astronomer Edwin Hubble's dramatic 1925 discovery that some 'nebulae' like the Andromeda Nebula, are indeed distant galaxies of stars like our own Milky Way, separated from it (**53**). Our Galaxy is, in the picturesque phrase of the day, an 'island universe', one of many.

39.

Interstellar Nebulae

Stars, molecules, dust and gas

When William Herschel viewed the nebula in the Sword of Orion through his telescope, he described it as a 'fiery mist' that looked like the flame of a candle. What he saw is actually a small dent on the side of a Giant Molecular Cloud, most of it dark. Its shadowy fields of dust and gas conceal thousands of infant stars and newborn planets.

1

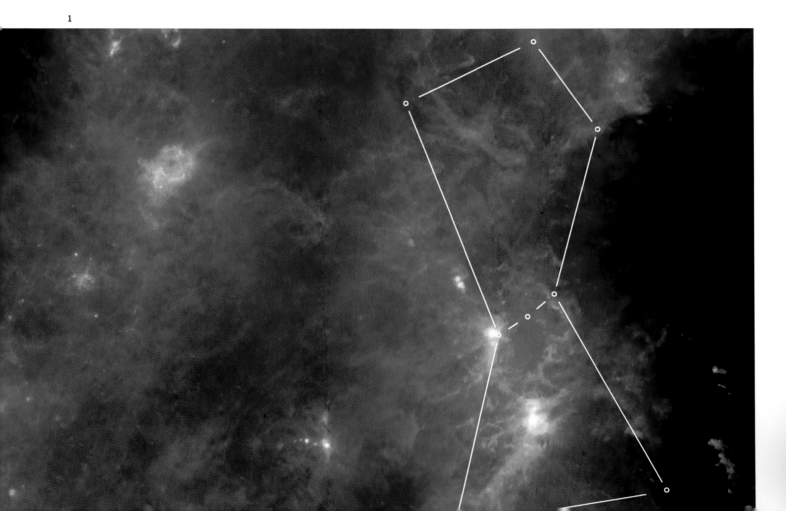

1 **The constellation Orion** in an infrared colour image made by the Infrared Astronomical Satellite (IRAS). The bright yellow region (lower right) is the Sword of Orion, containing the Great Orion Nebula (M42 and M43). Above it to the left is the nebulosity around the belt-star Zeta (ζ) Orionis, which contains the Horsehead Nebula (a small indentation on the right side). The whole area is glowing with infrared radiation emitted by interstellar dust, which emits no optical radiation and is thus invisible.

2 **Ainslie Common**, engineer and pioneering astrophotographer. Using his engineering skill, he built a telescope that could track a nebula smoothly and accurately enough for its image to be recorded on the photographic emulsions available in the 1880s, which required long exposure times.

3 **Ainslie Common's observatory** in Ealing near London contained a 90-cm reflecting telescope. In 1885 it was bought by Edward Crossley, a member of the British parliament, and ten years later was given by him to Lick Observatory, where it is still in use under the name 'the Crossley Reflector'.

4 **Henry Draper** made the first photograph of the Orion Nebula in 1880 in a 50-minute exposure.

Nicholas-Claude Fabri de Peiresc was a French aristocrat who trained as a lawyer and became a diplomat. He was also an amateur scientist. In 1610, when a friend acquired a telescope similar to Galileo's, Peiresc used it to discover a nebula (now known as the Orion Nebula) surrounding a star in the Sword of Orion. Galileo had already discovered three of the little cluster of four small stars (now known as the Trapezium) that illuminates the nebula but had not noticed the nebulosity around them. In 1656 the nebula was re-discovered by the Dutch scientist Christiaan Huygens, who published the first sketch of it. In Charles Messier's catalogue, the nebula was recorded as number 42.

On receiving a copy of Messier's catalogue, William Herschel turned his telescope to observe the Orion Nebula, which he described as 'an unformed fiery mist, the chaotic material of future suns'. Herschel thought that he detected changes in the overall shape and brightness of the nebula over the decades that he observed it. This claim was controversial, but potentially very significant, as it suggested that the nebula was not a large, distant object, but small and relatively close to the Earth. If the nebula was very large, its individual parts would not be able to communicate with each other quickly enough for their brightnesses to change at the same time. Nor could the nebula alter quickly if it was made of independent stars. Herschel's son, John Herschel, made a special point of mapping the nebula very carefully in 1826 so that any changes could be proved in future observations. A decade later he remapped it and found that his father had been mistaken about the changes: he saw no differences between the new map and the one he had made ten years earlier. John Herschel's findings were confirmed when the first photographs of the Orion Nebula were made independently in 1880–83 by two pioneers of astronomical photography: the American amateur astronomer Henry Draper, and the English retired businessman Ainslie Common, who photographed the nebula to test a new technique. Their images proved that the brightness of the Orion Nebula was unchanging.

Because the Orion Nebula is so bright, it has always been a natural target for people testing new technology. In 1864 William Huggins deployed his new astronomical spectroscope on the brightest nebulae. Instead of the continuous rainbow spectrum of colours that would indicate that the light from the nebula was similar to sunlight, and therefore made by stars, Huggins saw the individual spectral lines of a glowing gas. Astronomers had speculated that all misty 'nebulae' would eventually be found to be a mass of stars like the Andromeda galaxy or the Pleiades. Huggins's result showed that this was not true: some nebulae like the Orion Nebula are truly gaseous, although they may contain a few individual stars.

Huggins was especially puzzled by a strong green line in the nebula's spectrum. Between 1880 and 1889, Huggins and his wife Margaret persistently tried to photograph the spectrum, and discovered that the strong green line did not have a wavelength that coincided with any known terrestrial element. The Hugginses incorrectly attributed the line to a new cosmic element, which they called 'nebulium' (**49**). The light represented by the mysterious line gives the nebula a green colour when it is viewed by the naked eye, although photography or the more modern electronic detectors show the nebula as reddish, because the technological processes emphasize the red light generated by burning hydrogen more than the human eye does.

The Orion Nebula contains not only gas, but also copious amounts of dust. One especially dusty region shows as a protrusion of dark material into a bright nebula near to the Orion Nebula that is illuminated by the star Sigma (σ) Orionis. The dark nebula and its distinctive horse-head shape were discovered in 1888 by Williamina Fleming. She had emigrated to the USA in 1878 from her native Scotland, and after her husband abandoned her with a young child, supported herself by working as a maid in the home of Edward Pickering, the director of the Harvard College Observatory. Eventually Pickering asked her to take on clerical work at the observatory, and then promoted her to scientific assistant. The story is that Pickering exhorted his male assistants to achieve more with the admonition 'My maid could do better!' and then found that, indeed, she could. Fleming discovered the Horsehead Nebula as a dark indentation recorded on a photograph taken by Edward's brother, William Pickering.

The Orion Nebula and the Horsehead Nebula are just two features in a vast complex of dust, gas and stars that eventually became known as the Orion Giant Molecular Cloud. It is 1,500 light years across and covers the entire constellation of Orion. The bright object that Herschel and the Hugginses had seen is only a hollow dent on the surface of the enormous cloud, illuminated by the four Trapezium stars.

The concept of 'giant molecular clouds' (GMC) was proposed in the 1970s by radio astronomers who mapped the radio emission from molecules inside the Orion cloud. The radio emission was first detected in 1963, when Sander Weinreb of the National Radio Astronomy Observatory at Green Bank, West Virginia, and MIT physicist and engineer Alan

6

5

5   Orion Nebula In a modern view of the Orion Nebula, obtained from a mosaic of Hubble Space Telescope pictures, the nebula shows as a hollow cavity, like the white flesh of an apple left after the first bite through the apple's red surface. We look under the curve of its upper left hand wall to the Trapezium stars inside it: the stars that have blown this bubble into the surface of a larger nebula of interstellar material that stretches behind and to the right and now excites the inner surface, making it visible.

6   Common's photograph Ainslie Common made the first recognizable photograph of the Orion Nebula, and modestly wrote: 'Although some of the finer details are lost in the enlargement, sufficient remains to show we are approaching a time when photography will give us the means of recording, in its inimitable way, the shape of a nebula and the relative brightness of the different parts in a better manner than the most careful drawing.'

Barrett identified the presence of the hydroxyl molecule (OH) in the cloud. Over the next few years, ammonia (NH_3), water (H_2O), formaldehyde (H_2CO) and carbon monoxide (CO) were also detected. All these chemicals are commonly found on Earth, where they would normally be broken down by light from the Sun, but in the Orion GMC the massive dark clouds had shielded the molecules from the destructive effects of starlight, allowing them to survive in large quantities. These molecules also play a key part in the evolution of the cloud, and end up in the by-products of the formation of stars – comets and planets (**50**) – as 'seeds' that develop, it seems, into the chemical building-blocks of life (**65**).

How do astronomers know that there are stars inside the dark cloud? The dust of the Orion GMC can be very dense and hides the light of most of the stars inside it. Infrared radiation can penetrate the dust, and following the development of the technology in the mid 1960s, astronomers have been able to detect infrared radiation from the hidden stars. Pioneers like Eric Becklin, Gerry Neugebauer, Frank Low and Douglas Kleinmann discovered individual sources of infrared by laboriously scanning the cloud with single detectors in raster scans, just like the survey of the sky that William Herschel made with his conventional telescope (**38**). These early infrared readings were constantly hampered by the abundant infrared radiation emanating from the Earth itself, but in the 1980s the US military declassified a range of infrared detectors that had been developed during the Cold War to identify the warm engines of approaching vehicles and rockets, and which were capable of taking pictures. Astronomers eagerly exploited their ability to see warm objects in the sky, using them not only in ground-based telescopes but also in satellites, such as IRAS (InfraRed Astronomy Satellite, launched 1983), ISO (Infrared Space Observatory, 1995) and SIRTF (the Space Infrared Telescope Facility, 2003, later renamed the Spitzer Space Telescope). The satellite-mounted detectors operate in the cool of space, avoiding interference from the Earth's infrared radiation.

These developments have made it possible to see that the Orion GMC, and others like it, contain thousands of stars, which recently formed inside the cloud. Approximately two thirds of these stars are orbited by planet-forming discs – new solar systems in their infancy (**50**). Herschel's prescient description of the Orion Nebula as 'the chaotic material of future suns' could be extended in modern times to 'the chaotic material of future suns, planets and life itself'.

40

Star Clusters

Nebulae resolved

1

With the invention of the telescope, certain individual stars and nebulae that had been known since antiquity were revealed to be dense clusters of stars. As they mapped these clusters and elucidated their properties, astronomers began to comprehend the astounding size of our Galaxy and to understand how stars age and, eventually, die.

1 **The Pleiades** Amateur astronomer Nick Szymanek made this picture of the Pleiades open star cluster. It is clearly visible to the naked eye and this picture fairly represents the way it looks through a good telescope. The stars in the cluster are so close that they seem to be connected; all of them formed at the same time, so they are all the same age. The blue nebula is a dust cloud that the Pleiades have encountered by chance and lit up.

2 **Omega (ω) Centauri** is a globular star cluster. It is bright enough to be visible with the unaided eye, and is so compact that it was originally confused with a star and given a star's name.

2

Some star clusters have been known since antiquity. The Pleiades and the Hyades are readily visible in the constellation of Taurus and are mentioned in the *Iliad* and the Old Testament. The Praesepe star cluster, known as 'the Beehive', was described by the Greek poet Aratus (3rd century BCE) and by the astronomer Hipparchus (2nd century BCE). When Galileo turned his telescope on the sky in 1609–10 he looked at the Praesepe and discovered that it was a star cluster: 'The nebula called Praesepe is not one star alone, it is a mass of more than 40 small stars.'

Praesepe, the Pleiades and the Hyades are all star clusters in our Galaxy, loose conglomerations of perhaps a few hundred or a few thousand stars. Other star clusters are more densely packed and spherical, containing hundreds

of thousands or even millions of stars, orbiting around and through our Galaxy. These dense, mobile clusters are called 'globular clusters'. The discovery of the first globular cluster was reportedly made in 1665 by the obscure German amateur astronomer Johann Abraham Ihle. In 1677, during an expedition to St Helena, Edmond Halley discovered one of the nearest and largest globular clusters, best visible from the southern hemisphere and known as Omega (ω) Centauri. Describing it as a 'lucid spot or cloud', Halley could not resolve the cluster into individual stars with his telescope.

William and Caroline Herschel found many 'nebulae' during their 'sweeps' of the sky, and with their superior telescopes identified some of them as star clusters. William classified their shapes, labelling some of them as 'globular'. His son John

3

the keys that unlocked the mystery of how stars evolve.

Star clusters allow astronomers to develop techniques for comparing different stars, because they eliminate many of the usual variables. All the stars in a given cluster are at the same distance from the Earth, so the light of each is dimmed to the same extent by distance. Assuming that all the stars were formed at the same time from a single gas cloud, they are of the same age and composition, but have different masses. On the basis of these assumptions, the Danish astronomer Ejnar Hertzsprung was able to discover between 1907 and 1911 that the brightnesses of most of the stars in a cluster correlate to their temperatures. He found that in both the Pleiades and the Hyades the brighter stars were hotter (blue dwarfs) and the fainter ones were cooler (red dwarfs). Henry Norris Russell noticed the same correspondence in nearby stars (**34**). It transpired that the hot, bright stars were the more massive ones and the cool, dim ones less massive. In old clusters the bright blue stars were missing, and there were bright red stars instead (red giants and supergiants): the blue stars became bright red stars as they aged. In even older clusters many of the red giants had died and become white dwarfs, neutron stars or black holes (**35**, **36**, **37**).

Putting all the different clusters in order of age, the situation became clear. The bright, massive stars aged quicker than the dim, less massive ones, becoming red giants and then dying faster than the others. Massive stars have more hydrogen fuel than less massive ones but they burn it much faster, and consequently run out of energy and die earlier, just as a profligate millionaire may go bankrupt more quickly than a poor miser.

The internal structures of the different kinds of stars were first calculated by Arthur Stanley Eddington and James Jeans

Herschel noticed that the globular clusters were not uniformly distributed across the sky, but were concentrated towards the constellation Sagittarius. In 1909 the Swedish astronomer Karl Bohlin deduced from this that the globular clusters surround the centre of the Galaxy, which also lies in the direction of Sagittarius. The American astronomer Harlow Shapley used Henrietta Leavitt's method for determining the distance of Cepheid variable stars (**43**) to map the three-dimensional distribution of globular clusters. In 1918 he estimated the distance to the centre of the globular cluster system as some 60,000 light years, and the diameter of the whole system of globular clusters as 300,000 light years. Although these values are twice the currently accepted estimates, Shapley had discovered that our Galaxy is astoundingly large – much larger than Galileo and the Herschels had ever imagined.

Straggly, irregular-looking star clusters congregate in the plane of our Galaxy near its spiral arms. The early observers could not know that these clusters would be

3 **Harlow Shapley** in his office at Harvard University, seated at his famous rotating desk. Polygonal, nearly circular, the desk could be rotated about a central pedestal. Shapley arranged papers around the edge, according to the topics on which he was working, and could pull the papers on a given topic around to his chair when he was ready to work on them.

4 **Ejnar Hertzsprung** Danish by birth, Hertzsprung worked in Leiden in the Netherlands on topics in astronomy and astrophysics, and co-invented the Hertzsprung–Russell diagram, a potent diagnostic tool to determine the age and other properties of the stars in clusters.

5 **Star cluster NGC 265** The stars in star cluster NGC 265, which lies in the Small Magellanic Cloud, show contrasting jewel-like colours. Most of the stars are white, like diamonds, but a few of the more massive stars in the cluster have become red supergiants and shine brightly, with a deep ruby-red colour.

4

6 **Harvard Observatory, *c.* 1917** Female astronomers compiled vast amounts of data about millions of stars without the aid of modern information technology, providing the groundwork on which modern astrophysics was founded. The women, nearly all of whom were university graduates, worked together in a large room on the top floor of the building in the background, separated from the men out of propriety. Left to right: Ida E. Woods, Evelyn F. Leland, Florence Cushman, Grace R. Brooks, Mary H. Vann, Henrietta S. Leavitt, Mollie E. O'Reilly, Mabel A. Gill, Alta M. Carpenter, Annie J. Cannon, Dorothy W. Block and Arville D. Walker. Pickering is standing at the right.

between 1916 and 1924. In the 1950s and 1960s, with the advent of the first electronic computers, it became possible to link these calculations together to track how stars changed from one kind to another as they aged. The results could then be used to date stars in different star clusters, each cluster consisting of an array of stars of all possible masses, but each star having reached a different stage in its life history.

For decades, star clusters were the main way of verifying theoretical calculations about the size, composition and life cycles of stars. There was no way to see past the surface of a star in order to study what was really happening inside it. This changed in the 1970s with the discovery of helioseismology and solar neutrinos (**47**), which proved that these calculations had been remarkably accurate. Star clusters were the ancient keys to a modern problem, keys with which astronomers were first able to unlock the opaque outer layers of the stars in order to discover their secrets.

5

6

41

Supernovae

Origins of the stardust from which we are made

Before the 16th century, stars were thought to be fixed and eternal. We now know that stars are in constant flux, undergoing a cycle of birth, death and rebirth. The roadside observations of a 16th-century nobleman, and the striking data from modern telescopes bear vivid witness to supernovae, the explosive collapses of dying stars. These events also generated virtually all the basic elements that make up the Earth.

1

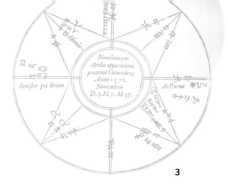

One winter's evening in 1572, the Danish nobleman Tycho Brahe was returning home in his carriage when his curiosity was piqued by the excitement of group of people standing beside the road, marvelling at something unusual in the night sky. Brahe halted his carriage to see for himself. It looked like a new star. Excited by this sensational phenomenon, not quite believing the evidence of his own eyes, he asked all who passed to confirm what he had seen.

When he reached home, Brahe began an extensive programme of observing the 'new star' or *nova stella* ('nova', for short). Brahe and astronomers Michael Mästlin of Tubingen and Thomas Digges of Kent all measured the position of the nova by noting that it lay on the intersection of the lines joining certain pairs of stars. They showed that the new star did not move through space as a comet would, nor did its position in the sky change when the Earth's rotation changed the position of the observer. This proved that the nova was a long way from Earth, among the fixed stars. Brahe wrote: 'I conclude, therefore, that this star is not some kind of comet or a fiery meteor, whether these be generated below the Moon or above the Moon, but that it is a star shining in the firmament itself – one that has never been seen since the beginning of the world.'

This momentous discovery challenged the ancient Christian and classical concept that the stars were eternal and unchanging. In the cosmology of philosophers such as Aristotle, the orbits of the Moon and Sun marked the boundary between the changeable and permanent parts of the Universe. The Moon had phases, the Sun had spots, but the stars were the same forever. The Earth was imperfect and harboured misery, disease and sin, but the stars were perfect and pure, the home in heaven for the blessed saints. Brahe's new star was irrefutable evidence that this world-view was wrong.

The star was in fact a supernova. This word was coined by Fritz Zwicky in 1931 when he discovered that some new stars were much brighter than others. These new stars were releasing colossal amounts of energy, their brightnesses rivalling the light from their surrounding galaxies.

A supernova is not the birth of a new star, but the destruction of an old one. Stars are in constant balance

1 **Ghost of a supernova** The tattered remnant of Cassiopeia A, a Type II supernova in the Milky Way. Astronomers calculate that Cassiopeia A exploded in 1667, making it the youngest supernova in our galaxy, but unlike Brahe's supernova, it went unnoticed until 1947.

2 **Tycho Brahe** The Danish nobleman must have been a striking figure, with his corpulent body, resplendent red mustache and beard, and false nose of copper (some say gold), having lost his real one in a duel.

3 **Horoscope** Tycho Brahe based this horoscope on the time of appearance of the supernova of 1572. Brahe was sceptical about astrology but was obliged by King Frederik of Denmark to cast horoscopes for the royal family. He linked the supernova with an approaching 'great mutation' into fire, including the decline of the Catholic religion.

4 **Uraniborg** The fame brought to Brahe by the supernova gained him royal sponsorship, with which he built Uraniborg, the first modern observatory.

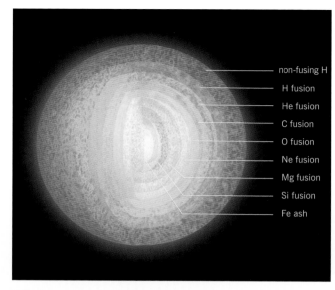

non-fusing H
H fusion
He fusion
C fusion
O fusion
Ne fusion
Mg fusion
Si fusion
Fe ash

5 (i)

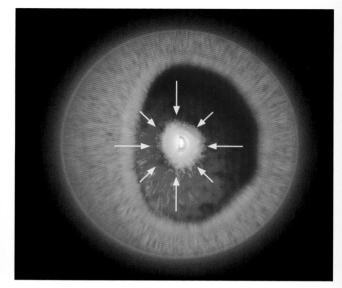

5 (ii)

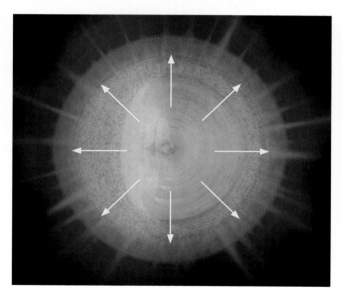

5 (iii)

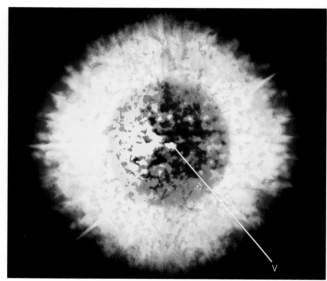

5 (iv)

between two opposing forces: the downward force of gravity and the upward force of the pressure that is generated by the heat and density within the star. Nothing turns off the downward force of gravity, but the amount of upward pressure depends on the energy generated by nuclear processes within the star. Stars contain a finite amount of nuclear fuel, and eventually the fuel gives out. The exhausted star may stabilize as a white dwarf, but some stars are too massive to stabilize. If the internal pressure can no longer support the structure of the massive star, it collapses like a house of cards from which a supporting card has been flicked away. The energy of the star is released as the infalling material crashes together near the

star's centre; this energy causes the infall to bounce and the star explodes. When it explodes, it becomes visible as a 'new star', in the same place where the old star had been, too dim and feeble to be noticed.

The explosion causes the old star to disintegrate, though not completely. It may leave a black hole (**36**) or a neutron star (**35**). The body of the star, including all the elements made inside it by nuclear fusion (**46**), is dispersed into space as it disintegrates. Eventually this material mixes with interstellar gas, and congeals to form new stars and planets. This happened in own Solar System before our present Sun was formed, and elements from ancient supernovae constitute the physical

BRAHE'S NEW STAR

On the 11th day of November in the evening after sunset, I was contemplating the stars in a clear sky. I noticed that a new and unusual star, surpassing the other stars in brilliancy, was shining almost directly above my head; and since I had, from boyhood, known all the stars of the heavens perfectly, it was quite evident to me that there had never been any star in that place of the sky, even the smallest, to say nothing of a star so conspicuous and bright as this. I was so astonished at this sight that I was not ashamed to doubt the trustworthiness of my own eyes. But when I observed that others, on having the place pointed out to them, could see that there was really a star there, I had no further doubts.

Tycho Brahe, 1572

5 **Type II supernova** A Type II supernova takes place in a lone, massive star. As the star ages, releasing nuclear energy, it builds up an onion-like structure of shells containing different elements (i). When the central core runs out of nuclear fuel, it collapses (ii). The outer shells of the star pick up the energy released by the collapse (iii), fragment and speed outwards (iv).

6 **The new star** In his book *De nova stella*, Brahe sketched the stars of the constellation of Cassiopeia to help other people locate the supernova of 1572. He labelled the stars A–H and (inset) identified in Latin where they were in relation to the constellation figure (A at the head, F at the foot). The supernova (the brightest star, I) is identified as *Nova Stella*, 'the New Star'.

7 **Brahe's supernova remnant** Where Brahe saw his supernova explode more than four centuries ago, X-ray telescopes now show a near-spherical bubble blown by the explosion in the interstellar gas. In this colour-coded picture from the Chandra X-ray Observatory, particularly hot regions (blue) on the periphery appear where exploding bits of the star collide with previously undisturbed gas. The green and red fingers are stellar debris.

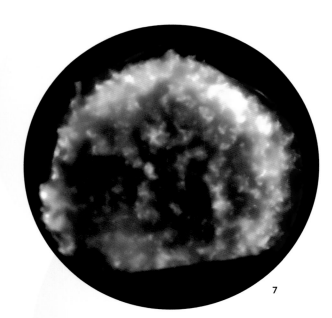

7

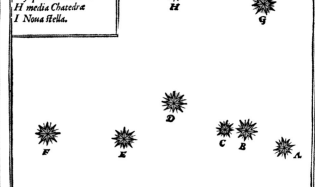

A caput Cassiopeæ
B pectus Schedir.
C Cingulum
D flexura ad Ilia
E Genu
F Pes
G suprema Cathedra
H media Chatedra
I Noua Stella.

6

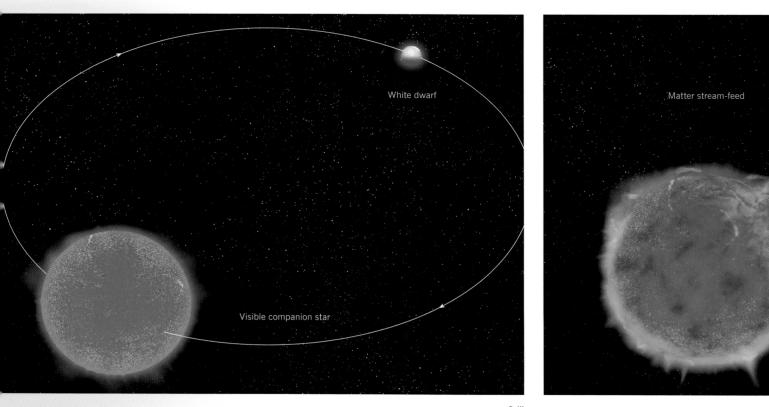

White dwarf

Matter stream-feed

Visible companion star

8 (i)

makeup of our own planet and everything on it, including our own bodies. As planetary scientist Carl Sagan put it in a 1980 television programme: 'We are made of star stuff'.

Brahe's supernova was a so-called Type Ia supernova: its progenitor star was a white dwarf that for some reason became too massive, probably by feeding on material donated by a close companion star. When the supernova exploded, the companion star was released like a stone from a slingshot. The companion star released by Brahe's supernova was discovered in 2004 by a team led by Spanish astronomer Maria Pilar Ruiz-Lapuente. She used the William Herschel Telescope on La Palma in the Canary Islands to identify the star, and later confirmed her discovery by observing the star more closely with the Hubble Space Telescope (HST). 'Here we have identified a clear path: the feeding star is similar to our Sun, slightly more aged,' Ruiz-Lapuente reported. 'The high speed of the star called our attention to it.'

In the case of Type II supernovae, the progenitor stars are actually too massive to support themselves from the outset. Massive stars, those more than eight times the mass of our Sun, are ticking time-bombs; all of them eventually collapse

and become Type II supernovae. Stars are long-lived, so even in a galaxy the size of the Milky Way where there are a lot of stars, there is on average only one visible supernova every 50 years. But if you watch 50 galaxies, you will find one supernova exploding every year, and if you watch 1,000 galaxies you will find a couple of dozen every week. This approach makes it possible to discover supernovae so that powerful instruments like the HST can study large numbers of them. Astronomers working on the Supernova Cosmology Project and the High-Z Supernova Search Team use large ground-based telescopes and powerful automatic image analysis systems to search fields crowded with galaxies that lie just ahead along the path of the HST. When the teams locate supernovae in these galaxies, they can direct the HST to study them in detail, discovering more of the secrets of these explosive dying stars (**63**).

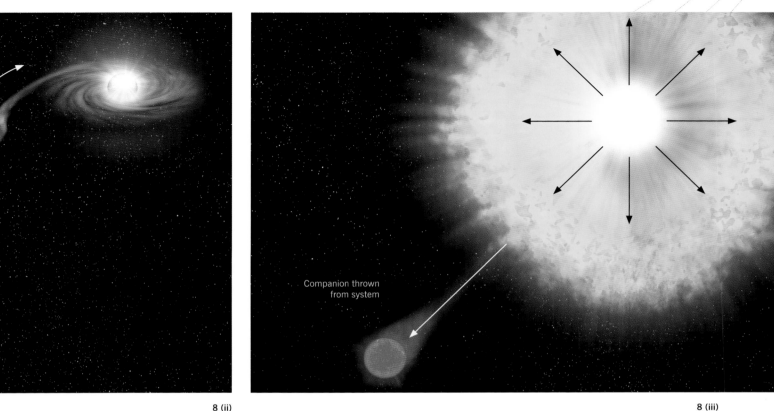

8 (ii)

8 (iii)

Companion thrown
from system

8 **Type Ia supernova** Supernovae like Brahe's begin as binary star systems in
which a small white dwarf star orbits together with a larger companion star
(i). The white dwarf begins to feed on material from the companion star (ii).
Eventually the white dwarf consumes so much of the companion star that
it can no longer support its own mass. It explodes as a supernova, ejecting
the companion star from the system (iii).

9 **The William Herschel Telescope** Ruiz-Lapuente's team used the
William Herschel Telescope, located on La Palma in the Canary Islands,
to study the remnants from Brahe's supernova. The telescope represents
a considerable technological advance on the primitive mural quadrant used
by Brahe.

10 **Maria Pilar Ruiz-Lapuente** In 2004 the Spanish astronomer identified
the companion star that was ejected when Brahe's supernova exploded.

10

42

Supernova 1987A

The whisper and the vision

1

Most supernovae are in distant galaxies and therefore rather faint. The appearance of a bright, nearby supernova like Tycho Brahe's 'new star' is always an occasion for excitement, and a source of valuable astronomical information. The most recent supernova that could be seen on Earth with the unaided eye was called SN 1987A (indicating that it was the first supernova of 1987). In 2005 astronomers studying the supernova witnessed an astoundingly beautiful and unprecedented event.

2

1 **Supernova 1987A in the Large Magellanic Cloud** SN 1987A is the very bright star in the middle right. The massive nebula in the left half of the picture is the Tarantula Nebula, full of bright, massive, relatively young stars similar to the one that exploded.

2 **SN 1987A and its precursor** This 'before and after' pair of pictures made by David Malin with the Anglo-Australian Telescope was used to identify the star that exploded in 1987. The cross-shape on the image of the supernova was generated by structures that support the camera within the telescope, which are usually regarded as a nuisance by astronomers, but in this case helped to locate the massively overexposed supernova image on the fainter background stars.

3 **SuperKamiokande** Kamiokande (Kamioka Nucleon Decay Experiment) was a neutrino detector that detected neutrinos from SN 1987A. In 1996 it was superseded by a larger detector called SuperKamiokande (shown here), a cylindrical tank, 40 m tall and 40 m in diameter, filled with 50 million litres of purified water. Hemispherical photomultiplier tubes lining its inside wall catch flashes of light produced when neutrinos are absorbed by the water. Technical staff access the photomultipliers by boat after draining the water.

Supernova 1987A was discovered by Ian Shelton at 05:40 GMT on 24 February 1987 at Las Campanas Observatory in Chile. He had taken a photograph of the Large Magellanic Cloud (**54**) and developed it before going to bed. To his surprise there was a black spot on the photograph, which at first he thought was some sort of blemish. Then he realized that it was actually a bright star where none was indicated on the charts. The spot was a supernova.

Shelton wanted to share his discovery with someone and went to another telescope to talk to his colleagues. A fellow astronomer, Oscar Duhalde, said that he had seen the star earlier that night with his own eyes, while strolling about outside during a break, but when he returned to the telescope another researcher immediately harassed him about some problem with the equipment and he had forgotten to mention the strange object. Later it transpired that the supernova had in fact been photographed the day before by astronomer Rob McNaught in Australia, but he put off looking at his photographs and thus failed to discover the nova before Shelton.

Shelton's discovery made it possible to identify the first neutrino particles from a supernova. When a Type II supernova collapses, its protons and electrons are jammed so close together that they merge and form neutrons. This nuclear

3

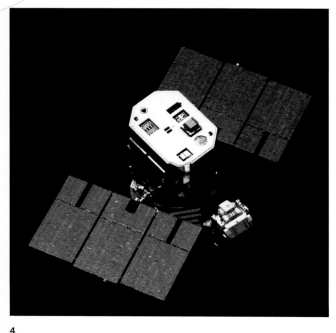

4

5

6

reaction produces huge numbers of elusive particles called neutrinos. When SN 1987A exploded, scientists in Gifu, Japan, and an American team operating in a salt mine in Ohio happened to be using neutrino detectors to collect particles from the Sun (**47**). Upon learning of Shelton's discovery, the Japanese scientists searched their computer files and discovered an unusual burst of eleven neutrinos when the supernova collapsed at 07.35 GMT on 23 February, a day before the outrushing explosion had made the supernova bright enough to be noticed. The Ohio team then searched their records and reported that their detector also had recorded a burst of eight neutrinos at the same time. Although both figures were mere whispers compared with the enormous scale of the explosion, the neutrinos from SN 1987A were the first from outside the Solar System to be discovered, providing remarkable insight into the conditions in the hitherto unseen interior of a supernova explosion.

Neutrinos are by-products of the nuclear reactions that build up the heavier elements from hydrogen and helium inside stars (**46**). One of the main types of nuclei made in a Type II supernova is called nickel-56, which eventually

becomes cobalt-56 and finally iron-56 through the process of radioactive decay, emitting gamma rays in the process. At first these gamma rays are absorbed by the expanding body of the exploding star. But eventually the explosion thins out and the gamma rays can escape. Gamma rays from SN 1987A were discovered by a satellite-mounted gamma-ray detector called Solar Max, which had been designed to study the Sun. It was only by luck that the satellite was in orbit and operational at the time the supernova went off and that its design let it detect gamma rays from a completely different celestial object.

As SN 1987A faded away it became easier to see the immediate neighbourhood around it, and astronomers realized that something else about the star was interesting. In 1989 Joe Wampler discovered that the spectral lines of a small nebula were visible in the supernova's spectrum. The spectral lines were confirmed by an orbiting satellite called IUE (the International Ultraviolet Explorer). The next year the Hubble Space Telescope (HST) found that the spectral lines came from a ring that surrounded the supernova, and, in 1994, discovered that the ring was in fact a symmetrical, hollow, three-dimensional bipolar structure, shaped like two glass tumblers

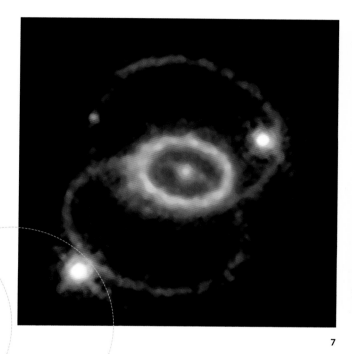

7

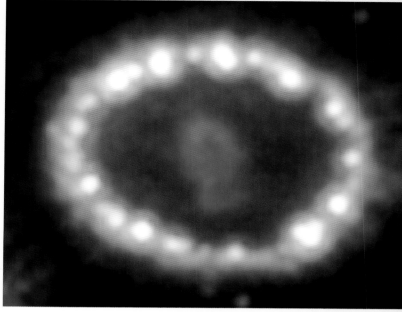

8

set bottom to bottom. This small nebula had been produced by the star that exploded as SN 1987A during previous phases of its life, some 20,000 years before the explosion witnessed by Shelton. In 1998 the HST was able to see that the exploding supernova was about to crash into this smaller nebula. By 2005 the whole of the central ring was involved in the crash, and the nebula lit up like a celestial firework display.

4   Solar Max In the MMU (Manned Maneuvering Unit, below right), astronaut George D. Nelson carefully approaches the Solar Max satellite to bring it back to the Space Shuttle for repair. The successful repair in 1984 (the first ever of a satellite in orbit) extended the life of Solar Max so that it happened to be in space and able to detect gamma rays from the supernova of 1987.

5   SN 1987A in the Large Magellanic Cloud In this picture from the HST, glittering young massive stars and wisps of gas surround SN 1987A several years after it exploded. The supernova occurred in the middle of the bright pink ring at centre, part of the nebula ejected by the star before it exploded.

6   Joe Wampler first inferred the existence of the star's ejected nebula from a spectrum made with the European Southern Observatory (ESO) telescopes.

7   Rings around SN 1987A The outflowing remains of the exploded star lie at the very centre of the central ellipse in this picture made by the HST in 1994. The three other rings are from a nebula created by the star before it exploded. The supernova lies at the centre of the small, bright central ring; the larger rings are in front of and behind the supernova. The whole structure is shaped like a pair of tumblers, held base to base and tilted at an angle to the line of sight.

8   SN 1987A in 2003 Sixteen years after the explosion, the body of the star has expanded halfway out to the bright ring. Shock waves extend outside the visible areas of the explosion and have reached the bright central ring, exciting it so it lights up like a glowing string of pearls.

43

Cepheid Variable Stars

Stars' heartbeats that measure the Universe

Cepheid variable stars have a regular 'heartbeat'. Thanks to the work of pioneering astronomer Henrietta Leavitt, this stellar heartbeat makes it possible to measure the size of the Universe.

1

Cepheid Variable Stars

2

VIRGO — 7,500,000 — H+K — 750 MILES PER SECOND

URSA MAJOR — 100,000,000 — 9,300 MILES PER SECOND

CORONA BOREALIS — 130,000,000 — 13,400 MILES PER SECOND

BOOTES — 230,000,000 — 24,400 MILES PER SECOND

HYDRA — 350,000,000 — 38,000 MILES PER SECOND

3

1 **RS Puppis** A dusty reflection nebula surrounds the Cepheid variable star RS Puppis, which is hidden behind the dark central stripe. As the star pulsates over a period of 41.4 days, its changes in brightness travel throughout the nebula. This allows astronomers to calculate the precise distance to the star as 6,500 light years, since they know the fixed speed at which light travels.

2 **Henrietta Leavitt** Hired by Pickering as a 'computer' at Harvard Observatory, Leavitt discovered the period–luminosity law for Cepheids, on which basis astronomers still measure the distance scale of the Universe.

3 **The expanding Universe** A historic image from the Mt Wilson Observatory using obsolete terminology shows how a distant galaxy ('cluster nebula') looks small because it is far away. At the same time the spectrum of its light (central strip in each right-hand picture) shows a 'redshift' that indicates how fast the galaxy is speeding away from us. Each galaxy spectrum has two gaps labelled H & K, which move relative to the short vertical bars either side; the displacement of these gaps indicates the 'redshift'. The vertical bars are spectra of a stationary gas in the telescope, which anchors each galaxy spectrum. The distance of each galaxy has been estimated using Henrietta Leavitt's period–luminosity relation.

In 1784 English astronomer John Goodricke, who discovered that Algol was a variable star (**33**), discovered a second one, the star Delta (δ) Cephei. This star pulsates, expanding and contracting like a heart beating, the regular changes in size causing its brightness to increase and fade between magnitude 3.6 and 4.4. Goodricke estimated the period of its brightness cycle as 128 hours and 45 minutes: 5.36634 days.

Delta (δ) Cephei became the prototype for similar pulsating variable stars, which were given the generic name of 'Cepheids'; other examples have periods that range from a few days to about a hundred days. These stars are heat engines: they the convert heat energy made in their interiors into mechanical motion that moves their outer layers up and down. In the 1930s the English astronomer Arthur Stanley Eddington explained how this worked in theory; the details were confirmed by the Russian astronomer S. A. Zhevakin the 1950s.

Essentially, Cepheid stars contain a valve mechanism. The valve is closed when star is small, causing pressure to build up and the star to expand. When the valve opens, heat and pressure are allowed to escape and the star shrinks back to its initial size. The 'valve' is a layer of ionized helium in the upper layers of the star. When the star is smallest, the helium is opaque and traps radiation. As the radiation increases, it lifts the ionized layer so that the helium recombines and becomes transparent, releasing the radiation and causing the outer surface of the star to fall again.

Cepheid variable stars are fascinating in and of themselves, but are especially important in astronomy because they are 'standard candles' that can be used to measure the distances of other galaxies. Normally the brightness of a star as it appears in the sky is only a very rough indicator of its distance, as stars vary enormously in brightness from one to another, according to their sizes and temperatures. But the special qualities of Cepheid stars make it possible to estimate their intrinsic brightnesses precisely. One of the main projects of the Hubble Space Telescope (HST), led by American astronomer Wendy Freedman, was to discover and measure the brightness of Cepheid variable stars in external galaxies in order to find their distances from the Earth. The ultimate goal of Freedman's team was to measure the size of the Universe. Henrietta Leavitt's discovery made this possible.

Leavitt was born in Lancaster, Massachusetts, in 1868 and died too early, of cancer, in 1921. She studied at the Society for the Collegiate Instruction of Women (later Radcliffe College and now part of Harvard University) and in her final year took an astronomy course, which enthused her about the subject. An illness left her deaf, but following her recovery she was still hired as a research assistant at Harvard College Observatory at a salary of 30 cents an hour, working for its director Edward Pickering. It was not a time when female scientists were encouraged to carry out independent research, although Leavitt

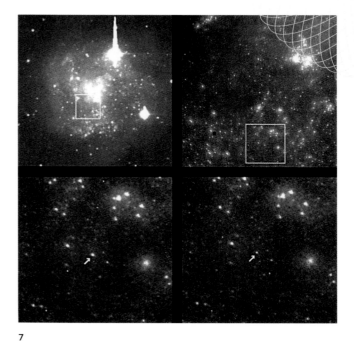

7

4   Harlow Shapley used Leavitt's period–luminosity relation to measure the distances of clusters of stars and determine the size of the Milky Way Galaxy.

5   Ejnar Hertzsprung calibrated Leavitt's period–luminosity relation by finding the distance to a few nearby Cepheids and used it to prove that the Small Magellanic Cloud was a galaxy outside the Milky Way system.

6   Wendy Freedman, of the Observatories of the Carnegie Institution of Washington, led the international team of nearly 30 astronomers that carried out the HST Key Project on the Hubble diagram.

was undoubtedly of the highest intellect and capability. In her work for Pickering she discovered 2,400 new variable stars, doubling the number then known. 1,800 of them were found on photographs of the Magellanic Clouds taken at the Boyden Observatory at Arequipa, Peru. Some proved to be Cepheids.

In 1908 Leavitt discovered that the brighter Cepheids in the Magellanic Clouds took longer to complete their brightness cycles. Leavitt did not know that the Clouds were galaxies separate from the Milky Way (**54**) but she did reason that all the Cepheids in each Cloud were at the same distance from Earth, so their different brightnesses were not an illusion caused by their various distances, and must somehow relate to the period over which they changed their light output. Leavitt's discovery became known as the period–luminosity relation; it demonstrated that Cepheids were standard candles that

could be used to measure distances by comparing apparent and intrinsic brightness. Good standard candles in astronomy are highly luminous (for observation at great distance) and reliably recognizable. Cepheids are indeed bright and they are easy to pick out in the sky because of the particular way that they vary. Good standard candles are also consistent in their brightness; once you have determined the period of a Cepheid, this Cepheid is the same average brightness as others of the same period. You anchor the relationship by locating Cepheids in nearby clusters of stars whose distance you can find by other methods, such as surveyors' triangulation techniques.

After her untimely death, Leavitt's work made it possible for modern astronomers to measure the size of the Universe and discover the exact location of objects within it. Astronomers Ejnar Hertzsprung and Harlow Shapley found that the Magellanic Clouds were outside the Milky Way galaxy, at distances currently estimated to be 160,000 light-years from Earth (**54**). The Andromeda galaxy – the nearest galaxy of the same size as ours – is 2.5 million light years from Earth. Wendy Freedman's team has used Cepheid variable stars to measure the distance of 31 galaxies, out to distances of 70 million light years. All distance calibrators for more distant galaxies, such as supernovae (**63**) are now referenced against this scale.

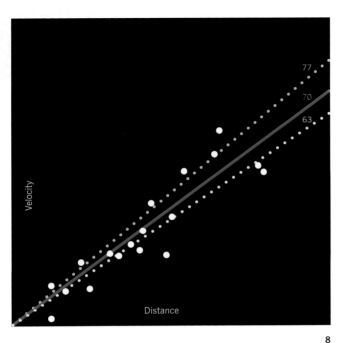

8

7 **Cepheid variable stars in the galaxy IC 4182** This relatively nearby irregular galaxy produced a supernova of Type Ia in 1937; astronomers intensively study it to pin down its distance, so that similar supernovae can be used as distance indicators for other galaxies. In the first two images (top row), the HST homed in on Cepheid variable stars in the galaxy; one is shown in the bottom row at its maximum and minimum brightness.

8 **Hubble Diagram for Cepheids** The plot of the distance of a galaxy versus the velocity at which it speeds away from us in the expansion of the Universe is called a Hubble Diagram, after Edwin Hubble who first used it. Making this diagram as accurate as possible was one of the key reasons to launch the HST. Wendy Freeman's Key Project used Cepheid variable stars to calibrate the expansion rate. The central line is the best fit to the data, the lines either side are drawn at ± 10%.

9 **A Cepheid variable star in the galaxy M100** M100 is a 'grand design' spiral galaxy in which the HST has discovered over twenty Cepheid variable stars, one shown here at three stages of its light cycle. The galaxy's central area has a bright bar from which two spiral arms emerge symmetrically, and thick dust clouds. The central area was magnified when positioned on one of the four CCD detectors in the HST camera, then demagnified when stitched together with the other three images; this process is responsible for the curious stepped appearance of many HST pictures.

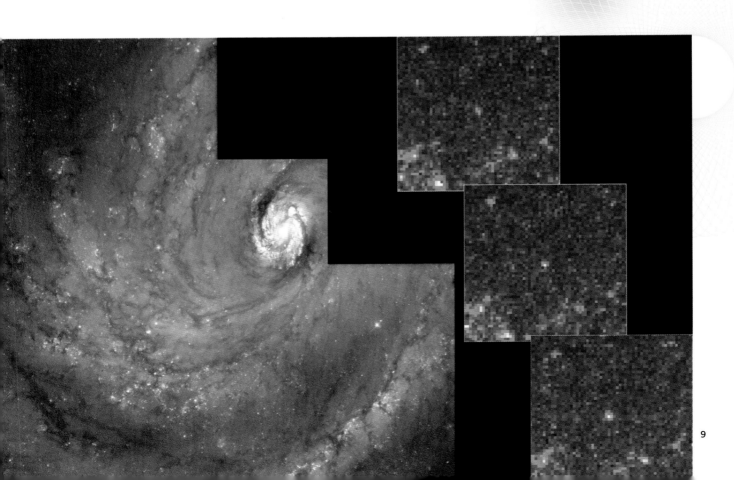

9

44

Exoplanets

Other worlds beyond ours

1

Until a few years ago astronomers knew of only one planetary system in the Universe: our own Solar System. This changed in the 1990s with the discovery of a number of large, Jupiter-sized planets orbiting central stars. Yet none of these exoplanetary systems seems to have formed in the same way as our own Solar System.

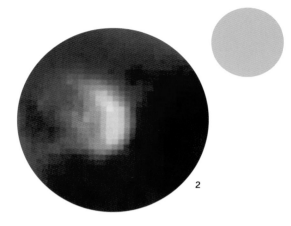

For 3,000 years philosophers had been theoretically convinced that there were other worlds like our own in existence. More recently, astronomers realized that planetary systems are a necessary consequence of the formation of stars. But until the last decade of the 20th century no actual planets had been found orbiting other stars.

In 1992 Aleksander Wolszczan, a Polish-born American radio astronomer, was timing the rapidly rotating pulsar PSR1257+12 (located at a distance of about 1,000 light years in the constellation Virgo). He noticed that its pulses arrived alternately earlier and later than expected. Wolszczan's interpretation was that the pulsar was being pulled nearer to and further from Earth by three Earth- and Moon-sized planets in orbit around it. Wolszczan had discovered the first planets outside the Solar System, which are collectively called 'exoplanets'. But Wolszczan's planetary system exists in circumstances completely different from our own: it consists of the remains of a supernova explosion, and is also a second-generation planetary system, formed not at the birth of a star but at its demise. The system's central star is a neutron star, a star as unlike the Sun as it is possible to get.

In 1995 two teams of astronomers discovered three planetary systems orbiting stars that were much more like our own Sun. Two Geneva Observatory astronomers, Michel Mayor and Didier Queloz, made the first discovery. In April 1994 they had embarked on a programme to detect any changes in the speed of nearby stars that could be due to the gravitational pull of Jupiter-sized planets. The analogy of our own Solar System guided the search. Although we typically say that the planets orbit around the central Sun, this is actually a simplification. In fact the Sun and the planets orbit around their common centre of mass. Because the Sun is so much more massive than any planets, the Solar System's centre of mass is actually inside it, so the Sun scarcely moves in its orbit: its motion of 13 m/second is not much faster than world sprinting records and on an astronomical scale is more of a slow quiver than an orbit. This delicate motion of the Sun is largely caused by the gravitational effect of the most massive planet in the Solar System, Jupiter. The main period of the Sun's motion is therefore the same as the orbital period of Jupiter: twelve years.

To find signs of this 'quivering' motion in other stars, Mayor and Queloz developed a spectrograph capable of detecting such small oscillations. They then drew up a list of 142 bright, nearby solar-type (Sun-like) stars that had been selected because they showed, at coarser accuracy, no large velocity changes that would suggest they were members of double-star

1 **Artist's impression of an exoplanet** Space artist David Hardy wrote of his picture: '*The Millennium Planet* depicts a planet of the star Tau (τ) Boötis, a huge, bluish gas giant, bigger than Jupiter...I postulated a close moon, which would be tidally disrupted, giving rise to tectonic and volcanic features – hence the glows on its night side. There may also be active aurorae around the poles of the gas giant.'

2 **Protoplanetary disc in the Orion Nebula** A newly born star glows orange at its centre, its light reddened by the nearby dust and made blue in the light reflected from the far side of the disc.

3 **Michel Mayor** of the Geneva Observatory, co-discoverer of 51 Pegasi, the first exoplanet discovered in orbit round a solar-type star. The system is shown in the artist's impression behind Mayor's head (see Fig. 7).

AN INFINITY OF WORLDS

For there is a single general space, a single vast immensity which we may freely call Void; in it are innumerable globes like this one on which we live and grow.

Giordano Bruno, *On the Infinite Universe and Worlds*, 1584

4

6

5

systems. It seems that planets only survive in orbit around single stars: astronomers calculate that a planet in a double-star system would loop in complicated orbits among the two stars, and in a relatively short time would be ejected from the system.

Despite their careful planning, Mayor and Queloz had not expected to be successful so quickly. Their original programme was essentially a career decision, since they could only be sure that they had found a twin of Jupiter by observing at least two periods of the duration of Jupiter's orbit – amounting to 24 years in total! But within only eighteen months they had discovered their first planet, orbiting the star 51 Pegasi, 45 light years distant. Even more surprisingly, the star's oscillation was much greater than expected, and its period was very much shorter – 4.2293 days! The planet, 51 Peg B, is roughly the same mass as Jupiter. Its orbit is close to circular and its orbital period of four days indicates that the planet lies much closer to its sun than does Jupiter to our Sun, in fact, closer than any planet in our Solar System. The distance of 51 Peg B to its parent star is only 1/20 of the Earth–Sun distance.

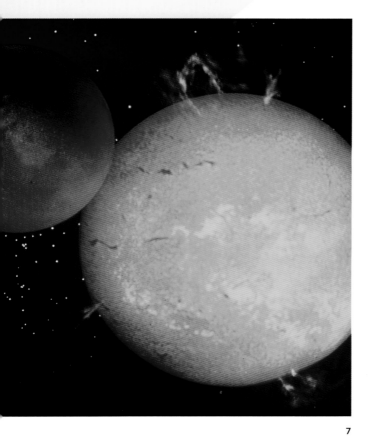

4 The 1.93 m telescope at Haute-Provence used to discover 51 Pegasi's planet.

5 Exoplanet HD209458b This exoplanet was the first that had been detected by looking for its orbital effect on its parent star; it then was seen in transit across the face of the star. It is about 2/3 of the mass of Jupiter but slightly bigger, because it is bloated by the heat of its star. The HST observed its spectrum against the light of the star and established that the planet is surrounded by an exosphere of escaping hydrogen gas.

6 Geoff Marcy of the University of California Berkeley, leader of the team that has found the largest number of exoplanets so far.

7 51 Pegasi b SETI Institute astronomer Seth Shostak drew this artist's impression of the planet orbiting 51 Pegasi. The planet is informally named Bellerophon.

7

As news of the discovery spread around the world, Mayor and Queloz's observations were swiftly confirmed during a brief four-day observing run with the powerful Lick Observatory telescope conducted by Geoffrey Marcy of San Francisco University, Paul Butler of the University of California, and a team from the High Altitude Observatory and the Harvard Smithsonian Center for Astrophysics. Like Mayor and Queloz, Marcy and Butler had been monitoring solar-type stars for changes of speed variations indicating the presence of jupiters, but 51 Pegasi was not on their original observing list. After hearing of the Swiss discovery, and altering their expectations about the orbital periods of exoplanets from decades to days, the team quickly found further examples, some of them in archives of previous observations that they had not yet examined closely because they thought there was no rush.

Another technique that has been used to discover exoplanets is to look for a 'winking star': a star whose light is periodically dimmed a little by the transit of a planet across its face. This has the potential to detect planets smaller than jupiters, in fact, 'earths'. In February 2009, the French space satellite COROT, launched to find transiting exoplanets, found a planet – cumbersomely listed as COROT-Exo-7b – which is only 1.7 times the diameter of the Earth, and perhaps five times the Earth's mass. With an orbital period of only twenty hours the planet is very close to its sun, and its surface has a temperature of over 1,000 °C, but apart from its lava-like surface, it is the most Earth-like exoplanet yet discovered. All in all, nearly 400 exoplanets have been found since the mid-1990s.

Due to technical constraints, only very specific kinds of exoplanets have been discovered thus far, representing a minuscule proportion of the probable total. Most of the stars that are currently known to have exoplanetary systems are near to Earth – up to about 1,000 light years away – and therefore relatively bright; you need a lot of light to measure accurately the small radial velocity shifts produced by jupiters or the small winks produced by earths. Nearly all these exoplanetary systems consist of a single large Jupiter-sized planet orbiting close to a central star, because the largest planets are easiest

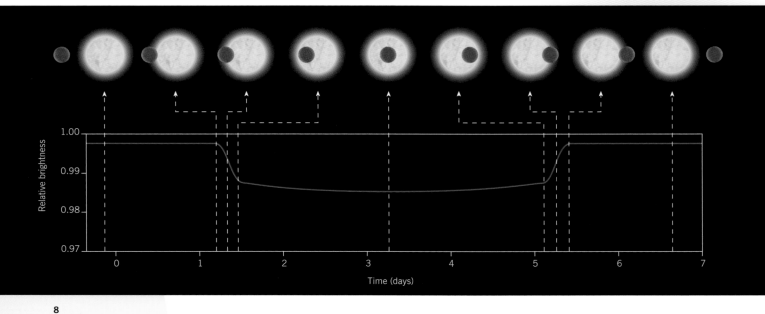

8

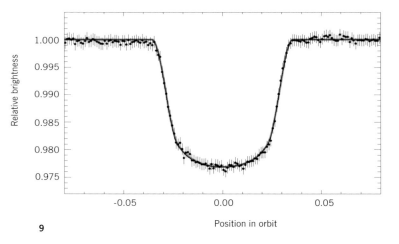

9

to discover with the presently available equipment, which has only been deployed for a few years, although it seems likely that Earth-sized planets also exist, perhaps in the same systems.

These new systems of planets, large and close to their central stars, contradicted astronomers' theories about the formation of our own Solar System. Astronomers had believed that planets as near to their suns as most of the ones discovered recently would be small terrestrial planets, resembling Mercury. Large planets were not supposed to be so close to their suns, but were expected to orbit the colder outer reaches of a planetary system, like Jupiter, Saturn, Neptune and Uranus, which retain their gases because they are massive and cold (**27**). The explanation for the apparent reversal seems to be that new-born planets

migrate in towards their sun, the bigger ones swallowing up smaller inner ones. In fact this phenomenon had been discovered and investigated theoretically in 1978 by astronomers based around the University of Cambridge (such as Douglas Lin, John Papaloizou, Peter Goldreich and Scott Tremaine) without its significance for planetary formation being recognized. If this phenomenon is routine during the formation of planetary systems, something exceptional – it is not known what – must have happened in our own Solar System to stop the migration of Jupiter. This was good for us: otherwise our Earth probably would not have survived.

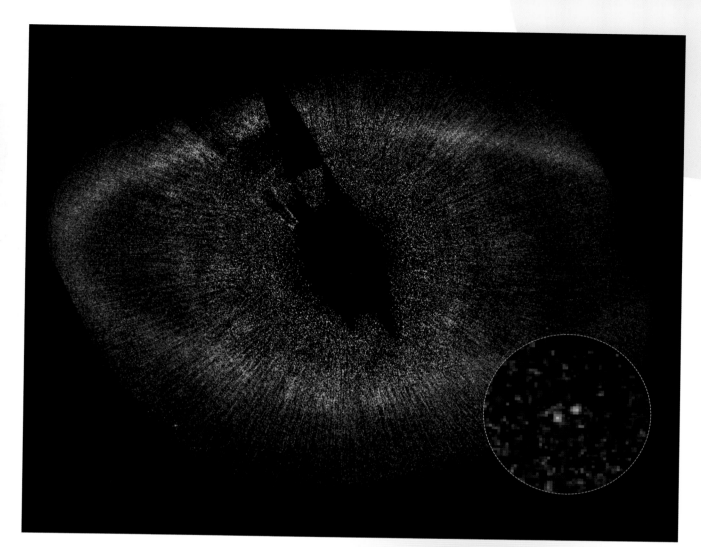

10

8 **Transit** An opaque planet passes across the face of its star, causing light from the star to dip. Features of the light curve, such as the duration of the transit, show how big the planet is.

9 **Corot-Exo-1b** In 2007 the French space satellite COROT detected the exoplanet as it passed across an otherwise unremarkable solar-type star in the constellation Monoceros. The planet is a very hot gas giant, with a radius equal to 1.78 times and a mass equal to 1.3 times those of Jupiter. It circles its star every 1.5 days in an orbit that is almost edge-on to us, which made it easy for COROT to spot the transit.

10 **Fomalhaut's dusty disc and planet b** The dusty disc, which is similar to the Kuiper Belt, was discovered in the 1980s in orbit around the star by the IRAS satellite and imaged by the HST in 2004. It contains what appears to be a planet, shown inset as it moved in its orbit between 2004 and 2006.

11 **HR 8799 and the 'first family'** The three planets orbiting the solar-type star HR 8799 were discovered by their gravitational pull on it. The Gemini Observatory telescope was able to image two of them (circled) by hiding the light from the star; they are seven to ten times the mass of Jupiter.

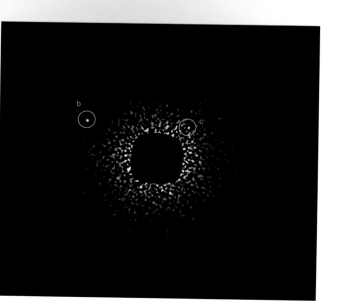

11

45

The Energy of the Sun and Stars

Discovery of nuclear fusion

1

How does the Sun shine, and how long has it been shining? Could the fuel ever run out? These questions have perplexed scientists since they first began to comprehend the awesome size and age of the Solar System. The secrets of the Sun's power system have – for better or for worse – made nuclear technology possible on Earth.

As soon as the scale of the Solar System and the distances of the stars became apparent, it became clear that the amount of heat and light emitted by the Sun and stars was extraordinarily great. When astronomers calculated the mass of the Sun using Newton's laws of gravity, it too was enormous. If the Sun is a kind of a normal fire, providing power by chemical means (for example, by chemically combining carbon and oxygen to make carbon dioxide, as happens when wood or coal is burnt), there is a lot of fuel available to feed the fire that we see. But for how long could the fuel last?

Presumably the Sun is at least as old as the Earth – the one depends on the other. In 1650, to determine the age of the Earth, Archbishop James Ussher published an analysis of the chronology of events in the Bible, which was reproduced in the standard edition of the Bible used in England for centuries and so became widely accepted. He set the date of the creation of the Earth at 4004 BCE. If the Sun really had been created along with the Earth just 6,000 years ago and chemical energy was the source of its power, the amount of material consumed as the Sun shone was only a small fraction of its total mass.

But in the 19th century British geologists such as Charles Lyell and John Phillips began to estimate that the Earth was actually millions of years old, based on calculations of how long it would take for sedimentary rocks to be laid down from sea deposits or for rocks to be eroded away (20).

This created a problem for the idea that chemical energy was the source of the Sun's power – a conventional fire or chemical reaction would not be able to continue burning for millions of years. The discrepancy was made worse when calculations by physicist Lord Kelvin and biologist Thomas Huxley suggested that the Earth was actually hundreds of millions of years old. Even this seemed short, considering

1 **The Sun** The only star whose surface we can view in any detail, the Sun has a surface temperature of 6,000 °K. In this ultraviolet image from the SOHO satellite, prominences leap and arch above the surface at temperatures even ten times higher. The radiation and the motions of the outer part of the Sun are driven by nuclear energy generated deep in its inner core.

2 *Pegwell Bay: a recollection of October 5th, 1858* (**William Dyce, 1858**) The chalk cliffs of the coast in the south-east of England show strata that were laid down over many millions of years. In the upper left of the painting is Comet Donati; both it and the Sun, setting from view beyond the horizon, must have been shining for at least as long. Dyce (who depicts himself standing at the foot of the cliff at the extreme right) contrasts their vast ages with the brief lifespans of the people who comb the beach for small treasures and paint on fragile canvas, oblivious of the evidence of the near-eternal things around them.

the length of time needed for the evolutionary processes that Charles Darwin envisaged in 1859 to produce the variety of living species on Earth. The German physicist Hermann von Helmholtz and the Canadian astronomer Simon Newcomb came up with another idea at the end of the 19th century, suggesting that the Sun's energy was supplied not by chemical energy but by gravitational contraction. They calculated that the Sun's age could be consistent with the Earth's if this was indeed the case.

But in the 20th century scientists began to realize that the Earth was much older even than hundreds of millions of years. After Henri Becquerel and Marie and Pierre Curie discovered radioactivity, the British physicist Lord Rutherford developed a technique for using radioactive decay to measure the age of rocks, by which a young American chemist, Bertram B. Boltwood, discovered that some rocks were as much as 1–2 billion years old. How could the Sun keep shining for such a length of time?

The answer was discovered by two physicists from the University of Göttingen, Fritz Houtermans and Robert d'Escourt Atkinson. As they passed their summer holiday in 1927 on a walking tour they discussed the problem of the source of the Sun's energy. They knew about the physical conditions inside the Sun from the work of the British astrophysicist Arthur Stanley Eddington: the high density and high temperature inside the Sun created high pressure, which countered the force of gravity that was drawing the material of its body tightly together. Atkinson knew of Einstein's formula for converting mass to energy, $E = mc^2$, and understood that what was then called 'atomic transmutation' was possible. The atoms (or, as we now know, their broken-down nuclei) in the centres of stars and the Sun are frequently colliding together because of the high densities and high temperatures. If the collisions transform some atoms from one kind to another, losing mass in the process, atomic (we would now say 'nuclear') energy would be produced. This might be the source of the Sun's energy. 'Let's

just work the thing out, shall we?' said Houtermans. 'How could it happen in the Sun?'

The two young scientists discovered how the fusion of light elements into heavier ones could fuel the Sun. When Atkinson later learned that the Sun was mainly hydrogen, he realized that the source of its energy was specifically the conversion of four hydrogen atoms to one helium atom. A helium atom is 0.7% lighter than four hydrogen atoms. This tiny amount of excess mass, multiplied many times over by the astonishing number of hydrogen atoms present in the Sun and the frequency of their collisions, provides the Sun's energy. 400 million tonnes of mass disappears from the Sun every second and is transformed into solar energy. This attrition rate can be kept up for billions of years.

Houtermans and Atkinson's work was followed up in 1939 by the German physicist Hans Bethe. He discovered the exact process that enables hydrogen fusion in the Sun. It is called the CNO cycle, since the hydrogen nuclei are fused together in successive stages, with carbon, nitrogen and oxygen as intermediate steps. Bethe was awarded the Nobel Prize in 1967 'for his contributions to the theory of nuclear reactions, especially his discoveries concerning the energy production in stars'. The details of what happens inside the Sun have since been confirmed with astonishing accuracy by the detection of neutrinos, tiny particles given out in its nuclear processes, which travel from the Sun's interior and have been detected on Earth using specially built neutrino detectors (**47**).

In the decades after Bethe's discoveries, physicists from many nations were organized into programmes that attempted to reproduce on Earth what happens in the stars, finding ways to release nuclear energy slowly in a reactor or suddenly in a bomb. For better or for worse, many of these projects have been successful. The discovery of the energy source of the Sun and stars may well prove to be the most momentous secret of the Universe ever uncovered.

3 **Hermann von Helmholtz** calculated the age of the Sun.

4 **Richard d'E. Atkinson** co-discoverer of nuclear fusion in the Sun.

5 **Simon Newcomb** director of the US Nautical Almanac Office.

6 **Madame Marie Skłodowska Curie** investigated radio activity and received two Nobel Prizes, but paid with her life when she died of radiation poisoning.

7 **Hans Bethe** Nobel laureate for his contributions to the theory of nuclear reactions, especially concerning the energy production in stars.

8 **Ernest Rutherford** British physicist, awarded the Nobel Prize for chemistry, and known for his work on atomic and nuclear physics.

9 **Ivy Mike** The first hydrogen bomb (tested by the USA on 1 November 1952), was powered by a device that fused deuterium and tritium to helium in a nuclear reaction similar to the one that powers the Sun.

HOUTERMANS FAILS TO IMPRESS

That evening after we had finished our essay, I went for a walk with a pretty girl. As soon as it grew dark, the stars came out, one after another, in all their splendour. 'Don't they sparkle beautifully?' cried my companion. But I simply stuck out my chest and said proudly: 'I've known since yesterday why it is they sparkle.' She didn't seem the least moved by this statement. Perhaps she didn't believe it. At that moment, probably, she felt no interest in the matter whatever.
Fritz Houtermans, in Robert Jungk's
Brighter than a Thousand Suns, 1958

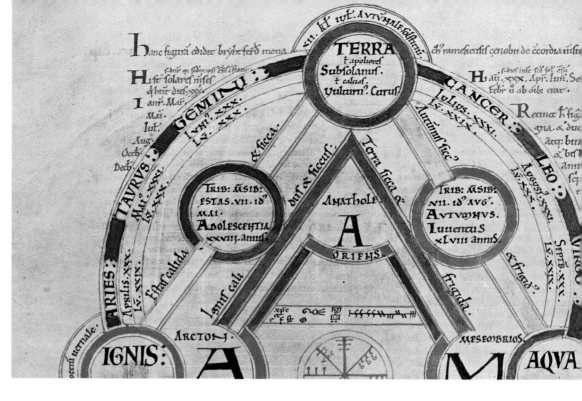

46

The Origin of the Elements

Making star stuff

All of the elements in the Universe – including the building blocks of our own bodies – are made inside stars.

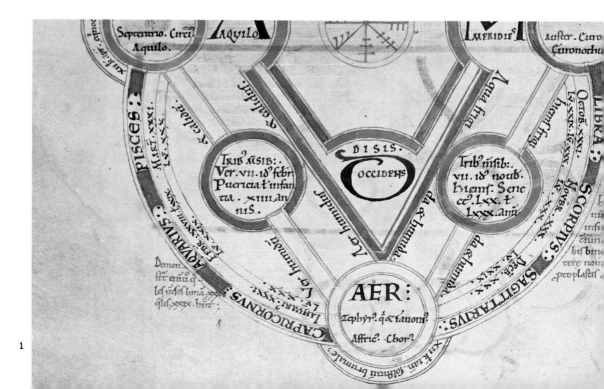

1

1 **Four elements** Schematic diagram linking the Aristotelian elements, the directions, the seasons, and the zodiac in a medieval English illuminated manuscript, now at St John's College, Oxford, *c.* 1110. From the top clockwise: earth (east), water (south), air (west), fire (north).

2 **Henry Norris Russell** American astrophysicist from Princeton University.

2

W e are bits of stellar matter that got cold
by accident, bits of a star gone wrong.
Arthur Stanley Eddington, 1932

Classical philosophers thought that every substance in the Universe was made of four basic elements – earth, water, air and fire – but in the 18th and 19th centuries, chemists realized that all materials were made of molecules, and molecules themselves were made of atoms in fixed arrangements (**28**). In the 20th century the actual physical make-up of atoms was discovered. Each atom consists of electrons orbiting a nucleus, itself made of protons and neutrons. The number of electrons in an atom is equal to the number of protons in its nucleus. The number of neutrons in an atom is roughly equal to the number of protons, but can differ from atom to atom of the same element: atoms with these variations are called 'isotopes'.

There are about 100 chemical elements. The atoms of each element are distinguished by the number of protons they have (or equivalently electrons, since their numbers are the same). Changes in the arrangement of the electrons produce light; astronomers can see the light with a spectroscope, and, in general, the clearer the spectral signature of a particular element in a celestial body like a star, the more of that element is present in the star – its 'abundance'.

In her 1925 doctoral thesis, Harvard astronomer Cecilia Payne (later Payne-Gaposchkin) suggested that hydrogen was the most abundant element in the Sun. Reviewing her thesis, the influential Princeton astronomer Henry Norris Russell dismissed the idea, but changed his mind in 1929. 70 other elements have since been identified in the solar spectrum. 71% of the Sun's mass is hydrogen, 27.1% is helium, together comprising 99.9% of the number of atoms in the Sun. The remaining 0.1% are (in order of abundance): oxygen, carbon, nitrogen, magnesium, silicon and neon. Payne-Gaposchkin had discovered that the abundance of elements in the stars was broadly the same as in the Sun, so astronomers could treat the Sun as representative of other stars.

After Fritz Houtermans and Robert Atkinson discovered in 1927 that the stars generate energy by nuclear reactions (**45**), astrophysicists were able to address the question of where the elements came from and why some are more abundant in the Universe than others. In 1939 Hans Bethe showed how hydrogen transformed into helium via a cycle involving carbon, nitrogen and oxygen. This explained one source of helium – but where did the hydrogen, carbon, nitrogen and oxygen come from in the first place? According to a 1948 paper known as 'αβγ' (alpha-beta-gamma) after its authors Ralph Alpher, Hans Bethe and George Gamow, the hydrogen was made in the explosion (later known as the Big Bang) that started the Universe. Elements were built up in stages from the simplest element, hydrogen, by successively adding neutrons one at a time to make heavier and heavier nuclei, and additional helium.

The theory failed because it could not create elements heavier than lithium. There is no stable atom with four protons and four nucleons, so when you get to eight nucleons, the nucleus decays spontaneously back to seven. Armagh Observatory director Ernst Öpik and astrophysicist Edwin Salpeter found a better explanation in 1951–52: carbon, with twelve protons and twelve neutrons, is made when three helium nuclei (each with four protons and four neutrons) collide at the same time inside a star.

In 1953 cosmologist and nuclear physicist Fred Hoyle discovered the exact nuclear reactions involved in this process, although some of the bold assumptions he made to tie up the loose ends in his theory were thought by many to be crazy. He persuaded physicist Ward Whaling in the Kellogg Radiation Laboratory at the California Institute of Technology to measure some of the properties of nuclei and find if they had the values that Hoyle needed to make his theory work. They did.

3

4

5

Hoyle was the first to establish the currently accepted explanation for how elements are made inside stars. Although some of his colleagues said that Hoyle's motivation was to find an alternative to the idea that the elements originated in the Big Bang, he had started work on the topic long before he proposed the rival Steady State theory (59). In studying the way that stars evolve he had simply asked himself the question 'What would be the very last of the nuclear reactions that take place in stars, instead of the first reactions, which had so far occupied the attention of astronomers?'

The detailed scheme by which the elements are made in stars was summarized by Margaret Burbidge, Geoffrey Burbidge, William Fowler and Fred Hoyle in 1957. Their seminal paper is known by astronomers as B^{2}FH ('B-squared FH'), from the authors' initials and is one of the most frequently referenced papers in astronomy. The Burbidges provided the astronomical expertise for the paper and Fowler and Hoyle contributed the nuclear physics. In 1983 Fowler received the Nobel Prize 'for his theoretical and experimental studies of the nuclear reactions of importance in the formation of the chemical elements in the Universe'; it is not clear why the Nobel Prize committee did not also honour Hoyle.

The B^{2}FH paper started from Houtermans, Atkinson and Bethe's discovery that stars make helium by 'burning' hydrogen, which is the most simple and abundant element in the Universe, but pushed the idea further, proposing that the burning process goes on to convert helium to carbon, carbon to oxygen, oxygen to neon, then silicon and finally iron. When a star explodes as a supernova the explosion irradiates elements such as carbon and oxygen in the body of the star, producing additional elements that the hydrogen burning process cannot make on its own.

Other elements are produced on the surfaces of red-giant stars. Dramatic proof of this was provided in 1952 by Mt Wilson astronomer Paul Merrill, who discovered the spectral lines of the element technetium in some red giants. Technetium is radioactive and even its longest-lived isotope decays relatively quickly – in a matter only of a million years. Since red giants are much older than this, the technetium must have been made recently on the surface of the star.

The elements made in stars are dispersed into interstellar space by supernova explosions, stellar winds and the formation of planetary nebulae. There they mix with hydrogen gas and form clouds that may eventually condense into stars. These are the origins of all the chemical elements that make up the Earth and all that is on it, including ourselves.

3 **Ralph Alpher** American physicist, who as a student gave flesh to the theory that the chemical elements were created in the Big Bang.

4 **George Gamow** Russian-born American atomic physicist, cosmologist and popular-science writer, known for his development of Big Bang theory.

5 **Fred Hoyle** Iconoclastic British physicist, known for his work on the structure of stars, the origin of the elements and nuclear astrophysics.

6 **Margaret Burbidge, Geoffrey Burbidge, William Fowler and Fred Hoyle** Known collectively as 'B-^{2} FH', the team admires a model steam train in a light moment of relaxation from their work on the origin of the elements in stars.

7 **N49 supernova remnant in the Large Magellanic Cloud** Supernovae explode the contents of their parent stars into space and spread newly made atoms of the heavier elements into interstellar hydrogen. This picture was taken by the Hubble Space Telescope.

7

47. Inside the Sun

Whispers and rings

We cannot look at the interior of the Sun directly, but over the past century we have developed ingenious ways to 'see' what happens inside it. Astronomers first captured neutrinos, incredibly small particles generated inside the Sun, by burying a vast quantity of dry-cleaning fluid in a gold mine. Studies of solar quakes showed that the Sun rings like a gigantic bell and provided clues to its interior makeup.

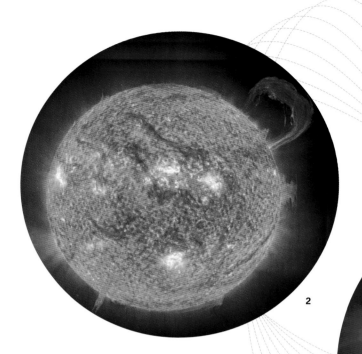

1 **Solar prominences** (red) are most readily seen as they stand above the granulated surface of the Sun at its edge (or 'limb').

2 **Image from the SOHO satellite** Giant solar prominences ride on magnetic field lines that arch above sunspots. The magnetic fields are generated under the Sun's surface as its interior layers rotate against each other.

3 **Solar activity** The SOHO satellite reveals the Sun's turbulent surface, its sunspots linked by arching loops, and the magnetic streaking of its atmosphere.

Astronomers who wanted to understand what happens inside the Sun faced one big problem: none of its internal workings could be seen because the Sun is completely opaque. At first only the Sun's surface characteristics and its global properties – such as its diameter and the amount of energy that it radiates – could be determined by direct observation. However, we now know what happens inside the Sun thanks to three lines of astronomical enquiry. Ingenious mathematical calculations built up a theoretical picture of the Sun's interior. This picture was verified by enormous efforts capture a tiny quantity of the neutrino particles that the Sun produces and later by measuring the sound waves generated by motions inside the Sun.

Our understanding of the inner workings of the Sun is the result of one of the great feats of modern mathematical reasoning. From the 1920s astronomers knew the physical conditions inside the Sun by calculation and from the 1930s they knew that nuclear reactions were the source of the Sun's energy (**45**). In the 1950s they had begun to understand the way that stars evolve in relation to one another from observations of star clusters (**40**). At the same time that the calculations became progressively more precise, the

way that they fitted with the observations built up astronomers' confidence in their theoretical knowledge of the Sun's interior.

The physics suggested that the Sun would be an abundant source of neutrinos. Neutrinos are small nuclear particles, whose existence was suggested by the Austrian-Swiss physicist Wolfgang Pauli in 1930 to explain some details of nuclear reactions, and they were first seen experimentally in 1956.

Neutrinos are made in the nuclear reactions inside the Sun and carry off small parcels of energy. The numbers of neutrinos given off by the Sun is immense – about ten billion pass through every square centimetre of the Earth every second – but they are whisper-quiet and can travel through a light-year (ten trillion

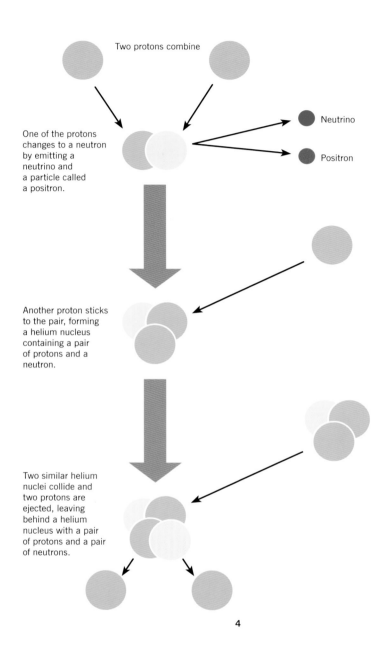

Two protons combine

One of the protons changes to a neutron by emitting a neutrino and a particle called a positron.

Neutrino

Positron

Another proton sticks to the pair, forming a helium nucleus containing a pair of protons and a neutron.

Two similar helium nuclei collide and two protons are ejected, leaving behind a helium nucleus with a pair of protons and a pair of neutrons.

4

4 **Proton–proton chain** This diagram shows how neutrinos are generated during nuclear reactions inside the Sun. The net result of this chain is that four protons (coloured red) combine to make a helium nucleus. Energy is released at each stage, and in the first stage some of that energy is released as a neutrino. If this neutrino can be captured on Earth, it constitutes direct evidence about the nuclear reactions in the Sun.

5 **Raymond Davis, Jr** swimming in the water surrounding the chemical tank in the Homestake Mine in 1971. The nearly 1.5 million litres of water, installed to shield the detector from background radiation, also provided welcome relief from the sweltering heat that the scientists endured during their twelve-hour shifts in the mine.

km) of material without interacting with it in any detectable way. Neutrinos travel as fast as light so that it takes only eight minutes for them to reach the Earth. Despite the astonishing speed and elusiveness of solar neutrinos, it is possible to build detectors that do catch a few.

The first solar neutrino detector was built by Brookhaven National Laboratory physicist Raymond Davis, Jr following technical suggestions from the notorious Italian-born physicist Bruno Pontecorvo (who later defected to join the Soviet nuclear programme) and American physicist Luis Alvarez. Three American nuclear astrophysicists – William Fowler, Alistair Cameron and John Bahcall – had insisted that it was practical to try to catch neutrinos, as the vast numbers constantly released by the Sun increased the likelihood that a few might be captured. In the bowels of the Homestake Gold Mine, in Lead, South Dakota, deep enough underground to avoid interference from cosmic rays, Davis installed a tank containing 615 tonnes of carbon tetrachloride, a solvent normally used for dry cleaning. Some solar neutrinos were captured on the chlorine atoms in the solution and converted to argon atoms. These argon atoms were flushed out of the tank every two months and counted.

The original estimates were that just seventeen argon atoms would be produced in the tank in the each extraction run, but in fact, in the first experiment in 1968, lasting six months, even fewer neutrinos were seen. As Davis repeated his experiment with improved equipment, the question became 'where are the missing neutrinos?' – this is known as the 'solar neutrino problem'. Another neutrino detector called Kamiokande, built and operated by Japanese astrophysicist Masatoshi Koshiba, confirmed in 1989 that Davis had detected neutrinos from the Sun. Unlike Davis's tank in the Homestake Gold Mine, Kamiokande could see where the neutrinos came from and Koshiba was able to prove that the neutrinos it captured really came from the Sun and that there were indeed lots missing.

At first, some physicists thought that the discrepancy between Davis's result and theory had arisen because astronomers' standard calculations relating to the solar interior were flawed. The calculations relied on knowing the density, composition and temperature inside the Sun, and these parameters were not accurately enough known. But the astronomers rejected this explanation, in part because they had found an independent way to look inside the Sun. The approach they used was called 'helioseismology'.

Helioseismology is the study of oscillations in the body of the Sun, which resemble earthquakes studied by

Inside the Sun

seismologists on Earth. In the general turmoil of motion
of hot material in the Sun's interior, the Sun generates
sound waves whose resonances travel across the body of
the Sun, causing its surface to oscillate up and down. The
Sun rings, like a bell quietly singing as it is brushed by a
succession of impacts from a stream of sand grains.

Cal Tech physicist Robert Leighton discovered the surface
oscillations of the Sun in 1960, and measured the oscillation
periods at about five minutes. In the 1970s UCLA physicist
Roger Ulrich suggested that the duration, frequency and tone of
these oscillations could provide clues to the composition of the
Sun's interior. Ulrich pointed out that the frequencies at which
the Sun rings depend on the time it takes sound to cross the
Sun. This in turn depends on the composition, temperature
and density structure of the solar interior. The sound waves thus
carry information about the interior of the Sun to the surface

5

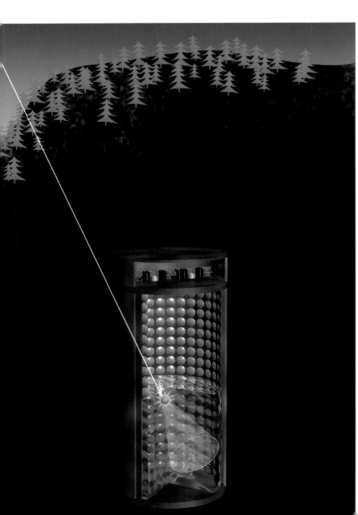

6

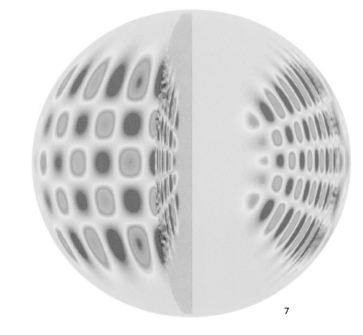

7

6 **Neutrinos from the Sun** (out of the picture, upper left) penetrate the
atmosphere; most of them zoom right through the Earth and continue into
space. However, in the Kamiokande Neutrino Observatory, a few interact
with water in a chamber and generate a burst of 'Cherenkov light' which
is caught by some of the 1,000 detectors in the walls of the chamber and
used to measure the energy and direction of incoming neutrinos.

7 **A computer-generated picture of the Sun's surface** shows a sound wave
resonating in the Sun's interior. In red and blue zones the surface of the Sun
is oscillating up and down. Each oscillation lasts for 340.613 seconds (just
over five minutes), giving an amazingly accurate measurement of the speeds
at which sound travels through the Sun.

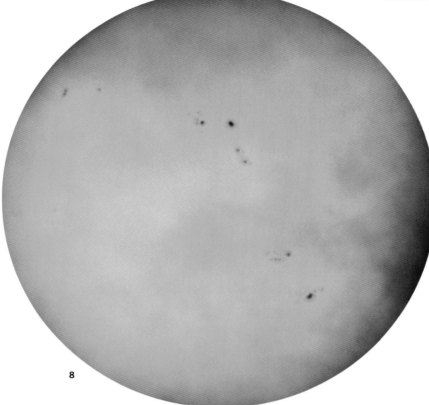

8

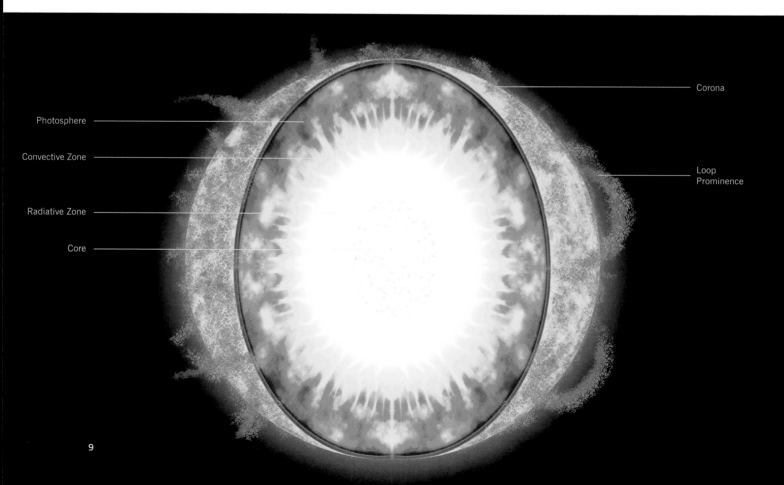

Corona

Photosphere

Convective Zone

Loop
Prominence

Radiative Zone

Core

9

HELIOSEISMOLOGY

There's not the smallest orb which thou behold'st
But in his motion like an angel sings,
William Shakespeare, *The Merchant of Venice*

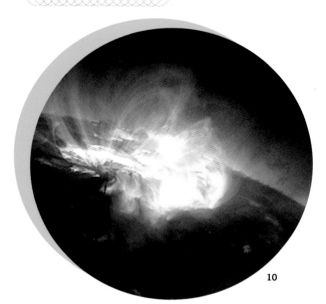

10

8 The surface of the Sun The outer layers of the sun are opaque, but by calculating the amount of heat that the Sun radiates in relation to its size, astronomers were able to make deductions about what goes on inside.

9 Inside the Sun A cutaway diagram of the Sun shows its core, surrounded by the layers of its body, up to the surface ('photosphere') and atmosphere ('corona').

10 Magnetic explosion The magnetic fields generated by friction between the different interior layers of the sun burst explosively through the surface, pushing aside the surface gases and funnelling solar surface material into the Sun's atmosphere.

where it can be seen, just as the oscillations of earthquakes carry information about the interior structure of the Earth.

Individual Earth-based telescopes had a great limitation: they could not observe the Sun after it disappeared daily below the horizon at night. Astronomers therefore set up networks of ground-based solar telescopes around the world to measure the frequencies of solar oscillations more accurately – the networks have names like GONG (Global Oscillation Network Group of the US National Solar Observatory), BiSON (Birmingham Solar Oscillations Network), and HiDHN (High Degree Helioseismology Network) – but intermittent cloud still interfered with their observations. The Solar and Heliospheric Observatory (SOHO) satellite, a joint project involving the European Space Agency and NASA, avoided even this limitation. It has been staring at the Sun continuously from space since its launch in 1995.

The comprehensive observations of the SOHO satellite provided new data on the temperature inside the Sun, and the way that its interior rotates slower than its surface layers, generating a hot layer inside the Sun that is the ultimate

cause of sunspots and prominences on its surface. SOHO also proved that the standard calculations that had been used to measure sound-speed at various depths in the interior of the Sun were 99.9% accurate. Clearly, Davis's 'missing neutrinos' were not the result of a miscalculation of solar conditions. Assuming that astrophysicists knew about the inside of the Sun and nuclear physicists knew how many neutrinos that would create, physicists now began to suspect that something happened to neutrinos after they had left the Sun: some of them did not make it across space to the Earth. This explanation was first proposed by Pontecorvo only a year after Davis first found the solar neutrino discrepancy in 1968.

Neutrinos come in three different kinds, or 'flavours'. They can oscillate from one 'flavour' to another as they travel for eight minutes across the distance between the Sun and the Earth. Davis's detector had been designed only to capture neutrinos of the original 'flavour' generated deep within the Sun. By the time the neutrinos arrived on Earth, many of them had acquired a different flavour through oscillation and could not be detected by Davis's device. Terrestrial experimental evidence for this explanation was discovered in the early years of the 21st century, most significantly by the Japanese Kamiokande detector.

Astronomers were proud that their meticulous work on the Sun had led to a new discovery about particle physics. The importance of this work was justly recognised by the award of the Nobel Prize in 2002 to Masatoshi Koshiba and Raymond Davis 'for pioneering contributions to astrophysics, in particular for the detection of cosmic neutrinos'.

The Crab Nebula

A supernova remnant

1

Chinese imperial astrologers and Native Americans in what is now the south-western United States recorded the appearance of a 'new star' in 1054. 1,000 years later, Knut Lundmark and Edwin Hubble realized that these early observers had witnessed the birth of the Crab Nebula, a magnificent supernova remnant that has intrigued astronomers since it was first mapped in the 18th century.

2

3

1   **The Crab Nebula** An ellipsoidal network of red and orange filaments is what remains of the star that exploded in 1054. The ellipsoid encloses a luminous white glow from electrons trapped inside. They are injected into the nebula by the pulsar at the centre of this supernova remnant.

2   **The sketch that named the Crab.** Made with Lord Rosse's 6-foot reflector in Ireland, the picture looks at first sight little like the nebula that we see on photographs, except that the mottling is reminiscent of the network of filaments.

3   **The 'Leviathan of Parsonstown'** Lord Rosse's massive 6-ft telescope at Parsonstown was slung on ropes and chains between two masonry walls. It could track for a limited time while viewing an object that was crossing the meridian (due south). Ladders gave access to the eyepieces. Lieutenant-Colonel Harry J. Watson, the project leader of the telescope, is seated on a ladder at left.

Like the planet Uranus, the Crab Nebula was found as a result of a systematic whole-sky survey, conducted by an 18th-century British doctor, John Bevis, who had an observatory near London. In 1745 he compiled his observations into an atlas called *Uranographia Britannica*, but the etched plates were costly to produce and the printer went bankrupt before printing it so only a few proof copies survive. On the map of the constellation Taurus, near the star Zeta (ζ) Tauri, Bevis drew a patch to represent a misty nebula that he had discovered.

The French astronomer Charles Messier used a copy of Bevis's atlas on his search for the predicted return of Halley's Comet in 1758. He found another comet, a new one, which passed through Taurus and drew his attention to the misty nebula. Comets and nebulae look much the same in a small telescope, so to avoid confusion Messier, who was known as the 'Ferret of the Comets', decided to make a list of known nebulae. The first item in Messier's catalogue was M1, Bevis's nebula, which became known as the 'Crab Nebula' after a weird sketch showed its 'claws'. The sketch had been made in the 1840s by William Parsons, the third Earl of Rosse, after viewing the nebula through his 'Six-Foot' telescope (the 'Leviathan of Parsonstown') at Birr Castle in Ireland. Modern pictures show the nebula as a generally oval shape of white light surrounded by a lacy network of filaments. The filaments are fragments of the body of an exploding star, and the white light comes from electrons spiralling around the star's magnetic field within the filaments.

The event that created the nebula was identified by the Swedish astronomer Knut Lundmark in 1931 as he listed the novae that had been recorded by Chinese astronomers and imperial historians. Chinese emperors maintained courts of astrologers who studied the sky in order to infer the future of affairs of state. Some of the signs that they recorded included 'guest stars' – temporary celestial phenomena, such as comets or novae. Generally if the 'guest star' does not move for several days or months relative to the other stars it is probably a nova. Number 31 on Lundmark's list was a 'guest star' of July 1054. He noted that M1 was at the same position.

Between 1913 and 1921, several astronomers (Vesto Slipher, Roscoe Sanford, Carl Lampland and John Duncan) individually and by various techniques measured the speed of expansion of the nebula. They saw that it was moving very

quickly, expanding outwards, growing in size as its filaments rushed continually outward from the centre of the explosion. The clincher in the identification of the nebula with the 'guest star' was American astronomer Edwin Hubble's realization that the filaments had begun expanding in 1054. Hubble pointed out the correspondence with the Chinese record, but his conclusion was overlooked until 1942 when it was revived by astronomers Jan Oort and Nicholas Mayall, and a Dutch orientalist, Jan Duyvendak. Since then, historical records of the nova of 1054 from Korea, Japan and Baghdad, have been identified, as well as rock paintings from Arizona and New Mexico.

As it changed over time, the Chinese astronomers had compared the brightness of the guest star of 1054 to other

4

5   6

celestial objects like Venus. This made it possible to draw a light curve of the nova and show that it was in fact a supernova, a stellar explosion in which nearly the whole star disintegrates. A supernova can leave behind a small stellar cinder – a neutron star (**41**). In 1968, while searching for the stellar remnant in the Crab Nebula with a radio telescope at Green Bank, West Virginia, radio astronomers David Staelin and Edward Reifenstein discovered a radio pulsar right in the middle of the nebula: a neutron star spinning on its axis at a rate of 30 rotations per second. This was a brilliant confirmation of the bold idea put forward 30 years earlier by Fritz Zwicky that supernovae produce neutron stars (**35**).

4 **Edwin Hubble** The Californian astronomer, best known as a cosmologist, realized that the supernova of 1054 had produced the Crab Nebula.

5 **Supernova of 1054** On a roof ledge of a cave (now partially collapsed) a member of the Early Pueblo people (also known as the Anasazi) living in Chaco Canyon, New Mexico, drew an image of the crescent moon and a bright star, signing the picture with a handprint. The star is thought to be an eyewitness representation of the supernova of 1054.

6 **Hsi and Ho** A late Ch'ing illustration to the *Book of Documents* shows the legendary astronomer-brothers receiving an imperial commission from emperor Yao to organize the calendar and pay respect to the celestial bodies. The Chinese imperial court relied on its cadre of astronomers to advise on potential threats and the auspicious moments to act.

7 **Crab Pulsar** X-rays have exposed this image of the Crab Pulsar (central star) from the Chandra satellite observatory. A disc of material swirls around the pulsar and a beam of light emanates from it along the axis of the disc, presumably along the rotational axis of the pulsar.

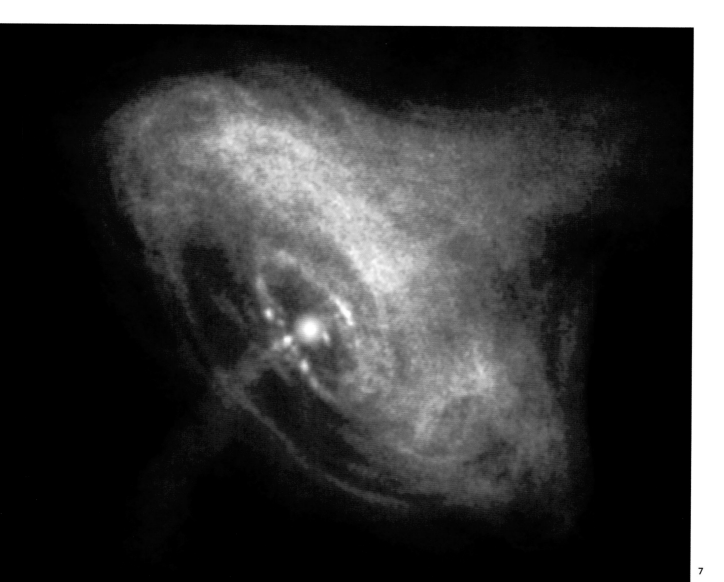

49. Planetary Nebulae
Looking into secret places

One August evening in 1864, English astronomer William Huggins used his spectroscope to examine a nebula that he thought was a collection of densely packed stars. To his surprise, the spectrum showed 'a single bright line only! ...The riddle of the nebulae was solved. The answer, which had come to us in the light itself, read: Not an aggregation of stars, but a luminous gas.' The nebula actually marked the quiet demise of a star much like the Sun, on its way to becoming a white dwarf.

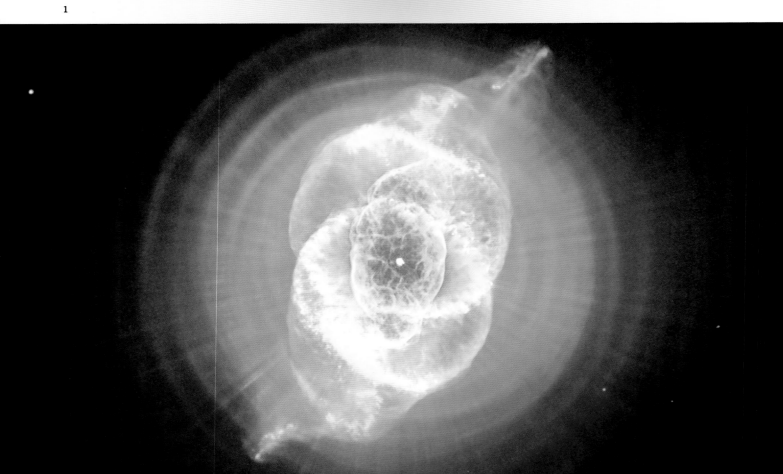

1 **Cat's Eye Nebula (NGC 6543)** The three-dimensional structure of the Cat's Eye planetary nebula is basically a bipolar structure lying within a series of spherical shells, all centred on a star turning from a red giant to a white dwarf. When it was a red giant, the star ejected some of its body into space at 1,500-year intervals, like a dog shaking dry its coat after a swim. Now the material is energized and made visible by the ultraviolet light radiating from the proto-white dwarf.

2 **Points of view** A bipolar planetary nebula consists of two equal and opposite lobe structures, perhaps throttled in the middle by some sort of disc. Depending on the point of view, the nebula can have strikingly different forms. Seen end-on, the nebula looks circular, the classic 'planetary' shape described by William Herschel. Seen sideways-on, the nebula looks symmetric, like a butterfly's wings. Seen at intermediate angles, it looks like overlapping arcs, circles and lobes.

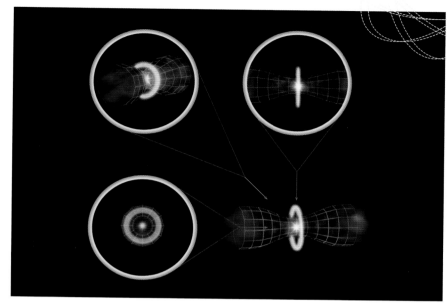

2

Planetary nebulae are shells of illuminated gas that surround old stars. They were discovered by the English astronomer William Herschel, who in 1785 referred to the 'planetary', or 'disc-like', appearance of the object we now call NGC 7009 – it reminded him of the appearance of the planet Uranus. He later discovered a similar nebula, NGC 6543 (also known as the Cat's Eye Nebula), which had a central star, describing it in 1790 as 'a most singular Phenomenon! A star…with a faint luminous atmosphere, of a circular form', adding that there could be 'no doubt of the evident connection between the atmosphere and the star.' Herschel speculated that planetary nebulae were 'generating stars' (that is, stars in the process of being born) and that 'the further condensation of the already much condensed luminous matter may be complete in time.' He was articulating what became known as the Nebular Hypothesis of the origin of stars (**50**), but it turned out that the 'planetary nebulae' he observed represented the deaths of stars, not their births. The name 'planetary nebula' has stuck, although these objects have nothing whatsoever to do with planetary formation.

William Huggins became a pioneer in stellar spectroscopy after selling the family business at the age of 30 to pursue his interest in astronomy. In August 1864 he examined the planetary nebula in the constellation Draco (now identified as NGC 6543) with his spectroscope. He saw a single green emission line, which was reminiscent of the spectral lines that he had seen in gas discharge tubes in his laboratory. He had discovered that planetary nebulae are made of low-density gas.

Later he identified other spectral lines in the nebula, some of which were generated by hydrogen.

In 1918, Lick Observatory astronomer Heber Curtis found that all planetary nebulae have central stars, which are visible if photographs are taken at sufficient depths inside the nebulae. The stars are all white-hot, emitting copious amounts of ultraviolet light, which turned out to be key to answering the question of why the planetary nebulae shine. The mechanism was discovered by the Dutch astronomer Herman Zanstra in the 1920s, while he was a post-doctoral fellow. He showed that the amount of ultraviolet light emitted by the central star of a planetary nebula was identical to the amount emitted by the gaseous parts of the nebula: the energy from the star and the energy from the nebula matched. Every ultraviolet photon (light particle) emitted by the central star ionized one atom of hydrogen in the gas. When the ionized atom recombined, it emitted one photon of visible light.

Zanstra's theory accounted for the light from the most abundant element in the nebulae, namely hydrogen, but other spectral lines remained unidentified. Lick Observatory astronomers William Campbell and James Keeler identified some as being from helium soon after that element was discovered in the Sun in the late 19th century (**28**); Huggins attributed other lines to a new element that he called 'nebulium' (**39**). But Henry Norris Russell, the director of the Princeton Observatory, remarked on the failure to reproduce the 'nebulium' lines in the spectrum of any other material that had been investigated, concluding that they 'must be due

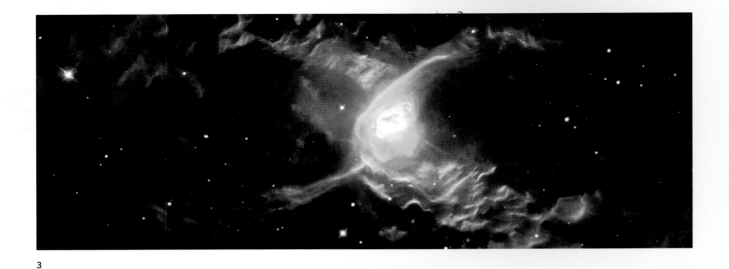

3

not to atoms of unknown kinds but to atoms of known kinds shining under unfamiliar conditions. The suggestion is tempting that the nebular lines may be emitted only in gas of very low density.' The American physicist and astronomer Ira Bowen confirmed Russell's guess in 1927, discovering that in certain circumstances in space, common elements such as oxygen and nitrogen can emit spectral lines that they do not generate in the laboratory. Such spectral lines are described as 'forbidden'. On Earth, oxygen atoms never get a chance to emit these spectral emissions because collisions with other atoms interrupt the process, but in space the time between collisions is longer and the atoms have sufficient time to release their radiation.

Although the central stars of planetary nebulae are very hot and bright, they look very faint to the naked eye because most of their energy is radiated as invisible ultraviolet. What has happened is that a star like the Sun has become a red giant.

At that stage it sheds its surface layers into the surrounding space. This exposes its central core, which shines as a hot star and causes the ejected gas to shine as a 'planetary nebula'. As time passes, the planetary nebula dissipates and the central star cools, making the slow transition to a white dwarf (**34**).

The curious shapes and colours of the gas in planetary nebulae inspire fanciful names, and motivate astronomers to take their pictures. Some planetary nebulae look circular and may be spherical. Others have complex shapes, often with some sort of symmetry like an hourglass (these are called 'bipolar nebulae'). Even some circular planetary nebulae like the Owl Nebula have double features (like the 'owl's eyes') that suggest they are bipolar nebulae seen nearly end on, like a hollow glass cylinder seen from one end. The explanation for this bipolar structure is still one of the undiscovered secrets of the Universe.

4

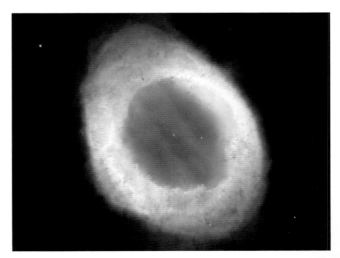

5

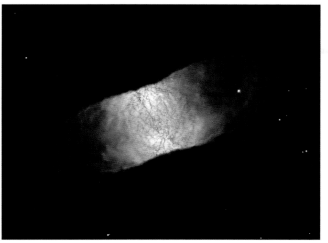

6

7

8

3 **Red Spider Nebula (NGC 6537)** Because the central star of this bipolar nebula is one of the hottest of the known white-dwarf stars, its central regions expand rapidly outwards and collide with the outer nebula.

4 **Little Ghost Nebula (NGC 6369)** A planetary nebula of the classic 'planetary' shape, precisely centred on the star that is evolving to become a white dwarf, the Little Ghost Nebula also has asymmetric extensions, evidence of some more irregular activity in the more distant past.

5 **Ring Nebula (M57)** is centred on a faint star, on its way to becoming a white dwarf and having a surface temperature of 120,000 °K. Like the Little Ghost Nebula, the symmetric 'planetary' shape is probably produced as we view down the axis of a three-dimensional barrel-like structure. The gradations of colour in the nebula illustrate the decrease in temperature of the gases away from the centre.

6 **Retina Nebula (IC 4406)** Intricate tendrils of material seen in silhouette against the nebula cause it to look like the back of an eye and give the nebula its name. The hot gas that emits the light by which the Hubble Space Telescope made this picture is surrounded by a larger complex of cold gas that can only be seen in radio telescopes. The dark lanes of gas and dust mark the boundaries where the outward expansion of the hot gas has trapped and compressed the colder areas.

7 **William Huggins** The British businessman and amateur Victorian astronomer spent considerable sums of money on a large telescope and equipment with which he discovered many spectral lines in nebulae.

8 **Ira Sprague Bowen,** American astronomer, director of both the Mt Wilson and Palomar Observatories in California.

9 **Egg Nebula (CRL 2688)** is a bipolar planetary nebula. Its central star has never been observed because it is surrounded by a particularly dense equatorial disc of gas and dust that, viewed from Earth, blocks the star's light. Along the axis of the disc is a succession of shells formed by a series of 'shrugs' from the hidden star, illuminated by beams of light and radiation shining through the disc. Throttled through the centre of the equatorial disc, the shells have generated the overall hourglass shape.

10 **Bug Nebula or Butterfly Nebula (NGC 6302)** Like the Egg Nebula, the Butterfly Nebula is a bipolar planetary nebula whose central star is hidden by an equatorial disc, which, unusually, contains significant quantitles of icy dust. It is a mystery how this cold material has survived for so long so near a star that appears to have a temperature in excess of 250,000 °K.

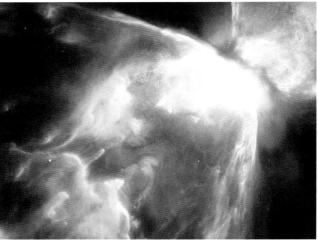

50

The Origin of the Stars and the Planets

The solar nebula and proplyds

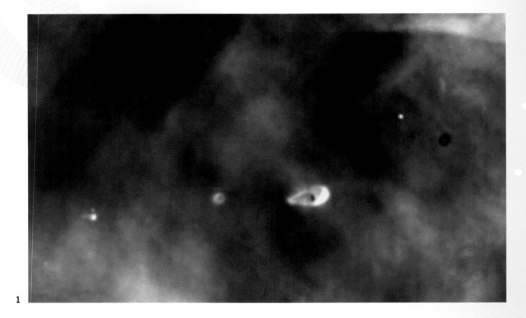

1

The discovery of the origin of the Sun and its Solar System, and of other stars and planetary systems, is a story of inspired guesses. The theoretical investigations of an international who's-who of astronomers and astrophysicists dating back over 300 years were gradually proved by observations with spacecraft and detectors. The Sun and the Solar System formed from the collapse of part of a very dense, very cold cloud of interstellar dust and gas. The main part of the cloud condensed to become the Sun; orbiting lumps gathered surrounding gas and dust and grew into planets.

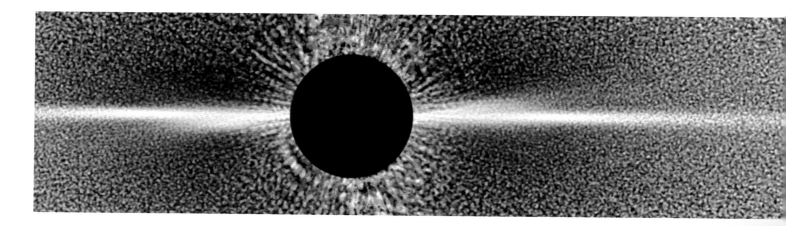

This explanation of the origin of planetary systems, first put forward by the Swedish scientist Emanuel Swedenborg in 1734 and by the Prussian philosopher Immanuel Kant in 1755, is called the 'nebular hypothesis'. The nebular hypothesis was given support as a theory by the French mathematician and astronomer Pierre-Simon Laplace in 1796. He had proved that the Solar System was stable, with the orbits of planets oscillating only by small amounts. The current shape of the Solar System, in which the planets all orbit in the same direction around a flat disc, reflects the way that it originally formed. In his book *Exposition du système du monde* Laplace put forward the idea that the planets had condensed out of a flat nebula that was whirling around the Sun, and that other stars and exoplanetary systems had formed in the same way.

The first embryonic exoplanetary system was discovered by Cal Tech astronomers Eric Becklin and Gerry Neugebauer in 1966. They used a newly developed infrared detector to make a laborious point-by-point scan of a region of the Orion Nebula, where they found a strong source of infrared radiation. Invisible to the unaided human eye, the infrared souce, which was named the 'BN Object' after the discoverers' initials, is the size of a planetary system. The infrared radiation comes from dust surrounding a newborn star. The dust traps the star's radiation and is heated to a temperature of about 700 °K, which causes the dust to radiate infrared especially intensely.

1 **Protoplanetary discs in the Orion Nebula** Some protoplanetary discs appear dark, their dust silhouetted against the nebula; others appear bright when their dust is illuminated by neighbouring bright stars. Eventually the dust will coalesce to form planets.

2 **Immanuel Kant** The great philosopher of the Enlightenment was also an early theorist of the 'nebular hypothesis' concerning the origins of planetary systems.

3 **Pierre-Simon Laplace** Laplace showed that the Solar System was stable and that its structure 'remembered' that it had formed in a rotating nebula.

4 **Gerry Neugebauer, Eric Becklin and their student Gareth Wynn-Williams** (left to right) display not only their 1960s fashion sense, but also their newly built infrared photometer, before mounting it on the telescope.

5 **Beta Pictoris** The bright light of the central star has been blocked out by the camera on the Hubble Space Telescope to reveal a disc of dust, seen edge-on and in orbit around the star. The second, inner disc, slightly tilted up to the right of the main disc, has been pulled into this position by an embryonic planet.

6 **Turbulence in the Orion Nebula** The curious cut-out shape of this photograph marks it as having been made by the HST. At the centre of the picture lie a number of proto-planetary discs, orbiting their newly born star.

7 **Kleinman–Low Nebula** The cluster of new, young stars embedded in a molecular cloud filled with dust photographed in infrared light by the Subaru Telescope. The area appears to be exploding as stellar wind is ejected at high speed from a new star 30 times the size of the Sun. In this turmoil, planets are forming.

8 **Chushiro Hayashi** The Japanese astrophysicist sketches the tracks of newly forming stars on his blackboard.

9 **Formation of a planetary system** A star coalesces from a cloud of dust and gas (i), which rotates (ii) and flattens (iii), (iv). Gas and ice in the inner planetary system are melted by the brightening star and blown away from the middle (v), to leave rocky planets in the inner regions and gas giant planets in the outer parts (vi). This is how our Solar System formed (vii).

The InfraRed Astronomy Satellite (IRAS) discovered further examples of protoplanetary systems in 1983, including discs of warmed dust grains orbiting the stars Zeta (ζ) Leporis, Vega and Beta (β) Pictoris. The disc of Beta Pictoris was photographed by Paul Kalas and Dave Jewitt in 1996, using a special camera attached to a small telescope. The clear skies at the high-altitude site in Hawaii concentrated the starlight into a small area, which was then blocked by a central obstruction so that the fainter light from the surrounding dust could be seen.

The first direct images of planetary systems were made with the Hubble Space Telescope HST in 1992 by Robert O'Dell of Rice University and his colleagues. The images showed dust discs silhouetted against the luminous background of the Orion Nebula and other nebulae. O'Dell's wife named the objects 'proplyds', a contraction of 'protoplanetary discs'. Their dust is concentrated in a disc rotating around a central star, just as Laplace visualized the solar nebula.

The way that the central mass of a proplyd contracts into a protostar was first calculated by Japanese astrophysicist Chushiro Hayashi in 1960. The contracting star system does not immediately settle down but goes through paroxysms,

during which it ejects a stellar wind in every direction and may also squirt jets of material from its poles. The material is ejected as a result of the rapid spin that the star has acquired as it contracts from the interstellar cloud, just as, at the end of her dance routine, an ice skater will spin faster as she brings her arms closer to her body. The ice skater then brakes by grinding her skates into the ice but also by throwing her arms out in a final flourish; the star brakes by throwing off material. This quickly clears the neighbourhood around the newly formed star. But before the surrounding nebula is all cleared away, it starts to form planets, creating solid lumps that are too compact and massive to be scoured away by the material ejected from the newborn star.

The nebula surrounding the condensing star is mostly made up of hydrogen and helium, but also contains dust grains that were made in old stars. Some of the material in the gas cloud assembles into molecules, which condense as ice on the surface of the dust grains. Still other material condenses into crystals. When the star switches on its nuclear reactions, it radiates energy (**45**), which melts the ice from the dust and blows away more gas from the inner, warmer parts of the solar nebula.

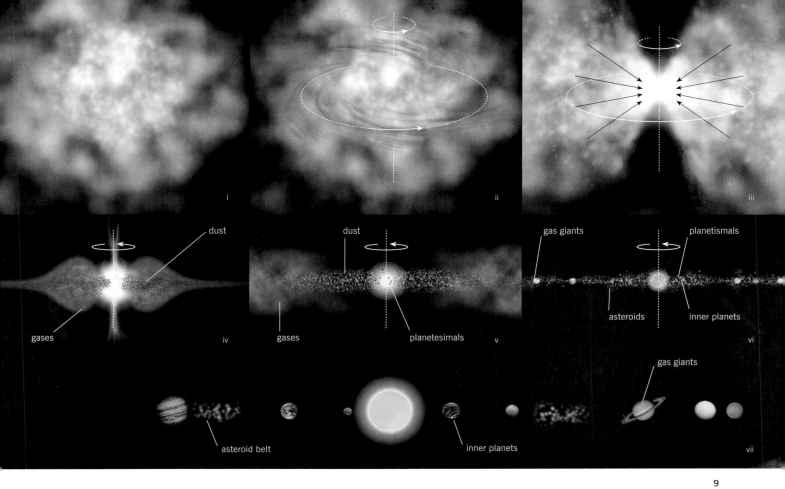

The dust that is left collides during its revolution around the star and sticks together into larger lumps. According to the calculations proposed in 1969 by the Russian theorist Viktor Safronov, and in 1973 by Cal Tech theorists Peter Goldreich and William Ward, the disc then condenses further, and forms lumps of about a kilometre in diameter, which are called 'planetesimals'. The gravity of the planetesimals is high enough to attract others; a large planetesimal is better able to do this, and the more it grows, the more its gravity increases, making it grow even faster. This process is called 'accretion'.

Our inner Solar System once contained about 100 protoplanets formed by accretion. The accretion process stops when the protoplanets have emptied their immediate neighbourhood of raw materials. Fragments that are left over are called chondrites; occasionally some fall to Earth as a type of meteorite (15). Astronomers can estimate their ages by looking at the radioactive elements that are trapped in the meteorite. This is the main way that astronomers establish the age of the planets, the Solar System and the Sun. It is a surprisingly precise and accurate known number: 4.555 billion years old.

The larger gas planets in the outer Solar System formed by the same process, although material in the more distant part of the gas cloud is less affected by the heat of the star. Jupiter grew the fastest, eventually becoming so massive that its gravity disturbed the orbits of the inner proto-planets, some of which jaywalked across the paths of the others and collided. These collisions caused some protoplanets to aggregate with Mercury, Venus, Mars, and the predecessors of the Earth and the Moon. Other collisions shattered protoplanets, creating the most common type of asteroid and meteorite. The fragments of shattered protoplanets rained down in the 'heavy bombardment' and made the large craters on the Moon and Mercury (22). One impact produced neither small pieces nor one single planet, but a 'twin planet': the Earth and the Moon (21).

As you look up to the Moon and planets on a clear night, it is easy to imagine – as Laplace did – the solar nebula rotating around the Sun, driving the planets in their orbits; the irregularities in its rotation triggering the collisions that created the Moon.

51

Interstellar Dust

Curtains of diamonds and graphite

Imagine a cathedral, with a shaft of sunlight shining through the window. Specks of dust are floating in the sunlight. Imagine the cathedral cleaned so scrupulously that there is only one speck of dust inside it. This represents the density of grains of dust that float in interstellar space, oxygen- and carbon-rich material expelled from supernovae and the interiors of red-giant stars. Interstellar space is indeed filled with stardust.

1

2

3

1 **North America Nebula** (NGC 7000) is an emission nebula (left), shaped like the continent of North America. The 'Gulf of Mexico' is a dark cloud that appears to separate the North America Nebula from the Pelican Nebula (IC 5070 and IC 5067, right), but actually lies in front of both, which are connected.

2 **Robert Trumpler** Swiss-born American astronomer, known for his work on star clusters.

3 **Edward Barnard** American photographer-turned-astronomer, who made a number of discoveries by application of photography to astronomy and compiled a catalogue of dark nebulae, now known by his name.

4 **Thackeray's globules** IC 2944 is a star-forming region in Centaurus. Silhouetted on the nebula are dense, opaque clouds of interstellar dust, first spotted by South African astronomer A. D. Thackeray in 1950. Thackeray's globules are shredded by intense ultraviolet radiation from the young, hot stars, and might disperse, rather than forming new stars themselves.

4

There is not much dust in space but there is a lot of space. The number of cathedral-sized volumes that stack one behind another in the line of sight to a star is very large, so individual dust grains can accumulate to form an opaque screen.

In 1847 the Prussian astronomer Wilhelm Struve, who was working at the Tartu Observatory in what is now Estonia, first proposed that something inhabited the space between the stars. He had discovered that the number of visible stars per unit volume in the Galaxy decreases with distance from the Sun, and therefore inferred that the light from distant stars was being absorbed by something in space. More evidence for the presence of this interstellar substance was discovered in 1909 by Dutch astronomer Jacobus Kapteyn, when he found that bluer stars moved across the sky more quickly than redder ones. Fast-moving stars are on average closer than slower ones, so Kapteyn concluded that the more distant stars were being reddened by a larger amount of interstellar dust, much as dust in the lower atmosphere of the Earth reddens the setting Sun. Similar work was carried out in 1930 by Robert Trumpler, then at the Lick Observatory, on clusters of stars (**40**): he found that the clusters with smaller diameters are more distant than larger ones, but still fainter than their distances alone would account for, because some of their light is absorbed by interstellar dust.

In the first two decades of the 20th century, the American astronomer Edward Emerson Barnard carried out a systematic programme to photograph our Galaxy. In his atlas of the Milky Way he identified distinct dark 'holes' in the star clouds. Astronomers since William Herschel had known of their existence, and for a long time thought they were true voids in the distribution of stars. But Barnard discovered that the holes were 'obscuring bodies nearer to us than the distant stars': dark clouds of unusually dense interstellar dust.

These dust clouds concentrate towards the plane of the Galaxy, which is why the Milky Way appears to be cleft along its central line when we view it edge-on from Earth (**03**). One of the most prominent clouds lies in the Southern Cross and is called the Coalsack. In the culture of some Australian aboriginal peoples, the Coalsack represents the head of an emu defined by the straggling form of the Milky Way between Crux and Scorpio, a unique 'constellation' made of dark dust clouds rather than stars.

Dust grains that lie near to bright stars can reflect starlight and form a 'reflection nebula'. There is a prominent example in the Pleiades star cluster, whose stars illuminate a dark cloud that they encountered as they coasted through space. The nature of this nebula, the first reflection nebula found, was

5

discovered by Lowell Observatory director Vesto Melvin Slipher in 1913, who observed the spectrum of the nebula and found that it was identical to the spectra of the brighter Pleiades stars.

The Solar System originally formed from of interstellar gas and dust. High temperatures destroyed most interstellar dust particles in the solar nebula, but some meteorites (known as carbonaceous chondrites) contain small particles, which differ in composition from the rest of the meteorite and are thought to be interstellar grains. These particles were convincingly identified in 1987 by University of Chicago physicists Ed Anders, John Wacker and Tang Ming, and Washington University physicist Ernst Zinner, who isolated interstellar diamond and silicon carbide in meteorites by dissolving the rest of the meteorite in acid, a method described as 'burning down the haystack to find the needle'.

The Ulysses spacecraft located interstellar grains in the Solar System by using a microphone to detect impacts by interplanetary dust particles. As it ventured beyond Jupiter in 1992, Ulysses encountered a higher than expected number of impacts coming from a particular direction in space, all of which hit the spacecraft at the same speed. They were from a stationary cloud of interstellar dust particles that the Solar System is passing through. Before this, it had been thought

that the solar wind would stop interstellar dust grains from entering the Solar System, but we now know that the larger grains manage to break through.

Some of these interstellar particles have been brought back to Earth by the Stardust spacecraft. Between 2000 and 2004 it deployed a sticky gel both in interplanetary space and in the vicinity of Comet Wild 2 (pronounced 'Vilt Two'); in January 2006 it successfully returned the gel to Utah and the material that had been collected in it was analysed. Most particles were from the comet but some have proved to be interplanetary and interstellar grains. The types of interstellar dust that have been discovered include tiny diamonds and larger graphite grains that are made in supernovae, as well as silicon carbide, aluminium oxide, spinel and titanium oxide grains made in the atmospheres of red-giant stars before they turn into planetary nebulae.

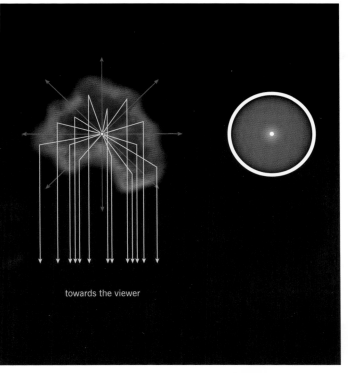

towards the viewer

6

7

IMMENSITIES

Until a person has thought out the stars and their interspaces, he has hardly learned that there are things much more terrible than monsters of shape, namely, monsters of space without known shape. Look, for instance, at those pieces of darkness in the Milky Way…You see that dark opening in it near the Swan? There is a still more remarkable one south of the equator, called the Coal Sack, as a sort of nickname that has a farcical force from its very inadequacy. In these our sight plunges quite beyond any twinkler we have yet visited.

Thomas Hardy, *Two on a Tower*, 1882

5 **The Coalsack** The prominent, broadly circular dark cloud of dust that obscures the light of the southern Milky Way at the edge of the constellation of the Southern Cross (right-hand quarter of the picture).

6 **Reflection nebula** If a cloud of dust surrounds a star, its red light escapes from the cloud but its blue light is scattered by dust grains and may be turned in our direction (left), making the nebula appear blue (right) and centred more or less on the star, perhaps with a redder scattering of light on the periphery.

7 **The Ulysses spacecraft encounters Comet Hyakutake** The spacecraft carried microphones to measure impacts of dust grains from comets, asteroids or interstellar space.

8 **Stardust** The spacecraft closed in on cometary dust grains from Comet Wild 2 (i). The comet ejects jets of dust as its surface melts in sunlight (ii). The spacecraft collected dust on panels coated with a sticky aerogel (iii). Small particles created tracks as they burrowed into the sticky substance (iv). The aerogel was returned to Earth for laboratory examination, where these microscopic images of dust grains were produced (v).

8 (i)

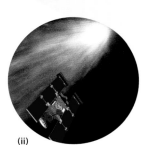

(ii)

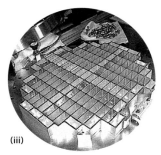

(iii)

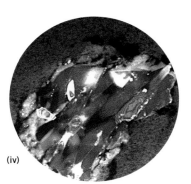

(iv)

(v)

Discoveries of the Universe
and its Galaxies

52. ● Hydrogen

The most abundant element in the Universe

1

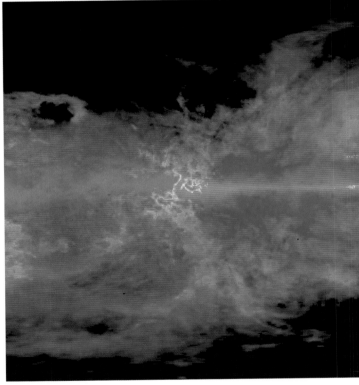

2

Hydrogen is the most abundant element in the Universe and the Galaxy. It was made in the Big Bang, and condensed to form large-scale gaseous structures, which in turn produced galaxies and stars. Our own Galaxy contains large amounts of hydrogen in stars and in interstellar space. Under the shadow of Nazi occupation in the Second World War, a group of Dutch astronomers worked in secret to identify the radio signature of this interstellar hydrogen. After the war, their discoveries would make it possible to map the entire Galaxy.

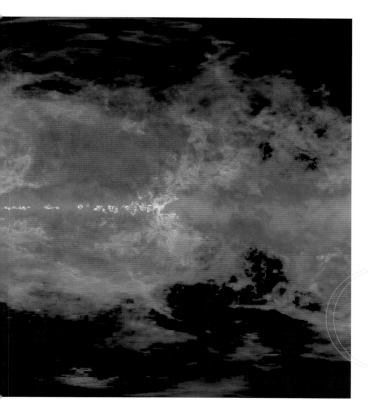

1 **Pillars of Creation** This iconic HST image is part of the Eagle Nebula (M16), showing three columns of dusty gas that are intruding into a hotter, less dense medium of mainly hydrogen gas. Stars are forming within each column, near the ends.

2 **Distribution of hydrogen in our Galaxy** The picture is oriented along the Milky Way (its axis lies horizontally along the centre of the picture). The image is colour-coded, with red meaning the hydrogen gas is moving towards us, magenta away from us, and green not moving much at all. The motions are a consequence of the rotation of the Galaxy.

3 **Hendrik van de Hulst** The Dutch radio astronomer calculated and predicted the radio emission at 21 cm wavelength from neutral hydrogen.

4 **Jan Oort** The Dutch astronomer had many interests, including the origin of comets and the rotation of the Galaxy.

There are two types of interstellar hydrogen in space. One type (called H II) is ionized hydrogen, which is present in areas where stars are forming and becomes visible to the naked eye when it is excited by the ultraviolet light from hot stars. By locating clouds of ionized hydrogen, William W. Morgan, Stewart Sharpless and Donald Osterbrock were able to plot the location of nearby nebulae in a 1951 map of the Galaxy. Their map was the first to hint at the Galaxy's spiral structure, but showed only the two nearest spiral arms; interstellar dust hid what lay beyond (**51**).

Outside the nebulae, away from any sources of ultraviolet light, are vast amounts of a cool invisible form of non-ionized hydrogen gas, which is called 'H I' or neutral hydrogen. The existence of neutral interstellar hydrogen was confirmed by radio astronomers in the 1950s, following a remarkable prediction made by a group, led by Dutch astronomer Jan Oort, who met in secret during the Nazi occupation of the Netherlands and whose research was only published after the Second World War ended in 1945.

When Germany invaded the Netherlands, Hendrik van de Hulst was an astronomy student studying under the astrophysicist M. Minnaert in Utrecht. After Minnaert was imprisoned in a detention camp for protesting the treatment of his Jewish academic colleagues, van de Hulst decided to make himself scarce. He moved to Leiden and studied under Jan Oort. Since observing with telescopes at night in the curfew was impractical and highly dangerous, Oort encouraged his astronomical students to concentrate on theoretical studies while the war was underway. Through Bart Bok, a Dutch astronomer working in the USA, copies of Grote Reber's papers on radio astronomy had been smuggled into the country and in the spring of 1944 Oort decided to hold a colloquium to discuss Reber's findings. Oort had the idea that if hydrogen showed a spectral emission it would be possible to use it to

5, 6

5 **The Pioneer plaque** The Pioneer 10 and 11 spacecraft carried plaques with pictorial messages intended for any extraterrestrial beings who found them, depicting the location of the Solar System, the trajectory of the spacecraft, and two people. The scale is indicated (top left) by an atom of hydrogen emitting 21 cm radio waves as it flips its spin axis.

6 **Edward Purcell** seated at the control table of a cyclotron in the 1950s.

7 **Whirlpool galaxy (M51)** Pink nebulae, blue clusters of stars, and lanes of dark clouds together delineate the spiral structure in this HST image.

8 **Whirlpool galaxy** sketched by Lord Rosse in 1845, the first 'nebula' to evidence spiral structure. The stars were numbered and keyed to measurements of their position that were intended to reveal, over time, the rotation of the spiral, but the galaxy proved to be so far away that no measurement of rotation was feasible.

9 **Our spiral Galaxy** A map of the distribution of neutral hydrogen in the Milky Way galaxy, prepared in 1958. In the middle of the diagram is the galactic centre, with the Sun near the top, where the two halves of the map touch. This was the decisive evidence that our Galaxy is a spiral.

see how the clouds of hydrogen were moving in the Galaxy and to infer where they were. Unlike visible light, radio waves could penetrate through interstellar dust to show the structure of the entire Galaxy, not just the neighbourhood of the Sun.

Although the occupying forces had banned most public gatherings, Oort arranged a meeting of the Astronomenclub (Dutch Astronomy Club), where van de Hulst presented calculations to support his theory that radio waves emitted by hydrogen have a distinct signature. Hydrogen atoms consist of an electron that is in orbit around a proton. Both the electron and the proton have a spin, and the axes of the spins can be parallel or antiparallel. Van de Hulst found that if the electron flips spontaneously from a parallel to an antiparallel spin, the hydrogen atom emits a pulse of radio waves that have a wavelength of 21 cm. Although these 'flips' happen only once every 11 million years in an individual atom, the number of hydrogen atoms in space is so large that a hydrogen cloud constantly produces measurable amounts of 21 cm radiation.

After the war ended, Oort and Lex Muller attempted to prove van de Hulst's theory by isolating the 21 cm radiation with a radio receiver, but suffered a setback when their equipment in Kootwijk was destroyed in a fire. The spectral

line from the hydrogen radio emission was finally discovered in 1951 by American radio astronomers Harold Ewen and Edward Purcell (later a Nobel laureate for his work on the fundamental physics of the hydrogen atom), using a radio telescope Purcell built at weekends with a grant of only $500. Their discovery was confirmed by Oort and Muller after they repaired their receiver, and subsequently reconfirmed by Australian radio astronomer Frank Kerr in Sydney. All three discoveries were eventually published together in the magazine *Nature*.

The first maps of the Galaxy, drawn according to Oort's vision, were made by van de Hulst, Muller and Oort in 1952, and correlated well with Morgan and Sharpless's earlier map of the local regions. In 1958, Oort, Frank Kerr and Gart Westerhout mapped the entire Galaxy by combining hydrogen radio emissions measured by Australian astronomers (who could see the southern part of the Galaxy) and Dutch astronomers (who could see the northern part). The map showed clearly that our Galaxy has spiral arms. Its structure is similar to the spiral galaxy M51, sketched in 1845 by William Parsons, the third Earl of Rosse (**53**). Our planet is a typical planet like others, our Sun is a typical star like others, and, it now appeared, our Galaxy is a typical spiral galaxy like others.

7

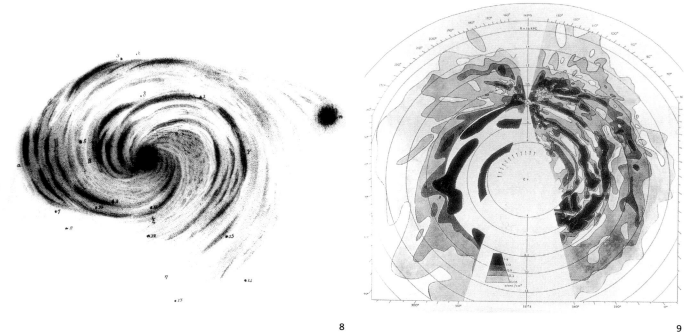

8

9

53

Galaxies

Ellipticals, spirals, mergers

1

  Galaxies are in constant motion, speeding outwards as the Universe expands, occasionally colliding with each other and changing shape. Thousands of distant galaxies – representing only a fraction of the total in the Universe – were captured in an astonishing set of images called the Hubble Deep Fields, which appear to show the frontier of the visible Universe.

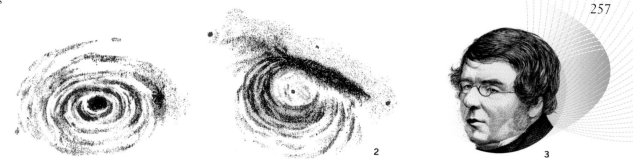

2

3

As they examined the sky between 1781 and 1847, William Herschel and his son John noticed that certain 'nebulae' in the regions of sky off the Milky Way displayed different degrees of central concentration, apparent flattening and mottling. A century later, astronomers realized that these strange 'nebulae' were actually outside our Milky Way system, and called them 'extra-galactic nebulae', later shortened to 'galaxies'. We now recognize them as star systems in their own right.

William Parsons, the third Earl of Rosse, saw the spiral structure of fourteen galaxies through his great telescope in 1845. In 1936, as advances in astronomical photography made it possible to record their complex shapes on film, Mt Wilson Observatory astronomer Edwin Hubble developed a classification system for galaxies that is still used today, describing them as spiral, elliptical or irregular. Elliptical galaxies are smooth and featureless. They can be spherical or ellipsoidal (the shape of a tangerine or an American football), but most are 'triaxial ellipsoids': lopsided balls. Spiral galaxies, like the Milky Way, are flat discs and have spiral arms. Lenticular galaxies are a bit like spiral galaxies that have a disc but no arms. Irregular

galaxies are either small, without a clear shape, or larger and appear to be two galaxies whose shapes have been disrupted as they pass close to each other.

Astronomers believe that elliptical galaxies have been formed by the collision of spiral galaxies, and that most larger galaxies have been formed by the merger of smaller ones. At the present time, a small galaxy is being absorbed by our own Milky Way. The Sagittarius dwarf elliptical galaxy was discovered in 1994 by Cambridge astronomers Rodrigo Ibata, Mike Irwin and Gerry Gilmore, who noted an excess of faint stars grouped just above the plane of our Galaxy at a distance of about 70,000 light years from Earth. The Milky Way has disrupted the structure of the smaller galaxy, breaking it down into a stream of stars, which loop in orbit over the pole of our Galaxy and are merging with it. Our Galaxy has grown by several such mergers in the past, but if it were to merge with another comparable spiral, it seems likely that, in the melee, both of them would lose their shapes and form an elliptical galaxy. Astronomers James Binney and Scott Tremaine discovered in 1987 that the Milky Way galaxy will collide

1 **Bode's spiral galaxy (M81)** This spiral galaxy is seen at an oblique angle. It has a delicate spiral that run in towards the centre, delineated among the areas of bright stars by dust lanes. The upper part is nearer the tilt; the dark clouds are clearer there, seen in silhouette against the brighter central regions.

2 **Spiral galaxies illustrated by Lord Rosse**

3 **William Parsons, the third Earl of Rosse** Mathematician, astronomer, country gentleman and builder in 1847 of the then largest telescope in the world (see also Chapter 48, Fig. 3).

4 **Edwin Hubble** at the eyepiece of the guide telescope, holding in his hand the controls of the Schmidt Telescope on Mt Palomar (now known as the Oschin Telescope), in 1949, when it was still permissible to smoke in observatories.

4

5

NGC 1201   Type S0

NGC 2841   Type Sb

NGC 2811   Type Sa

NGC 3031 M81   Type Sb

NGC 488   Type Sab

NGC 628 M74   Type Sc

6   7

and merge with the Andromeda spiral galaxy M31 in about 2–5 billion years. The spiral arms of both galaxies will flail about and lose their identity; some stars will be ejected. Our Sun will be a red giant by that time, and will find itself in an elliptical galaxy or speeding through intergalactic space in its old age.

In addition to classifying the shapes of galaxies, Hubble also discovered how far they were from Earth and the speeds at which they were moving, using his observations of Cepheid variable stars in spiral galaxies to determine their distances (**43**). Working from measurements made by Vesto Slipher of the speed of 46 galaxies, in 1929 Hubble discovered that most of the galaxies were receding away from us at speeds proportional to their distances: distant galaxies tended to be travelling fastest. There was some scatter (in particular, the Andromeda galaxy is approaching us, which is why it will merge with our Galaxy soon), but, according to Hubble's calculations, a galaxy at a distance of 3 million light years was receding at 500 km/second on average. Modern calibrations show that Hubble's

figure was actually eight times too high, but the principle, which is called Hubble's Law, remains correct: the Universe of galaxies is expanding. It was the first indication that the Universe had exploded in a Big Bang (**59, 60**).

The most distant galaxies that have been well studied so far are radio galaxies (**59**) and the galaxies in the Hubble Deep Fields. In 1995 the Hubble Space Telescope (HST) stared at an almost empty area of the sky in the northern hemisphere for ten days to take the deepest picture ever before obtained, repeating the process in 1998 with a similar area in the southern sky to confirm that the Universe had the same appearance in all directions. In 2004 the HST made an even deeper image, known as the Hubble Ultra Deep Field, constructed over eleven days with several different types of cameras. This is the deepest (most sensitive) astronomical picture ever made. It covers an area only 1/100 the angular area of the Moon, yet contains the images of 10,000 galaxies, hinting at the astounding size of the Universe.

5   **The Sagittarius dwarf irregular galaxy** Not all galaxies are spectacular. Most are dwarf galaxies, composed of rather few stars, hard to identify. On this image the brighter stars, uniformly distributed over the picture, are foreground stars in our own Galaxy. The fuzzy patches, often red, are galaxies far in the distance. The individual stars of the Sagittarius dwarf elliptical galaxy are faint and distributed in the central zone of the picture, and lie in the middle ground.

6   **Spiral galaxies of various types,** photographed at the Mt Wilson and Palomar Observatories. The type (indicated by letters like S0, Sa, Sb) depends on the openness of the spiral arms.

7   **Arp 147** The galaxy on the left is a spiral galaxy seen almost edge-on, and it has just fallen down the axis of rotation of the galaxy on the right, passing straight through. It has made a splash in the target galaxy, and piled up its gas into a ring, triggering the formation of bursts of stars in massive clusters. The reddish area to the lower left of the blue ring is probably what was the central nucleus of the target galaxy, reddened with dust.

8   **Collision between two spiral galaxies** NGC 6050 and IC 1179 are members of the Hercules cluster of galaxies, linked by their swirling arms. In about 2–5 billion years our own Milky Way galaxy and the Andromeda galaxy will collide like this.

8

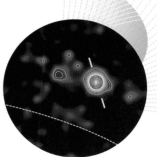
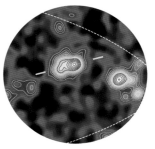

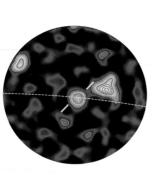
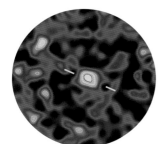

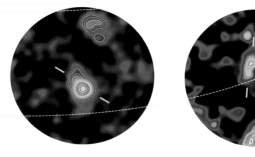

8 Hubble Deep Fields The HST homed in on one of the emptiest parts of the sky for about ten days in 2004. It recorded 3,000 images of which only a handful were stars in our own Milky Way (one of them the hard, whitish patch with a cross-like structure superimposed, seen to the right of centre). The rest are galaxies, many of the fainter ones extremely reddened by their high rate of motion away from us as the Universe expands.

9 Distant galaxies Galaxies that formed 12 billion years ago (the Universe formed 14 billion years ago) are shown in these false-colour pictures made by a camera called SCUBA mounted on the James Clerk Maxwell Telescope. The camera is sensitive to extremely infrared light and was able to image the galaxies because they are shrouded in thick dust, which, catching the light of the stars embedded inside, warmed up enough to radiate infrared.

9

Some of the galaxies in the Hubble Deep Fields are extraordinarily old, having formed perhaps 13 billion years ago, only about half a billion years after the Big Bang. However, these very ancient galaxies appear, paradoxically, very young in terms of their development. Because they are so far away, their image has been travelling to us unaltered for 13 billion years; we see them as they appeared approximately half a billion years after they formed, in what must be the closest practical approximation to eternal youth. The galaxies in the Hubble Deep Fields are therefore rather small compared to galaxies around us now; 14 billion years after the Big Bang, most galaxies around today have grown larger as they merged with others. The galaxies in the Hubble Ultra Deep Field are also less symmetrical than modern galaxies because they are younger, seen in gawky adolescence rather than settled, mature equanimity.

Just beyond the faintest galaxies in the Hubble Deep Fields lies an area where no galaxies are visible, which is called the Dark Ages of the Universe. Astronomers believe that galaxies actually exist in this region but are cloaked in dust that hides the light of the stars. The dust gets warm and emits infrared radiation, making the galaxies show up as faint infrared sources; several of these have been imaged by the SCUBA camera on the James Clerk Maxwell Telescope. They are so faint that not much will be discovered about their properties until the James Webb Space Telescope, the HST's successor, is launched, perhaps in 2013.

54

Magellanic Clouds

Our neighbour galaxies

The outline of the Milky Way as seen from the southern hemisphere is very irregular. Two pieces seems to have broken off; they are known as the Magellanic Clouds, and appear in the mythology of many peoples in Africa, Australia and South America. We now know that the 'clouds' are actually two separate galaxies that orbit together with our own.

2

The first known mention of the Magellanic Clouds is in the Book of Fixed Stars (964 CE) by the Persian astronomer Abd al-Rahman al-Sufi, who called the Large Magellanic Cloud al-Bakr ('the White Ox'). al-Sufi said that al-Bakr is invisible from northern Arab countries, but can be seen from the strait of Babd-el-Mandeb, which is the southern outlet of the Red Sea into the Indian Ocean. Europeans first saw the Clouds during early voyages of discovery to the southern seas, calling them the 'Cape Clouds' after the Cape of Good Hope. They were drawn with the Southern Cross on a 1516 star chart by an Italian navigator and spy, Andrea Corsali, who, travelling as a double agent for the de' Medici family on a secret Portuguese voyage to India, noted 'two clouds of reasonable bigness moving above the place of the pole, now rising, now falling, keeping their continual course in circular movement.'

Later the Clouds became associated with Ferdinand Magellan, the intrepid Portuguese captain who led the first voyage around the world from 1519 to 1522. Magellan never mentioned the Clouds himself, since he had been killed in the Philippines during the final months of the voyage, but members of his crew reported them in their accounts. Antonio Pigafetta was an Italian navigator who sailed with Magellan. After passing through what are now called the Straits of Magellan, he wrote: 'The Antarctic pole is not so covered with stars as the Arctic, for there are to be seen many small stars congregated together, which are like two clouds a little separated from one another and quite dim, in the midst of which there are one

3

1   **The Large Magellanic Cloud** The Cloud's location is indicated by a laser beam emitted at the Paranal station of the European Southern Observatory in Chile. The laser is used as a reference probe to correct distortions caused by the Earth's atmosphere in real time. The Small Magellanic Cloud is also visible near the centre; the gigantic tail of Comet McNaught fans over the horizon.

2   **Medieval Islamic astronomers at work** Abd al-Rahman al-Sufi and his colleagues and students probably assembled his star catalogue in an observatory similar to this one (Taqi al-Din's observatory in Istanbul, depicted in a 1577 illumination). Two astronomers (seemingly the leaders, upper right) use an astrolabe, timing their observations with the hourglasses on the table. Others are observing with quadrants and other instruments, drawing with a compass and adjusting a tripod, or consulting reference books drawn from the library.

3   **Magellanic Clouds** The opening page of (a contemporary copy of) the 1516 letter by Andrea Corsali to Giuliano de' Medici containing the earliest known sketch of the two Magellanic Clouds. This is also the first description of the Southern Cross as a cross (top, centre, straddling the circle).

or two stars'. Because they lie so close the south celestial pole, the Clouds were useful navigational aids for early explorers of the southern seas, just as Ursa Major, the Great Bear, is for northern sailors.

In the 17th century the Clouds were often called by their Latin names: Nubecula Major and Nubecula Minor (the Large and Small Magellanic Clouds, now abbreviated to LMC and SMC, respectively). John Herschel studied the two Clouds during his astronomical expedition to the Cape in 1834–38. He claimed to have seen a connection or star drift lying between the two Clouds, and catalogued 244 star clusters, double stars and similar objects in the SMC and 919 in the LMC.

The first suggestion that the Clouds were actually galaxies separate from our own Galaxy was made by Cleveland Abbe, an American astronomer who, unable to find a job in astronomy, had become a meteorologist. In 1867 he noticed that the Clouds contained a similarly large density of nebulae to the Milky Way. Abbe's theory was gradually confirmed by studies estimating the Clouds' distances from the Earth, conducted at the Boyden Observatory, which had been established by the Harvard College Observatory to study the skies of southern hemisphere, first at Arequipa, Peru (1889–1927) and then moving to Bloemfontein, South Africa. Henrietta Leavitt used data from these observatories to study variable stars in the Clouds and discovered the period–luminosity relationship (**43**).

These galaxies are some of our nearest neighbours – the largest two of about two dozen satellite galaxies to our Galaxy. The fact that they are separate from our Galaxy, but near enough for their contents to be seen in detail, makes them very useful to astronomers. For example, in 1987 a star in one of the Clouds exploded without warning as Supernova 1987A (**42**). The properties of the star – including the all-important fact of its distance from Earth – were already known to astronomers before the nova appeared; it was the first time it had been possible to study a supernova with such a complete case-history. As the Clouds cannot be observed very effectively from anywhere from the northern hemisphere, their study is a special priority for southern astronomers. Using the Parkes radio telescope in New South Wales, Australia, astronomers P. Wannier, G. T. Wrixon and Don Mathewson discovered the Magellanic Stream of hydrogen gas that links the two galaxies and our Galaxy. The gas was drawn out of the three galaxies by their mutual interaction by gravity as they orbited around each other. Even if the 'star drift' reported by Herschel did not stand the test of time, he was right in suspecting that the two Clouds were connected: all three galaxies exchange stars and gas.

Prima ego velivolis ambivi cursibus Orbem, Magellane novo te duce ducta freto.

4

5

6

7

8

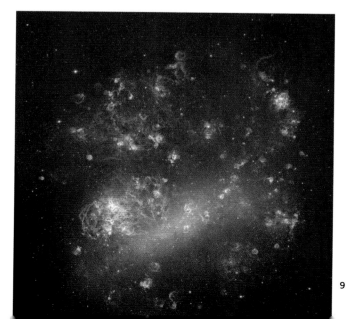

9

4 *Victoria* Magellan's ship, whose crew observed the Magellanic Clouds from the southern tip of South America as they circumnavigated the globe.

5 **Fernão de Magalhães** (known as Ferdinand Magellan), depicted in a 16th-century portrait by an unknown artist.

6 **John Herschel** The English astronomer, son of William Herschel, proposed that the Clouds were connected by a star trail.

7 **Parkes radio-telescope in New South Wales**

8 **Star cluster (NGC 1850) in the Large Magellanic Cloud** The LMC is so close that is possible to study its component stars with relative ease. Differences between its stars and the stars of our own Galaxy provide insights into how galaxies evolve differently.

9 **Large Magellanic Cloud** From the 'bar' of stars (lower half) protrude two sparse arms of red nebulae, one rising up at the left end of the bar, and another falling from the right end. The LMC is a barred spiral galaxy.

55.

Quasars

Active galaxies

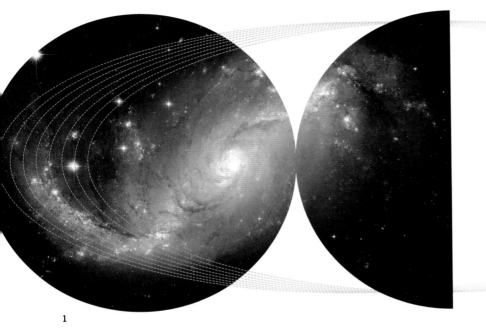

1

A quasar is an 'active galaxy': a galaxy that is exceptionally bright because it emits most of its energy from its nucleus. This energy is not starlight, but light and radio waves pouring from a massive central black hole. Quasars were discovered in the 1950s when astronomers tracking these strong radio emissions noticed a strange bright star that proved not to be a star at all.

3

4

5

1 **NGC 1672** A barred spiral galaxy with a Seyfert black-hole nucleus.

2 **Francis Graham-Smith** The British radio and optical astronomer was the thirteenth Astronomer Royal.

3 **The 200-inch telescope at Mt Palomar** This telescope founded the modern science of cosmology.

4 **The Cygnus A radio source** as photographed by the Palomar telescope. The original notion was that this irregular image showed a collision between two galaxies, but radio waves are now thought to be from a very active black-hole nucleus, which is one of the most powerful radio sources in the sky.

5 **Walter Baade** Baade photographed the Cygnus A radio source (Fig. 4), betting a bottle of whisky that it would prove to be two galaxies colliding.

The first known 'active galaxies' were certain spiral galaxies discovered by Carl Seyfert in 1943. He noticed that they had unusually bright nuclei and strong emission lines in their spectra, coming from gas that at times seemed to be moving very quickly. Only later was it clear that the gas was in orbit around something very massive. At the time, the galaxies were an unexplained curiosity called 'Seyfert galaxies'.

Then radio astronomers discovered that some galaxies emit radio waves. The first recorded radio galaxy was located on Grote Reber's 1939 radio map in the constellation Cygnus (**31**). In 1946 British physicist John Hey and his colleagues used military-surplus radar equipment to study this source, which they named Cygnus A. The source was very small, so some astronomers thought that it was a new kind of radio star. Others, including Thomas Gold and Fred Hoyle, argued that it was not a star, but an unknown object outside the Milky Way. 'Why . . . does not one find any identifiable visual object where those very near radio stars are supposed to be?' asked Gold. In 1951 he also pointed out that the 50 then known radio sources did not concentrate towards the Milky Way, but were uniformly scattered over the sky. It seemed likely that they were not stars, but galaxies.

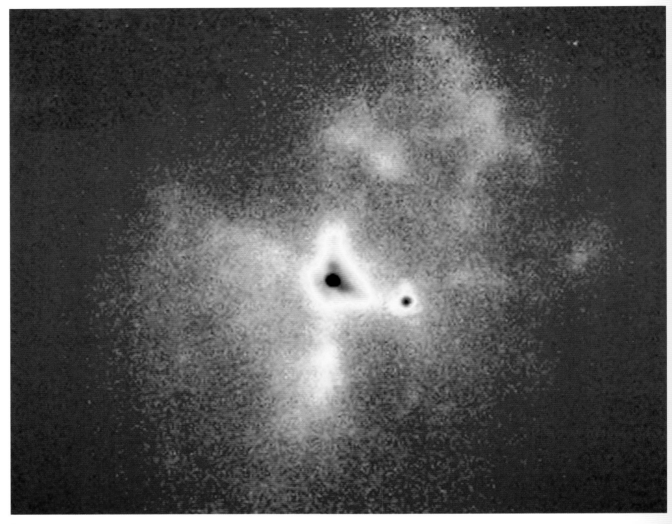

6

Later that same year, Cambridge radio astronomer Francis Graham-Smith measured the radio position of Cygnus A accurately enough to match it objects in the visible sky. Smith airmailed the position to astronomer Walter Baade at the California Institute of Technology, and in April 1952 Baade took two photographs of the area with the 200-inch Mt Palomar telescope. As he developed the images, a strange, bright object appeared at the centre of the plates. 'It showed signs of tidal distortion, gravitational pull between the two nuclei – I had never seen anything like it before. It was so much on my mind that while I was driving home for supper, I had to stop the car and think.' Baade concluded that Cygnus A was two galaxies in collision.

Discussing his discovery with a sceptical colleague, Ralph Minkowski, Baade bet a bottle of whisky that the spectrum of Cygnus A would contain emission lines from gas, energized by the shock of the collision. Minkowski took the spectrum with the Palomar telescope, found the evidence, and handed over the whisky to Baade. Reminiscing, Baade remarked that it was a small bottle and that Minkowski drank most of it in subsequent visits to Baade's office. In retrospect Minkowski might have kept it all for himself; although he saw the emission lines that Baade predicted, they had not been produced by a collision between two galaxies, but had been expelled from the massive black hole in Cygnus A (**56**).

As the technology improved from 1953 onwards, radio surveys of the northern and southern sky discovered several thousand galaxies that emit radio waves. In 1959 a comprehensive radio-survey catalogue called '3C' (as it was the third such catalogue made in Cambridge), identified a particularly strong radio source, which was designated '3C 273'. It lay in the band of the zodiac and from time to time the Moon

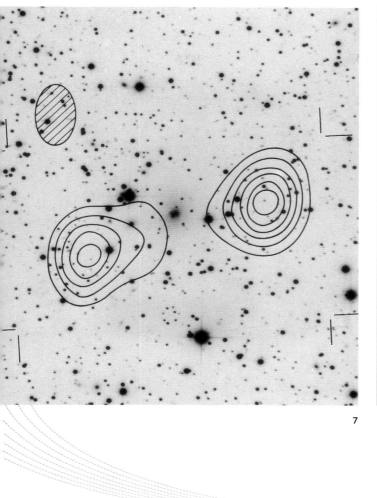

7

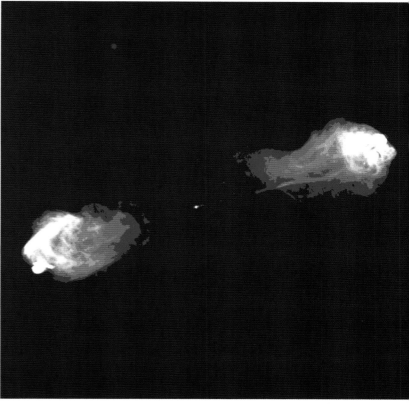

8

passed in front of it and occulted it. British radio astronomer Cyril Hazard used the newly built Parkes radio telescope in Australia to watch 3C 273 during a series of occultations in 1962, and was able to pin down its position. He also noticed that the radio source disappeared in two stages when the Moon passed in front of it, so it must have a binary structure. One of the two components of the radio source coincided with what looked like a thirteenth-magnitude star, which in photographs had a small 'wisp' or 'jet' attached to it. 3C 273 began to be described as a Quasi-Stellar Radio Source or Quasi-Stellar Object, later abbreviated as 'QSO' or 'quasar'.

Suspecting that the 'wisp' was a faint galaxy emitting radio waves, but knowing it was too faint to record an optical spectrum, Cal Tech astronomer Maarten Schmidt took a spectrum of the star in order to eliminate it as the radio source. It quickly became apparent that the 'star' associated with

6   **Cygnus A** photographed by the Keck Telescope shows an irregular structure that is generally oriented along the radio jets, all centred on an elliptical galaxy. No wonder that Baade and Minkowski were puzzled by its appearance in the first photograph (Fig. 4).

7   **Cygnus A** as observed by the Cambridge one-mile radio telescope and superimposed on a photograph of the sky, with the radio emission (contours) symmetrical on both sides of the fuzzy patch that is the optical galaxy seen by Baade and Minkowski.

8   **Cygnus A** in a false-colour image from the Very Large Array radio telescope. At such high resolution the radio source has a double jet, shooting from the nucleus, spanning half a million light years.

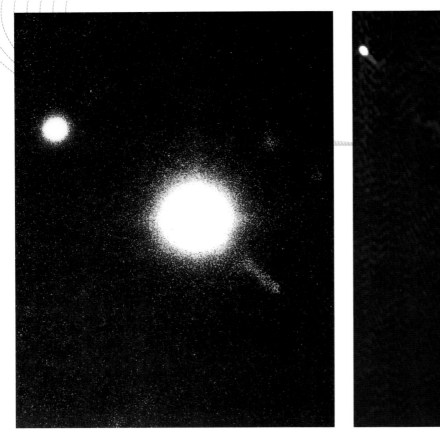

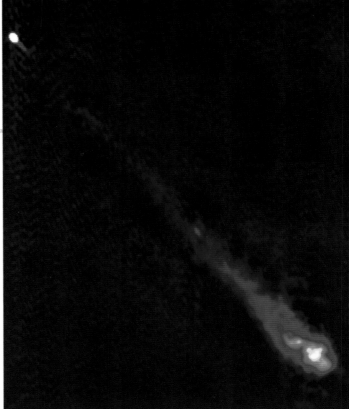

8

9

3C 273 was no ordinary star. The spectral emission lines in the star indicated hot gas was present, but neither Schmidt nor the world experts who subsequently examined the strange spectrum were able to interpret the lines. They had never before seen anything remotely like it.

Collaborating with Cyril Hazard in writing up his inconclusive findings about the 3C 273 source, Schmidt plotted the wavelengths of the spectral lines in a diagram. Suddenly he noticed that four of the lines formed a progression that reminded him of the spectrum of hydrogen – but with the wavelengths redshifted by a huge factor. When Schmidt adjusted the remaining spectral lines by the same factor, he was able to associate them with other known elements. But why was the star's spectrum so strongly redshifted? The only explanation that he could think of was that the 'star' was very distant, receding from us as a result of the

expansion of the Universe. Schmidt realized that the huge, unprecedented redshift was because the 'star' was a galaxy more distant than any yet been discovered. As the galaxy was visible from Earth, its brightness had to be a million times that of the Sun.

There was a final surprise: 3C 273 was a tiny galaxy. This became apparent in 1961 when Harlan Smith and Dorrit Hoffleit looked back through the archive of sky photographs at the Harvard College Observatory and saw that the brightness of 3C 273 had varied by large amounts on a time scale of only a few years, which meant that it could be no more than a few light years across (**30**), whereas a normal galaxy is many tens of thousands of light years in size. Incredibly bright, at incredible distances, incredibly small: this was the paradox of the quasars. Only later would astronomers discover the secret of these mysterious galaxies: they contained black holes (**56**).

10

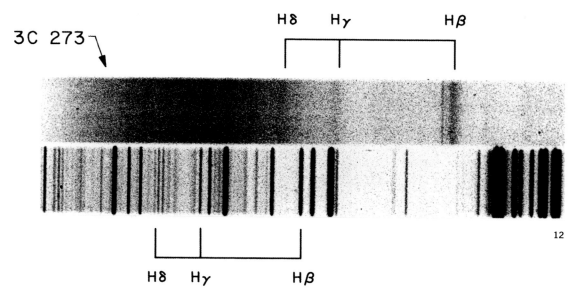

3C 273

Hδ   Hγ   Hβ

Hδ Hγ   Hβ

12

8 **3C 273** The main image, with 3C 273 taken by the 200-inch Palomar telescope, is of a point-like 'star', but there is a faint jet to one side (the third image is really a star accidentally near to the line of sight). The main 'star' is a quasi-stellar radio source (quasar).

9 **3C 273** A false-colour radio picture by the Merlin radio telescope shows a small nucleus (upper left, centred on the optical 'star') with a radio jet aligned along the same axis as the optical jet, shooting out from the quasar's black-hole nucleus.

10 **The Parkes radio telescope control desk in 1970** At the desk sits Fox Mason, who had an important role in maintaining contact with the Apollo 11 mission during its landing on the Moon. The telescope shows its engineering heritage from the late 1950s.

11 **Maarten Schmidt** The Dutch-born American astronomer made the key observation about quasars that proved they were at extreme distances.

12 **Spectrum of 3C 273** The spectrum of 3C 273 (top) has a few bright spectral 'lines' (marked here as Hβ, Hγ, Hδ). In the laboratory spectrum (below) the same lines are displaced from their position in 3C 273. This shift of 16% in the wavelength indicated to Schmidt the enormous distance of 3C 273 as it recedes at great speed in the expansion of the Universe.

56

Supermassive Black Holes

Monsters at the centres of galaxies

Nature makes black holes in two scenarios: in the aftermath of a supernova explosion or in the nucleus of an active galaxy. 'Active galactic nuclei' (AGNs) is the generic name for quasars, radio galaxies, Seyfert galaxies and the like. They are all supermassive black holes: black holes that are much more massive than stars, and lurk unseen in the centres of galaxies.

1

2

Although AGNs are astoundingly bright, they are tiny, perhaps only light hours in diameter (**55**). Gas in AGNs moves very quickly (as fast as 100,000 km/second) so there must be some kind of massive, compact structure around which the gas orbits. In fact, in 1964 Russian astronomers Yakov Borisovich Zeldovich and Igor Novikov calculated that if AGNs were not massive enough to generate a strong gravitational force, their intense radiation would blow them apart. The huge mass of AGNs was spectacularly confirmed in 1994 by a team led by Holland Ford, who used the Hubble Space Telescope (HST) to discover that the nucleus of the AGN known as M87 was 3 billion times the mass of the Sun.

Some AGNs (like 3C 273, **58**) have jets that are long, narrow and straight, and stay pointed in the same direction for millions of years. One explanation is that when the jet shoots out from a rotating body, along its axis, it acts as a stable gyroscope that maintains its own direction.

All these clues told astronomers Edwin Salpeter in the USA and Yakov Zeldovich in the USSR what the compact structure in quasars could be: a rotating, supermassive black hole. The key breakthrough came in 1969 when English astrophysicist Donald Lynden-Bell put clothes on the black hole theory. He argued that a quasar's energy came from frictional heating of a gaseous orbiting disc of material: the inner bits of the disc orbit faster than the outer bits and they scrape together. Astronomers had already seen evidence for the disc – the rapidly moving material – and found that its spectrum was just as Lynden-Bell predicted. So all AGNs have the same structure: a black hole, surrounded by a high-speed disc. Gas falls from the disc into the black hole, where it is compressed by intense pressure, heats and emits X-rays. The X-rays and friction heat the disc itself in turn, producing intense ultraviolet and optical light.

If AGNs are all the same, why are their appearances so different? This was explained by the 'unification model' of AGNs, put forth in 1984 by Robert Antonucci and Joseph Miller of the University of California, who proposed that the various types (quasars, radio galaxies, Seyfert galaxies and

1 **Centaurus A** In this image from the Chandra X-ray Observatory, there is a prominent X-ray jet from a supermassive back hole in the middle of the galaxy, hidden behind an obscuring dusty disc that passes across it. The main jet extends for 13,000 light years towards the upper left, with a shorter jet aimed in the opposite direction. The inner parts of the jets power enormous clouds of high-energy particles that balloon into the outer reaches of the galaxy. What look like 'stars' scattered over the picture are stellar black holes throughout the galaxy, which also emit X-rays.

2 **M87** This elliptical galaxy has a bright nucleus (upper left) from which shoots a fast jet 5,000 light years long. The jet seems to have stuttered, producing a succession of blobs. Superimposed on the starry main background are numerous spots, which are large, old globular star clusters.

3 **Donald Lynden-Bell** In 1969 the University of Cambridge theoretician provided the convincing argument that quasars were black holes, surrounded and powered by discs of material that was falling inwards.

4 **Joseph Miller** Lick Observatory astronomer, Robert Antonucci's teacher.

5 **Robert Antonucci** With Miller he 'unified' a diverse range of galaxy types, identifying them all as various orientations of the same basic black-hole structure.

3

4

5

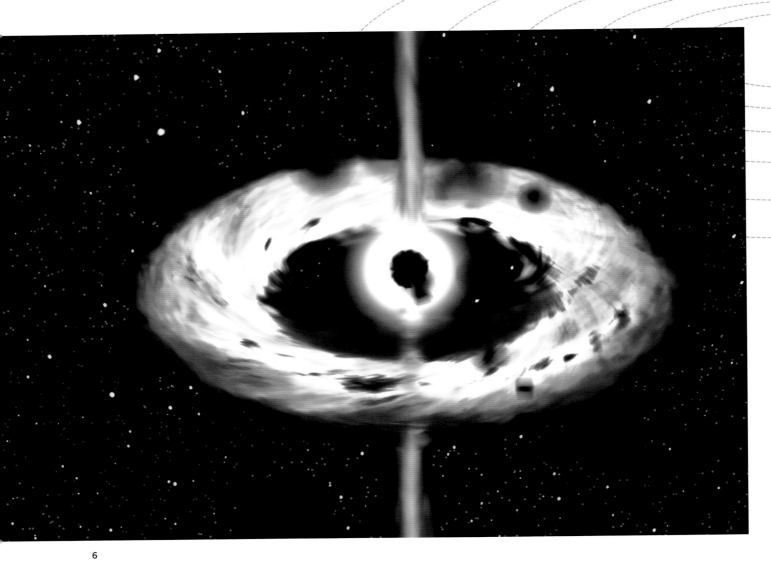

6

the like) differ only in the angle at which we happen to view the disc. Surrounding the black hole and the hot inner disc are the cooler outer parts of the disc: an opaque, rotating doughnut ('torus') of dust and thick gas, which has a radius of a few light years. If we happen to view an AGN edge-on, the torus obscures the inner parts, making it a Seyfert galaxy. However, if we see the AGN from a different angle, with the dust ring framing the nucleus, the inner parts of the AGN are revealed, including the blaze of light from near the nucleus, and we see a quasar.

The jets of material shooting out from some AGNs are paradoxically caused by material falling inward from the disc. So much gas falls in towards the black hole that some is turned around by the strong radiation of the black holes and ejected.

Astronomers think they now have a reliable basic picture of AGNs, and a recent discovery has solidly confirmed the blackhole theory. In 1995–99 Cambridge astronomer Andy Fabian used the Japanese satellite Asca to map the X-ray emission from the disc orbiting near the black hole in the AGN galaxy MCG-6-30-15. He found that the X-ray spectral lines from the disc had been bent into a unique shape, because of the combination of several effects of Special and General Relativity (**30**) generated by the black hole's powerful gravity. Even if black holes are necessarily permanently secrets in themselves, their surroundings give them away.

7

6 **The neighbourhood of a supermassive black hole** Within a rotating torus of dust and gas there is a supermassive black hole ejecting jets of material. Dark clouds of gas and dust float about in the general disruption nearby. This is the underlying concept of the unified theory of AGNs, whose individual appearances change as this structure is seen at different angles, and varies according to the strength of the black hole and the amount of material falling into it.

7 **MCG-6-30-15** The rotating circular disc of gas is seen obliquely in orbit around the supermassive black hole in the galaxy MCG-6-30-15 in a colour-coded simulation. There is an empty space in the middle of the disc where the gas falls into the black hole. In the central part of the disc, the red areas show where gravity is so strong that light escaping from this region loses energy. The image of the back side (upper half) of the circular disc is lifted up by the gravitational pull of the black hole as radiation passes over it.

8 **Andy Fabian** An astronomer at the University of Cambridge studying high-energy phenomena in galaxies and clusters of galaxies, he identified the conclusive evidence for active galaxies with images like Fig. 7.

8

57. The Black Hole in our Galaxy

A dormant monster

1

At the centre of our Milky Way galaxy, a cluster of a hundred stars is orbiting a mysterious object at incredibly high speeds. Although astronomers cannot see the object, they know its mass is over 2.5 million times the mass of the Sun. It seems to be a supermassive black hole that is invisible because it is sleeping.

1 **Central region of the Milky Way galaxy** Infrared light recorded by the Spitzer Space Telescope shows patchy nebulae (coloured red) in front of the vast numbers of more distant stars in the central bulge of the Milky Way. The bulge is centred on a bright, dense star cloud, which contains our Galaxy's black hole, among other spectacular cosmic phenomena.

2 **Sagittarius A** The radio picture of the galactic centre from the Very Large Array is complicated. The brightest (reddest, in this false-colour picture) circular feature is known as Sagittarius A East, thought to be a supernova remnant. Sagittarius A West is the orange and green extension to the right of Sagittarius A East. The threads and arcs in the top half of the picture

trace the Galaxy's magnetic field. Sagittarius A* (Sgr A*) – the yellow dot in the middle of the bright red area – is what we are seeking: the (not very) active nucleus of our Galaxy.

3 **The star cluster orbiting our Galaxy's black hole** Using infrared light to penetrate the dust that piles up in the direction of Sagittarius, astronomers at the European Southern Observatory imaged the individual stars near the galactic centre. By following the motions of the most central stars for nearly twenty years (Figs. 9, 10), astronomers have been able to determine the mass of the supermassive black hole.

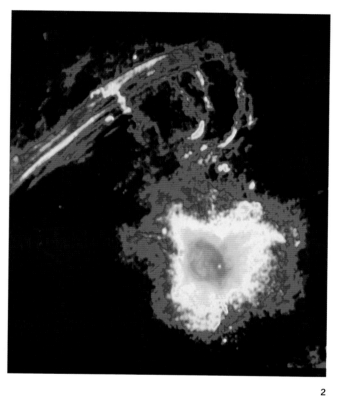

2

3

The maps that Karl Jansky and Grote Reber made of the radio sky in the 1930s showed that the strongest source of radio waves in the Milky Way came from the direction of Sagittarius (**31**). The centre of the Galaxy lies about 25,000 light years in this direction, but in visible light we can only see objects 1,000 light years away, because the interstellar space contains a 'smokescreen' of dust particles blown out by stars (**51**). Radio waves and infrared radiation can, however, penetrate through this smoke to reveal what lies at the centre of the Galaxy.

The directional discrimination of the earliest radio telescopes was only precise enough to show that there was a strong radio source in Sagittarius; the source was therefore named Sagittarius A, or Sgr A for short. It soon became clear that Sgr A was made up of several different objects, including

nebulae and supernova remnants (**48**). Radio astronomers discovered that Sgr A had two strong radio-emitting regions with distinct properties, which they named Sagittarius A West and Sagittarius A East. Sgr A East is probably a supernova remnant. But Sgr A West was a mystery. It has a complex spiral shape and it coincides with the highest density of stars in the Galaxy. Radio astronomers interpret it as the centre of our Galaxy; in 1959, when the International Astronomical Union agreed to set up a coordinate system for the Galaxy, Sgr A West was defined as the central point.

Because the structure of Sgr A West was so complex, it was difficult to study. In 1974 Bruce Balick and Robert Brown used the Green Bank 35-km radio link interferometer at the National Radio Astronomy Observatory to map Sgr A West

7

4   **Bob Brown** Cornell University radio astronomer.

5   **Bruce Balick** University of Washington astronomer interested in planetary nebulae.

6   **Reinhard Genzel** German infrared astronomer, builder of advanced space- and ground-based instruments.

7   **The galactic centre in X-rays** An image by the Chandra X-ray telescope shows hot gas (upper right, lower left) bursting out of the galactic centre (bright region at centre), as well as other X-ray sources nearby.

and discovered a bright point-like radio source at its centre. They concluded that the point-like source was in some way 'physically associated with the galactic center (in fact, defines the galactic centre).' Brown named the source Sagittarius A* (pronounced 'A-star' and abbreviated 'Sgr A*').

With the development of very large telescopes and detectors optimized for use in infrared light, it became possible to take pictures of the galactic centre. The pictures showed a region 30 light years wide filled with dust, stars and gas. Its very centre is surrounded by the so-called Circum-Nuclear Disc of dust. The Disc surrounds a cluster of stars, and Sgr A* lies at the centre of the cluster.

The motion of the stars in the central cluster has been studied by teams led by the German astronomer Reinhard Genzel and University of California astronomer Andrea Ghez.

Over the past decade they have repeatedly imaged the stars with telescopes at the European Southern Observatory in Chile and the Keck Telescope in Hawaii, using advanced image stabilization techniques to overcome the wobble of the star images caused by the Earth's moving atmosphere, so that very small changes in their position can be detected. The teams discovered that stars in the cluster are speeding around Sgr A* at velocities as high as 1,400 km/second. These measurements of the stars' motions make it possible to estimate the mass of the cluster, which is 2,600,000 times the mass of the Sun. As there are only about a hundred stars in the cluster, Sgr A* must be responsible for most of the mass, and the only object capable of being so small and yet so massive is a black hole. The stars are orbiting the supermassive black hole at the centre of our Galaxy.

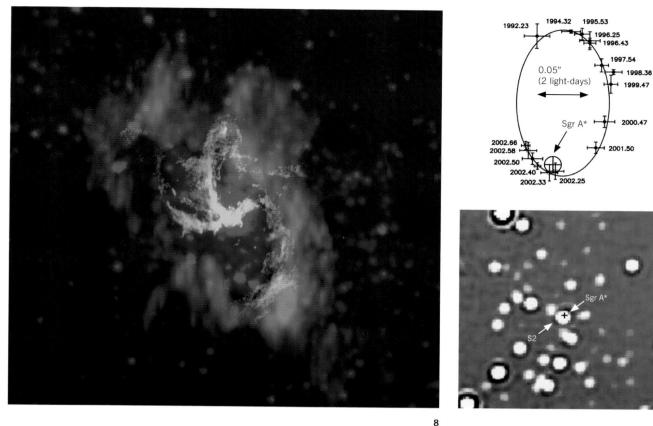

8

9

10

If we compare the central black hole of our Galaxy with a typical quasar or other Active Galactic Nucleus, it does not rank very high in brightness. Whereas the supermassive black holes in other AGNs can be seen on the far side of the Universe, it has been a struggle for astronomers to find our own, which does not rival, say, 3C 273 or M87 (**56**). One reason for this is that the mass of our Galaxy's black hole is nowhere near the top end of the range for supermassive black holes. Another is that not much matter falls on to it. There is a clear hole in its circum-nuclear disc. Collisions between clouds of material in the circum-nuclear disc causes some of its material to fall in towards Sgr A*, but only a small trickle of material goes on to dribble into the black hole. Our Galaxy's black hole remains asleep.

8 **Galactic centre** This image of the Milky Way's central region combines data from several different telescopes, including radio images from the Very Large Array (green and red, indicating cold and warm gas, respectively) as well as an infrared image from the Spitzer Space Telescope (blue). The red-coloured warm gas forms the hollow torus shape of the Circum-Nuclear Disc. It orbits around the three-armed structure that is Sgr A West, which has emptied the zone immediately around the black hole, starving it. This is why the centre of our Galaxy is a not-very-active galactic nucleus.

9 **Stars in orbit around the black hole in our Galaxy** A small star cluster surrounds the compact radio source Sgr A* at the centre of our Milky Way Galaxy, which is marked by a small cross. The stars are in orbit around the black hole. Star S2, for instance, has an orbital period of fifteen years; its successive positions are marked on the diagram, showing that the mass of Sgr A* is 2.6 million times the mass of our Sun, the key piece of evidence that it is a black hole.

10 **Enhanced image of the small star cluster** The star S2 is very close to the black hole Sgr A*.

58. Gamma-Ray Bursters

The biggest bangs since the Big Bang

1

Gamma-ray bursters are cosmic explosions of extraordinary energy, which were first discovered during the Cold War by American satellites designed to detect Soviet nuclear-weapons tests.

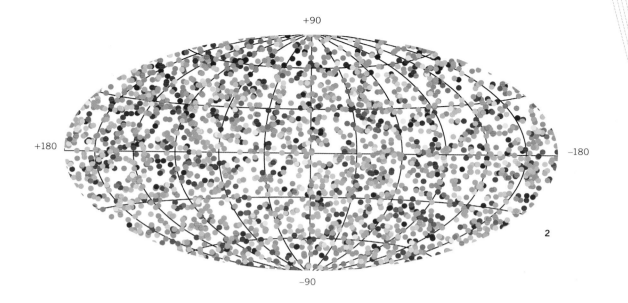

2

3

1   **Gamma-ray burster** In an artist's impression, a gamma-ray burster sends beams of light in opposite directions from a central explosion.

2   **2,704 gamma-ray bursts** recorded over nine years by the Burst and Transient Source Experiment (BATSE) on board the Compton Gamma-Ray Observatory. As they are uniformly distributed over the whole sky, they must be from the most distant galaxies.

3   **Ray Klebesadel** worked at Los Alamos National Laboratory as an electrical engineer and scientist on various projects, the specifics of which are still not publicly known, and discovered gamma-ray bursts.

The Partial Test Ban Treaty, signed by the USSR, the USA and the UK in October 1963, prohibited tests of nuclear weapons in the atmosphere or in space. To monitor whether the USSR was keeping to the terms of the treaty, the US Air Force launched a series of satellites, named Vela, to look for signs of illegal tests.

The satellites watched for the brief bursts of gamma rays that would signal a nuclear explosion. Almost immediately after the first satellites were launched in 1967, they began to detect short bursts of gamma rays, each lasting only about a second. The bursts were being detected on an almost-weekly basis; too frequent to be clandestine nuclear tests, they had to be coming from natural sources, but the phenomenon was completely unexpected and no one had any idea what had caused it.

As the Vela series of satellites became more sophisticated, they provided more information about the mysterious gamma-ray bursts. Vela 5 a and b (launched in 1969) and Vela 6 a and b (1970) were able to characterize their signatures, even though the bursts were so brief. In 1972 Ray Klebesadel and Ian Strong of the Los Alamos Scientific Laboratory studied secret records that revealed the direction from which the gamma rays had originated. They found that on several occasions the same burst had been detected by two satellites. (The Vela satellites operated in pairs so that they could detect nuclear tests on both sides of the globe; identical satellites called a and b travelled at opposite sides of a circular orbit 250,000 km in diameter.) The scientists found that on these occasions the gamma-ray bursts were not detected at exactly the same moment by both satellites; there was a one-second delay as the gamma rays, travelling at the speed of light, triggered first one satellite and then the other. This allowed Klebesadel and Strong to see from which direction they had come.

By 1973 Klebesadel and Strong were able to prove conclusively that the gamma-ray bursts were of cosmic origin. They had noticed that some bursts had been seen by all four satellites – both the Vela 5 and the Vela 6 pairs – and therefore could not have come from a single isolated source on the Earth's surface. Astronomers began to debate the possible cosmic sources of the random gamma-ray bursts, and Klebesadel and Strong were given permission to publish their discovery as a scientific paper, but were not allowed to report all the details of the satellites' instrumentation, as it could help other countries to tailor illegal weapons-tests to be undetectable. Because of this, some astronomers complained that they were not privy to all the relevant scientific background, but they had to put up with it.

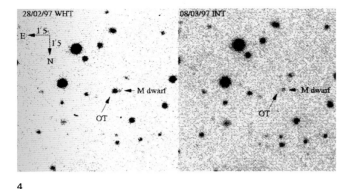

4

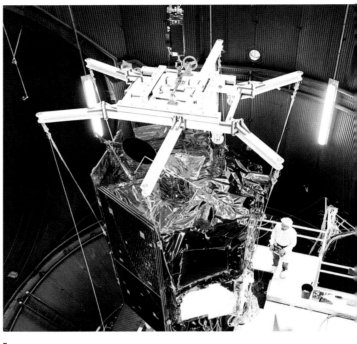

5

As more scientific satellites were deployed and gamma-ray bursts were discovered almost daily, the phenomenon was given a new name: a 'gamma-ray burster'. Before 1997 the best clue to the nature of bursters was that gamma-ray bursts came from all directions equally (isotropy). Some astronomers thought that the bursters might occupy a region surrounding the Solar System at a distance of up to a light year, the same region from which comets come. Others thought that bursters might occupy an extended halo around our Galaxy.

Alongside these theories for 'local' origins, some daring astronomers began to consider that the bursters might be 'cosmological' – distributed among the most distant galaxies, at distances of hundreds of millions of light years. But if bursters were really this distant, the energy they emitted had to be enormous.

In 1997 a key observation by an Italian-Dutch satellite called BeppoSAX finally solved the mystery of where the bursters were located. At 5 a.m. on 28 February, BeppoSAX detected a gamma-ray burst that was designated as GRB 970228 (gamma-ray bursts are numbered by date). The operations team in Rome had prepared for this and quickly rescheduled the satellite's

observing programme so that more accurate X-ray sensitive instruments could be deployed for follow-up observations. Eight hours later, they saw a new X-ray source in the same place, which quickly faded. It seemed probable that the gamma-ray burst and the X-ray emission had come from the same object. The team in Rome telephoned a colleague, Dutch astronomer Jan van Paradijs, whom they knew was scheduled for work on a large optical telescope. Less than a day later, using the William Herschel Telescope on La Palma in the Canary Islands, van Paradijs saw a faint optical source at the same position in the sky as the X-ray source. The optical burst was surrounded by a fuzzy patch, which van Paradijs found difficult to interpret. He passed the information on to colleagues at the Keck 10-m telescope on Mauna Kea, Hawaii, and the Hubble Space Telescope, who found that the fuzzy patch was a galaxy at an enormous distance. The 'cosmological' theory was correct: the gamma rays were being released as huge bursts of energy in galaxies far away from our own.

Gamma-ray bursters let loose as much energy as in a supernova – perhaps more – in a few seconds or less. Some gamma-ray bursters have subsequently proved to have spectral

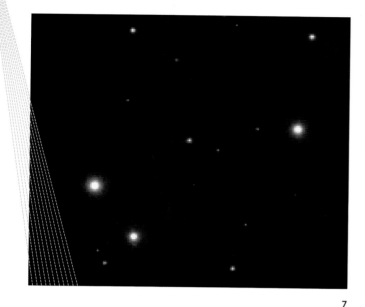

6

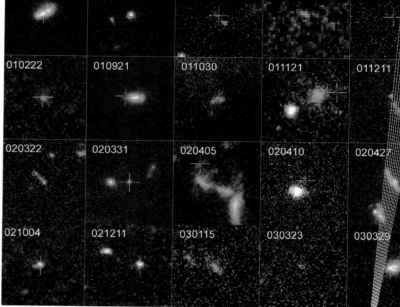

010222 010921 011030 011121 011211
020322 020331 020405 020410 020427
021004 021211 030115 030323 030329

7

8

properties that are the same as supernovae, and may in fact represent a type of supernova, which some astronomers refer to as 'naked supernovae'. For unknown reasons the gamma rays escape into space during the explosion, rather than being absorbed by the surrounding stellar debris as normally happens in supernovae.

Not all gamma-ray bursts are the same. Astronomers have gradually come to the conclusion that there is a second variety of burst that lasts for shorter periods of time than the ones that can be attributed to supernovae. Their cause is still a mystery; perhaps they are generated by the plunge of a neutron star into a black hole. It is rare for such things to happen in a galaxy, but, since gamma-ray bursts are so powerful and there are so many galaxies in the Universe, it is possible that astronomers could see it happening every day.

4 **GRB 970228 discovery image** The first burster to be identified with an optical afterglow is labelled here as OT ('optical transient'), together with a random star nearby, in images taken on 28 February 1997 (left) and on 8 March 1997 (right).

5 **BeppoSAX satellite** in its assembly bay.

6 **The brightest gamma-ray burster** The extremely luminous afterglow of GRB 080319B was imaged in 2008 by the X-ray telescope (left) and the optical/ultraviolet telescope (right) on NASA's Swift satellite. This was by far the brightest gamma-ray burst ever seen, 2.5 million times brighter than a supernova.

7 **Optical afterglow GRB 080319B** appears in the centre of this image from 'Pi of the Sky', a Polish group that monitors the sky for gamma-ray bursters and other short-lived optical sources. For about thirty seconds the burst would have been visible to the naked eye, even though the galaxy in which it occurred is at a distance of 7.5 billion light years.

8 **Host galaxies of gamma-ray bursters** The Hubble Space Telescope interrupts its prepared programme to image suitable gamma-ray bursters in order to identify their host galaxies, which have proved to be a motley collection.

The Evolving Universe

The past, the present and the future

Is the size of the Universe fixed and unchanging, or is it continually expanding? Has the Universe always existed, or did it have a discrete, explosive beginning in an event called the Big Bang? Throughout the 20th century, rival camps of physicists and astronomers fought over these questions about the Universe's past, present and future.

In the 17th century, Isaac Newton applied his theory of gravity to the Universe, on the assumption that the mass in it was uniform and static. Each mass particle – he thought of them as stars – attracted all the others and he realized that if this attraction continued ad infinitum eventually the Universe would collapse in on itself. Newton never resolved this difficulty, and thought it was proof of the existence of God, who prevented the collapse from happening.

In 1915 Albert Einstein discovered a new formulation of the theory of gravity, which came to be called General Relativity (**30**). When he tried to apply the theory to the entire Universe, he came to the same realization as Newton: a static Universe of galaxies was unstable. Einstein's solution was to invent the so-called Cosmological Constant, which acted as a repulsive force that held the Universe up.

That very year, Willem de Sitter discovered in the mathematics of General Relativity something that Einstein had overlooked: the Universe need not be static but could in fact be expanding. Around 1927 the Belgian astronomer Abbé Lemaître, visualizing the start of the Universe as an exploding atom, formulated theories of the expanding Universe that did not require the Cosmological Constant to stabilize it. In 1929 the American astronomer Edwin Hubble discovered that distant galaxies are moving back from us at a rate that is proportional to their distance, a phenomena that he expressed as Hubble's Law (**53**). In his 1927 paper Lemaître had implicitly predicted a linear velocity–distance relation of this kind. Hubble's Law showed that the Universe was indeed expanding.

The explosion of the giant atom was identified as the moment of creation of the Universe: the Big Bang. Some astronomers in the 1950s opposed this idea on philosophical or religious grounds, among them the Cambridge mathematicians Hermann Bondi and Thomas Gold, who wished to construct a theory in which there was no question of how the Universe originated; they were assisted in this by the physicist Fred Hoyle. After the failure of George Gamow's theory that the elements were created in the heat of the Big Bang, Hoyle formulated the more successful alternative explanation that chemical elements were generated inside stars (**46**).

The outcome of these objections was the Steady State theory, which held that the density and general arrangement of the Universe was always the same, rather than evolving. Since Hubble's Law showed that it was expanding, for the Steady State theory to be correct, something had to be filling

1 **Cambridge radio interferometer** The '4C Array' was a fixed array of radio antennae at Lords Bridge near Cambridge, incorporating a cylindrical reflector. It swept the sky as the Earth rotated, recording 4,844 celestial radio sources for what became known as the '4C' catalogue (1965–67).

2 **Willem de Sitter** Dutch mathematician, physicist and astronomer, famous for his research on General Relativity, the orbits of planets and satellites.

3 **Monsignor Georges Lemaître** Belgian Roman Catholic priest, physicist and astronomer who proposed the concept of the expanding universe.

4 **Thomas Gold** Austrian-born British-American astronomer.

5 **Hermann Bondi** Austrian-born British physicist and cosmologist.

6 **Fred Hoyle** Outspoken British physicist and cosmologist, also a popular-science and science-fiction author.

2

3

4

5

6

the new space at just the right rate. Hoyle therefore proposed that hydrogen was continuously and spontaneously created in the gaps that developed between the galaxies. In his arguments for the Steady State/Continuous Creation theory, Hoyle invented the derisory term 'Big Bang' to describe the rival theory of the origin of the Universe; to his surprise the term was taken up by his opponents without rancour.

The main difference between the two contrasting theories was that one insisted that the density and arrangement of the Universe had not changed and in the other insisted that it had. In the Big Bang theory, the Universe was denser in the past, with galaxies closer together. Astronomers have the equivalent of a time machine, as they can look far away and see things from which the light has taken a long time to travel; in effect, looking back in time. Optical telescopes were not able to see far enough to take advantage of this phenomena and investigate whether galaxies were more densely packed together in the past, but by the 1950s radio telescopes were discovering thousands of radio galaxies at large distances and the question could be addressed (**55**).

What they found divided radio astronomers into two camps, one led by Martin Ryle at the University of Cambridge and the other a loose alliance of research groups in Australia at the CSIRO Division of Radiophysics and the University of Sydney led by John Pawsey, Bernie Mills and Bruce Slee. It was a highly technical discussion. Ryle had invented a new technique called 'aperture synthesis interferometry' which was hard to get right, and the cosmological debate boiled down to arcane arguments about a concept called 'log N-log S'. This was a mathematical relationship that expressed the number, N, of radio galaxies of a given brightness, S. Fainter galaxies are generally further away than brighter ones, and since radio waves from the most distant galaxies, like light waves, take such a long time to reach Earth, if you find more of the fainter distant galaxies than expected, it means that the Universe was denser in the past, and must therefore have evolved.

The arguments about log N-log S became increasingly bitter. In 1955 Ryle's group published a catalogue called '2C', showing nearly 2,000 radio sources, with an overabundance of faint sources that suggested the large numbers of distant

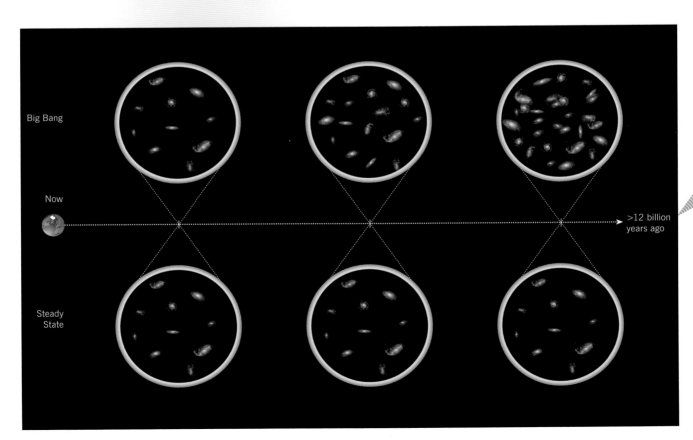

galaxies predicted by supporters of the Big Bang theory. But in Australia, Bernie Mills had just started making a survey with a radio telescope that was different from the equipment that had been used at Cambridge. Fred Hoyle wrote a worried letter to Mills asking if his results confirmed Ryle's. Even early on they did not, and by 1955 the Australian radio-source survey was showing a slight excess of faint sources but nothing like the huge numbers found by Ryle. The Australians privately expressed reservations about the 2C catalogue and suggested that most faint sources in the 2C catalogue were spurious instrumental effects artificially generated by the new technique. Hoyle was relieved – his Steady State Theory was still alive and kicking.

Eventually the Cambridge radio astronomers came to realize the validity of the criticisms and redesigned their equipment and analysis, producing the '3C' catalogue in 1959. Third parties like John Bolton, working with a group at Owens Valley, California, joined the arguments and pressed the observations towards their ultimate conclusion. When Ryle produced his '4C' catalogue in 1965, he invited Hoyle to a press conference

about it. Uninformed about what Ryle would say, Hoyle had to sit on the stage and listen to Ryle announce the irrefutable proof that the Steady State theory was wrong: there was indeed an excess of faint radio sources, which were at vast distances and therefore reflected an earlier, denser state of the Universe. Hoyle was humiliated in public by the stage-managed presentation of Ryle's new results but astronomers accepted the outcome: the Steady State theory, at least in its original form, was dead.

Ryle was right: the Universe has evolved. He was awarded the Nobel Prize in 1974 for 'pioneering research in radio astrophysics', both for his outstanding invention of the new aperture synthesis technique and for the important work that he had done with it.

8

7 Steady state or evolving? If the Universe has been expanding since the Big Bang (top) galaxies will appear more densely packed as astronomers look further into space and therefore further back in time. But in the Steady State model the Universe looks the same everywhere in space, so something must appear to fill in the gaps that are left by the expansion.

8 Martin Ryle British radio astronomer at the University of Cambridge, who discovered that the Universe is evolving.

9 The Cambridge 1-Mile telescope was built in 1964 and was the first to use the technique of aperture synthesis in radio astronomy, for which its lead scientist, Martin Ryle, was awarded the Nobel Prize.

10 4C catalogue From the dry list of radio sources, radio astronomers inferred that the Universe was different in the past than now.

60. The Cosmic Microwave Background
The after-glow of the Big Bang

1

 The cosmic microwave background (CMB) is the radiation left over

from the fireball of the Big Bang. Between 1946 and 1965 three groups

of people were tracking it down, all of them working from different

perspectives and unaware of each other's efforts. None of them realized

that the existence of the CMB was implied by a curious, unexplained

measurement made in 1938.

In the USA, George Gamow, Ralph Alpher and Robert Herman predicted the existence of the CMB in 1947 as part of their theory of the Big Bang (**46**). The hot radiation that had been created in the explosion and in the associated cosmic fireball cooled as the Universe expanded. Approximately 300,000 years after the Big Bang, when the Universe had expanded so much that it had become transparent, the cooling stopped and the radiation ranged throughout the Universe, forming the CMB. Although the temperature of the radiation had been 'frozen' when the cooling stopped, it gradually decreased as the Universe expanded. In 1948 Gamow calculated the temperature of the CMB radiation at 10 °K; Alpher and Herman got 5 °K. The calculations were remarkably close, considering the uncertainties.

Robert Dicke at Princeton University had his own variation on the Big Bang theory. He thought that the Universe might be oscillating, expanding in a Big Bang and then collapsing periodically to a Big Crunch. Each cycle set off the next Big Bang with an associated fireball. As in Gamow's theory, the radiation formed in the initial Big Bang phase subsequently cooled. Dicke's colleague James Peebles calculated the temperature of the radiation, and in 1964, joined by David Wilkinson, they built a so-called 'Dicke radiometer' to detect the radiation, modelled on equipment that had previously been used to measure the heat of the Moon and that, in 1946, had set an upper limit of 20 °K for the temperature of the CMB radiation.

1 **Cosmic microwave background** A map of the sky generated by the WMAP satellite, with red patches representing regions that are a bit hotter than average and blue indicating colder areas. This is, in effect, a picture of the irregularities on the cosmic fireball, with red patches being small holes and blue ones small hills, like dimpling on the surface of an orange.

2 **George Gamow** Russian-born American cosmologist who predicted the existence of the CMB.

3 **Robert Herman** American physicist and cosmologist who also worked in traffic flow, which he effectively invented while working for General Motors.

4 **Ralph Alpher** As Gamow's student, Alpher wrote his epoch-making dissertation on the creation of the elements in the Big Bang, and forever afterwards resented Gamow's joke in inviting Hans Bethe to join the publication's author list so that it could be called the 'Alpha-Beta-Gamma' ('αβγ') theory, as it belittled his own major role in its creation.

5 **Robert Dicke** American physicist, inventor of the lock-in amplifier used for sensitive radio applications, also known for a series of ingenious and beautiful experiments that he carried out to test General Relativity.

6 **David Wilkinson** American cosmologist who studied the CMB and for whom the Wilkinson Microwave Anisotropy Probe (WMAP) was named.

Meanwhile in Holmdel, New Jersey, physicists Arno Penzias and Robert Wilson were trying to identify all the sources of noise in a very sensitive receiver-antenna combination used by Bell Labs for early communication-satellite experiments. After they had accounted for all the instrumental effects, including the droppings from two pigeons that had decided to nest in the horn, they still found excess noise that they could not explain. Penzias and Wilson decided to see whether the noise was caused by radiation that had an astronomical origin. At Johns Hopkins University, Peebles gave a lecture mentioning Dicke's idea on the cosmic fireball, and his remark found its way to Penzias. The Holmdel and Princeton groups got together and decided in 1965 that what Penzias and Wilson had detected was the CMB radiation, and measured its temperature at 3 °K. The discovery was published and became a sensation because it disproved the Steady State theory of the Universe (**59**), which held that there had been no explosion to make the cosmic fireball, and therefore could be no residual radiation from one. Penzias and Wilson were awarded the Nobel Prize in 1978 'for their discovery of cosmic microwave background radiation'.

The discovery of the 3 °K radiation was anticipated in 1941 by Andrew McKellar, who used measurements that S. W. Adams made in 1938 of spectral lines from interstellar molecules called cyanogen to show that molecules in interstellar space were warmed to about 2.3 °K. The source of the warmth was only successfully identified as the CMB by N. J. Woolf and George Field after its discovery was announced in 1965.

Right from the start, measurements showed that the CMB was isothermal and isotropic. Isothermal means that the radiation has almost the same temperature everywhere

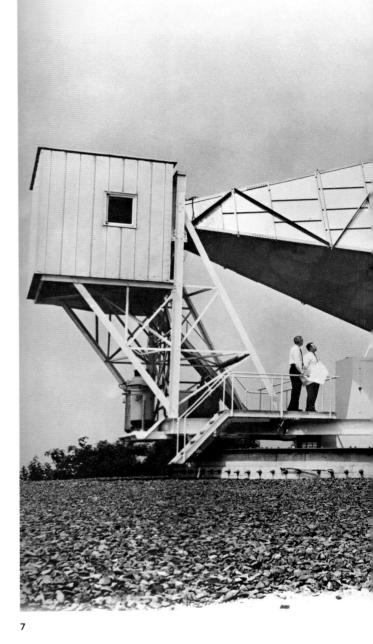

7

8

9

10

7 Horn Antenna at Holmdel, New Jersey Penzias and Wilson discovered the CMB with this equipment. The horn could be rotated around its horizontal axis, and around the horizon on its circular ground track. Radio waves entered the horn and were reflected along the axis and through a hole where they could be detected by equipment in the cabin.

8 Arno Penzias (right) and Robert Wilson The pair stand outside the Horn Antenna.

and must therefore have had a single origin. Isotropic means that it looks the same in every direction. Ever-more-sensitive measurements made from the ground and by equipment flown in high-altitude balloons failed to find any irregularities. By and large, there is no reason to think that any part of the Universe is different from any other, but it was expected that isotropy would break down at some level because nothing is completely uniform, especially when you get down to consideration of the quantum structure of matter and radiation. The Cosmic Background Explorer (COBE) spacecraft, launched in 1989, carried an instrument to search for these irregularities in the CMB. Led by Berkeley physicist George Smoot and NASA astrophysicist John Mather, the COBE team discovered that the CMB had shady patches at the level of 40 parts per million (actually much more uniform than the most perfect white paper). In 1992 Stephen Hawking described this as the 'discovery of the century, if not of all time', for which Smoot and Mather shared the Nobel Prize in 2006. The patches were mapped in more detail by the Wilkinson Microwave Anisotropy Probe (WMAP), launched in June 2001. In May 2009 the European Space Agency launched the Planck satellite in the hope of generating more accurate maps of the CMB, which will make it possible to calculate the properties of the Universe, including its age, size and density.

These irregularities (called 'anisotropies') formed in the first split seconds after the Big Bang, when gravity acted on the extremely small fluctuations in density of the matter. The irregularities developed into the major structures in the Universe, such as clusters of galaxies, which eventually formed everything from stars and planets to life-forms and even human beings.

9 **John Mather** American astrophysicist at the Goddard Space Flight Center in Maryland, who was the scientific leader on the COBE satellite team that first measured fluctuations in the CMB.

10 **George Smoot** American astronomer responsible for measuring the fluctuations in the CMB with the instruments on the COBE satellite.

11 **WMAP satellite** Its instruments are carefully shielded from the warmth of the Sun by a parasol.

61. Darkness at Night

The missing galaxies

1

The darkness at night hides surprising secrets. It tells us about our place among the stars and it shows that the Universe has not existed forever.

2

1 **Darkness** The stars relieve the darkness of the night sky.

2 **Jean-Philippe Loys de Chéseaux** Swiss astronomer, discoverer of nebulae and of the argument known as Olbers's Paradox.

3 **Forest of trees** No matter which direction you look, in a large forest your gaze ends on a tree trunk. In the same way, if the stars are infinite in number your gaze will always end on a star's surface.

3

There is obviously only one star – the Sun – in the immediate vicinity of our Solar System. It is actually rather rare for a star to be single like this; most stars have companions close by, in double or triple star systems, or in star groups or clusters. In science fiction (for instance, in the film *Star Wars*) scenes sometimes take place on a planet with two suns. There are doubts about whether such a planet could really exist (44) but if it does it would have complex day and night patterns. If the planet was orbiting a star that is part of a star cluster, with so many nearby stars scattered all over the sky, all of them shining as brightly as the Sun, it would never have a night as dark as ours, although it might have a brighter–less bright cycle.

Such scenes are imagined. But on Earth we do live in a star cluster of sorts: the Galaxy. At night we see its constituent stars, and they cast a dim light on the surface of the Earth. This is the reason why it is possible to see more on a starlit night (even if it is moonless) than on a cloudy one, as hunters and soldiers know well.

How bright is it at night? There is a theoretical argument that on Earth it ought to be as bright at night as it is by day, which developed progressively in the 18th to 19th centuries after it became clear that the Universe was not contained within crystal spheres, but extensive, with many stars beyond the edge of the Solar System; perhaps even stretching infinitely into the beyond. Edmond Halley was one of the first to consider this possibility in a paper published in 1721 under the name 'Of the infinity of the sphere of the fix'd stars' and to link it with the question of why the night sky was dark. The rigorous mathematical formulation of the problem was discovered by the Swiss astronomer Jean-Philippe Loys de Chéseaux in 1744.

Think of an infinite Universe of stars, uniformly distributed around the Earth in an infinite succession of thin spheres of the same thickness, each sphere getting larger and larger, like the shells of an onion: not real shells, as the crystal spheres were once thought to be, but imaginary boundaries that enable us to divide up infinity into tractable pieces. The number of stars in each shell is proportional to its volume, and increases in proportion to the square of the radius of each shell. But the light at the Earth from each star in each shell diminishes according to the same square law. As a result the total light at the Earth from each shell is the same. If there is an infinite number of shells, the light from all of them added together is infinite – there would be no dark sky at night.

Chéseaux realized that there is a limitation in this argument: the stars in the nearer shells would obscure some of the stars in the shells behind, preventing their light from reaching the Earth,

DARKNESS WOULD BE ABOLISHED

From innumerable stars a limitless sum total of radiations should be derived, by which darkness would be abolished from our skies and the 'intense inane [void]', glowing with the mingled beams of suns individually indistinguishable, would bewilder our feeble senses with its momentous splendour.

Agnes Clerke, *The System of the Stars*, 1890

5

4

4 **Wilhelm Olbers** German doctor and astronomer, discoverer of asteroids Pallas and Vesta, and popularizer of the paradox first formulated by Chéseaux.

5 **Hermann Bondi** Bondi gave Olbers's Paradox its contemporary significance in cosmology.

6 **The night sky over California redwoods** Outside the Milky Way, stars are separated enough for your line of sight to pass between them, into the darkness of the night sky beyond. If the galaxies were infinite in number and extent, every part of the sky would be at least as bright as the Milky Way.

just as in a forest we can see between the nearest trees, but, no matter which direction we turn, we cannot see the trunks of the most distant trees because they are blocked by nearer ones. Substitute 'star surface' for 'tree trunk' in this model, and no matter where we look in the night sky, our line of sight should end at the surface of a star. By this logic the entire dome of the sky should therefore be solidly ablaze with light, as bright as the surface of the Sun. Chéseaux estimated that we could see a distance of 3,000 trillion light years on average, which would number the visible stars in the trillions, making the whole sky contribute as much light altogether as 90,000 suns; his figures are broadly consistent with modern calculations.

Clearly night on Earth is nowhere near as bright as this. Chéseaux proposed two possible explanations for the problem. Perhaps the extent of the starry part of the Universe is finite and smaller than expected: there are regions where the stars are 'missing'. Alternatively, starlight is absorbed in space, diminishing the light from the more distant stars. Evaluated by

modern astronomers, the first explanation is along the right lines and the second less so.

For some reason Chéseaux's work was passed over by his coontemporaries, although many people had access to it, including Wilhelm Olbers, a Bremen doctor who had discovered two of the first asteroids (12). In 1823 he wrote a chapter for the *Berlin Observatory Yearbook*, which was translated into English three years later under the title 'On the Transparency of Space'. Without acknowledging Chéseaux's work – which was in Olbers's library, although it is possible that he hadn't read it – in the article he considered the question of why it is dark at night, and proposed Chéseaux's solutions. Olbers's article became well known, and although the problem had actually been formulated by Chéseaux (and others) nearly a century earlier, it is commonly known as 'Olbers's Paradox'.

One of Chéseaux's and Olbers's solutions, that the excess starlight is absorbed in space, raises a further problem, which was pointed out by John Herschel in 1848. If starlight is

absorbed by an object in space, the object must get hotter, until eventually it gets as hot as an average star's surface and begins to give out as much light as it absorbs. So suggesting that there is an absorber in space doesn't do much to solve the paradox.

Olber's Paradox is still relevant to the modern understanding of the Universe as made up of galaxies rather than individual stars. In 1960 the mathematician Hermann Bondi revived Olbers's Paradox, and listed Olbers's four major assumptions, substituting 'galaxies' for 'stars': (a) The Universe is uniform throughout space; (b) The Universe is unchanging in time; (c) There are no major systematic motions in space; (d) The laws of physics apply everywhere. Modern cosmology offers possible solutions to the paradox by attacking assumptions (a), (b) and (c). (a) The Universe is not infinite and therefore not uniform beyond a certain distance. The furthest that we can see is to the distance from which it has been possible for light to travel since the Universe formed; beyond that boundary we have no way of seeing anything in the Universe, including stars and galaxies. (b) In fact we cannot see even as far as this: light left the most distant regions of the Universe before any galaxies had formed in them; although these regions may contain galaxies, we will not be able to see them for millions or even billions of years. Both of these considerations mean that, compared with the assumptions of Olber's Paradox, lots of galaxies are 'missing' from the visible Universe. (c) Moreover, there is indeed systematic motion in space: galaxies are receding as the Universe expands (53), which makes the light from more distant galaxies progressively weaker than standard calculations (such as the inverse square law) would suggest, although this third objection is not as important as the first two. The night sky is dark because of missing galaxies, not because of small amounts of missing light from otherwise visible galaxies.

The banal observation that the night is dark reveals an astoundingly important secret: the Universe is not infinite. It is limited both in space and time, and it had a beginning.

Future Discoveries

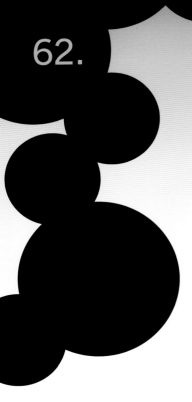

Dark Matter

A dark secret to uncover

Astronomers estimate that 80% of the material in the Universe is a substance called 'dark matter'. It is invisible to all currently available technology, and nearly everything about it is a secret yet to be uncovered, leading some scientists to question whether dark matter exists at all.

1

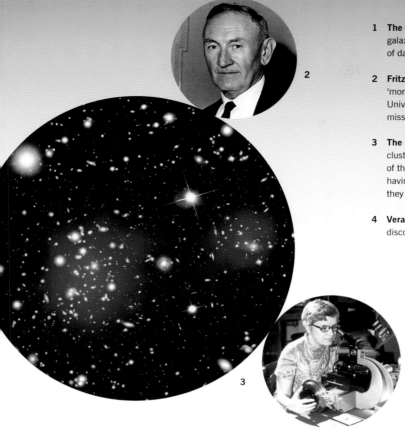

1 **The Coma Cluster of galaxies** One of the closest and richest of clusters of galaxies, the Coma Cluster alerted astronomer Fritz Zwicky to the existence of dark matter.

2 **Fritz Zwicky** Swiss-born American astronomer who, in a scheme he called 'morphological astronomy', attempted to account for everything in the Universe; this led him to the realization that large amounts of mass were missing.

3 **The Bullet Cluster** 1E 0657-56 is the result of a collision between two clusters of galaxies (at left and right). Dark matter is associated with each of them (blue tones), the dark matter and the galaxies in each cluster having passed through each other like marching bandsmen. The gas that they contained stuck together and lies between the two clusters.

4 **Vera Rubin** American astronomer, who studied the rotation of galaxies and discovered the effects of dark matter distributed in haloes around them.

Looking out at the broad view of the Universe, astronomers see a huge quantity of mass, which is distributed as stars and gas in galaxies. There is also a considerable amount of hydrogen and helium in intergalactic space, which constitutes clouds left over from the Big Bang; they have not turned into galaxies, they contain few or no stars and they do not shine. The only way this material makes its presence know to astronomers is by absorbing ultraviolet light from distant quasars: the light from the quasar penetrates the clouds as if they were pieces of meat skewered on a kebab, so when astronomers look at the quasar's ultraviolet spectrum, they see the gaps wherever the beam of light is passing through a cloud.

However, there is also a considerable amount of matter in the Universe that astronomers cannot see by any known method. It is generically called 'dark matter'.

Dark matter was discovered in the 1930s by the Swiss-born astronomer Fritz Zwicky of the California Institute of Technology, as he set out to provide a complete scheme that would encompass everything in the Universe. In order to establish the boundaries for his scheme, Zwicky began by measuring the mass of as many known space objects as he could. In 1933 he estimated the mass of a nearby cluster of galaxies in the constellation Coma by measuring each galaxy's speed of motion as the combined mass of the cluster pulled it along in orbit. The galaxies were moving much faster than he expected, which suggested that the mass of the cluster was also much larger than predicted. He had already estimated the mass of all the individual stars in the cluster on the basis of the light they emitted, but, judging from the speed of the orbiting galaxies, he estimated that the mass of the Coma Cluster as a whole was actually 400 times greater than the total mass of its stars. 399/400 of the mass was, as he put it, 'missing'.

Zwicky's discovery of dark matter in clusters of galaxies was not followed up by his colleagues – perhaps because he was so difficult to work with – but it was confirmed half a lifetime later on a smaller scale, within single galaxies, by astronomer Vera Rubin. After graduating from Vassar in 1948, she wanted to join the graduate astronomy programme at Princeton, but was told that it did not admit women, so she continued her career instead at Cornell and Georgetown Universities. She then took a position at the Carnegie Institution in Washington, where she worked with Kent Ford, who had developed equipment sensitive enough to measure the spectra of faint galaxies. Rubin and Ford used the new spectrograph to determine how spiral galaxies rotate, observing fainter and more distant regions of the galaxies than had previously been possible.

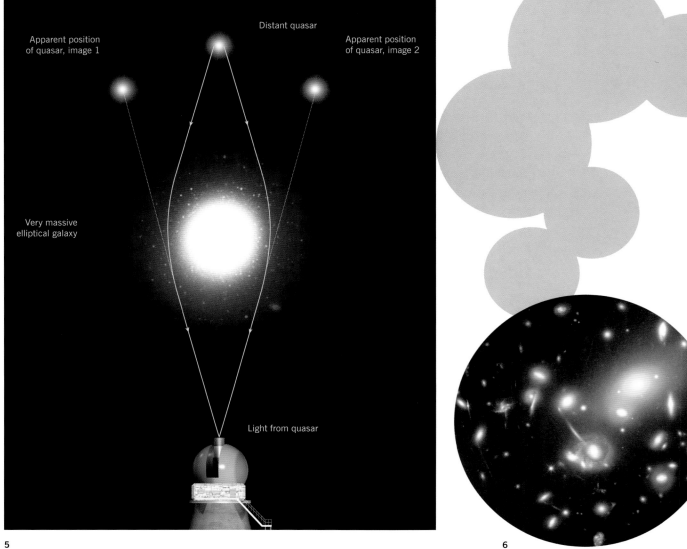

Distant quasar

Apparent position of quasar, image 1

Apparent position of quasar, image 2

Very massive elliptical galaxy

Light from quasar

5

6

7

5 **Gravitational lens** If a distant galaxy or quasar and a massive galaxy are accidentally aligned with the Earth, the galaxy bends the light of the more distant object like a lens, deflecting its position and magnifying its brightness.

6 **Abell 2218** The cluster of galaxies acts as a powerful lens, magnifying but also distorting all the galaxies lying behind the cluster core. The lensed galaxies (orange streaks) are all stretched and elongated. They may also be multiply imaged, and curve in arcs around the main mass-concentration in the cluster of galaxies.

7 **Westerbork Radio Synthesis Telescope** The Westerbork radio interferometer is located in a forest in the Netherlands.

8 **Boulby Mine** Maybe dark matter consists largely of a new heavy neutral particle, a 'neutralino' or 'WIMP'. If so, it might be detected in experiments like this one at the Boulby Mine of Cleveland Potash Ltd., on the edge of the Yorkshire moors, aimed at recording the rare events that would occur when a dark matter particle collides with ordinary terrestrial matter. Cosmic rays – which constantly batter Earth's surface – cannot pass through the kilometre of rock above. This protects the experiment from confusing particle bombardment.

Their expectation was that stars in the outer regions of a spiral galaxy would move more slowly than stars in the central regions (where most of a galaxy's star-mass is concentrated), just as the more distant planets in the Solar System rotate more slowly around the Sun than the inner planets. But to their surprise Rubin and Ford discovered that stars in the central and outer regions of each spiral galaxy were actually moving at the same speed. This means that there is unseen matter in each galaxy to give stars the extra pull along in their orbit, even in the outer regions of galaxies where there are few visible stars; in a typical spiral galaxy this accounts for up to ten times more mass than is immediately visible as stars. By 1975 Rubin had become convinced that 'What you see in a spiral galaxy is not what you get.' Although her findings were initially met with scepticism, with the construction of radio telescopes like the Westerbork Radio Synthesis Telescope in The Netherlands, which measured the rotation of hydrogen gas in spiral galaxies, the evidence became overwhelming.

In 1937 Zwicky had suggested another way to investigate the mass of galaxies. If a massive galaxy lies between the observer and a more distant galaxy, it acts as a 'gravitational lens', warping the surrounding space to magnify, distort and displace the image of the background galaxy, as predicted by Einstein's theory of General Relativity (31). Zwicky suggested that this effect would provide a way to probe a galaxy's mass, but acknowledged that no such lens was known. He did not live to see the first discovery of a gravitational lens in 1979 by Dennis Walsh, Bob Carswell and Ray Weymann. The team was using a small telescope at Kitt Peak Observatory to locate known quasars when they discovered a pair of identical quasars right next to each other; in fact, they were seeing two images of the same quasar produced by a gravitational lens galaxy. Astronomers have since discovered gravitational lensed images produced by clusters of galaxies, which they have used to implement Zwicky's suggestion.

Bringing together considerations like these, the latest estimate is that 5% of the matter in the Universe is made up of stars, 15% is made up of intergalactic gas clouds and 80% is made up of dark matter. The composition of dark matter is unknown. It might be some sort of unknown massive elementary particle, different models being known under various names such as axions or 'WIMPS' (Weakly Interacting Massive Particles). Laboratory searches for these particles are underway, which might produce the most important discoveries of the 21st century. If dark matter is indeed isolated in the laboratory, it will be a case – like the discovery of helium (28) – where a fundamental constituent of matter was first identified in the cosmos and only subsequently found on Earth. However, given that dark matter has still not been identified, some astronomers speculate that there may be something wrong with our theories of gravity and related calculations, or that dark matter is ordinary matter in some hard-to-see form. Either way, we know there is still a dark secret to uncover.

63. Dark Energy

On the threshold of a profound discovery

1

Imagine taking a region of space and removing all matter and radiation from it until the area is completely empty, much more so than ordinary space between planets or stars. The result is a 'vacuum', which has effects that physicists and astronomers call 'dark energy'. Dark energy is thought to account for nearly three-quarters of the energy in the Universe, but, as with dark matter, we have not yet discovered a way to see it.

3

2

1 **Cosmic web** The distribution of galaxies in space is accurately represented by this simulation. Threads of individual galaxies and clusters of thousands or millions of galaxies enclose voids, which are almost empty. Where the threads converge there are superclusters of galaxies, the largest structures in the Universe.

2 **Adam Riess** American astronomer, one of the leaders of the High-Z Supernova Search Team, now of the Higher-Z Supernova Search Team.

3 **Saul Perlmutter** American astronomer, head of the Supernova Cosmology Project at the University of California at Berkeley, poses against a backdrop of Supernova 1987A.

To common sense, the vacuum of space is nothing. To a scientist, the vacuum is not nothing: it is a physical state and it has an energy. In the absence of gravity, there is no way of measuring the energy of a state on an absolute scale; the best we can do is to compare energy differences. The strength of vacuum energy itself would be arbitrary. According to the theory of General Relativity, however, any form of energy has a gravitational effect, so the vacuum energy might be a crucial ingredient in the evolution of the Universe. The vacuum energy is known colloquially as 'dark energy.'

Dark energy is as important to the history of the Universe as dark matter (**62**). The Universe is expanding (**53, 59**),

but the galaxies and dark matter in the Universe mutually pull on one another and tend to slow down the expansion. This has been verified by the Hubble Space Telescope (HST), which in looking out into the distant Universe is looking back in time, because light takes billions of years to travel such great distances. The most distant galaxies, which reflect the state of earlier Universe, should be moving more quickly than nearby ones. To confirm this, in 1998–99 astronomers from the Supernova Cosmology Project and the High-Z Supernova Search Team used supernovae observed with the HST (**41**) to check the distances of remote galaxies and then, using the largest ground-based optical telescopes, measured how

ON THE COSMOLOGICAL CONSTANT

Much later, when I was discussing cosmological problems with Einstein, he remarked that the introduction of the cosmological term was the biggest blunder of his life.

George Gamow, *My World Line*, 1970

It's ugly. If you or I were making a Universe we wouldn't put it in.

James Peebles, 2007

4 Albert Einstein Was it a blunder or inspiration that caused Einstein to formulate General Relativity with a mathematical term that is the conceptual equivalent of dark energy?

5 Carlos Frenk British astrophysicist and cosmologist at the University of Durham, seen here lecturing about the Millennium Simulation.

fast the galaxies were moving. They discovered the opposite of what had been expected: the expansion of the Universe is speeding up, not slowing down. There is some progressive input of energy into the Universe, thought to be dark energy, but the nature and mechanism of its input is a mystery.

Dark energy has similar effects to something hypothesized 100 years ago by Albert Einstein. Since Einstein formulated his theory of General Relativity before the discovery of the expanding Universe, he applied it to a model of a static Universe. The force of gravity attracts all galaxies together. Einstein realized that if gravity was the only force there was, it would be impossible for the Universe to be static, because galaxies would continue to attract each other until they were packed together in an infinitely dense single cluster (**59**).

Einstein discovered that he needed something to stop the galaxies falling together. Quite arbitrarily, he added a term to his equations called the 'cosmological constant', symbolized by the Greek letter Λ (lambda), but when the expansion of the Universe was discovered by Edwin Hubble in 1929 (**53**), Einstein retracted the concept as a blunder. A cosmological constant has the tendency to cause galaxies to accelerate away from us. In a Universe with both matter and vacuum energy, there is a competition between the tendency of Λ to cause acceleration and the tendency of matter to cause deceleration. This has a big effect on 'the formation of structure', the expression that astronomers use to describe the way that the earliest irregularities in the material of the Big Bang grew. The denser bits drew in surrounding matter and grew to intergalactic-size clouds, which congealed into clusters of galaxies, condensing further into stars and planets.

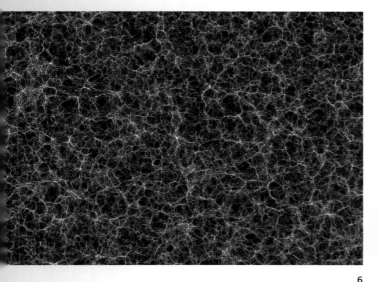

6

7

6   **The Millennium Simulation** mixes matter, dark matter, dark energy and
    their physical interactions and follows how they all evolve. The simulation
    starts with an only slightly inhomogeneous distribution of material in the
    Universe.

7   **Lumps and bump**s grow under the influence of gravity in the Millennium
    Simulation, to make the Universe as it is today.

8   **Pie chart** What do you need to cook to make the millennium pie? The most
    realistic simulation uses only 4% of normal matter, with 22% dark matter
    (their gravity draws together the largest lumps) but nearly three-quarters of
    the energy in the Universe has to be progressively released in the form of
    dark energy.

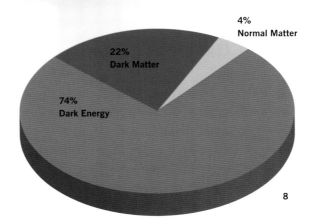

4%
Normal Matter

22%
Dark Matter

74%
Dark Energy

8

In a major calculation project called the Millennium
Simulation, astronomers of the University of Durham and the
Max Planck Institute for Astrophysics in Garching in Germany,
showed how the formation of a cluster of galaxies starts slowly,
because the fluctuations in Big Bang material are not very
pronounced; even 300,000 years after the Big Bang they had
an effect on the cosmic microwave background so small that it
is at the limit of astronomers' ability to measure (**60**). Gravity,
principally from dark matter, draws in surrounding material, but
the release of dark energy tends to stabilize the infall, which
peters out as the galaxies orbit one another. We can take as
an analogy what happens at a reception when a celebrity and
his entourage arrive. Other guests, who were formerly moving
quite slowly, are attracted by the celebrity and hasten towards
him, so that the irregularity in the density of figures around

him grows. But the closer they get to him the less confident
the guests become, starting to hang back and mill around one
another at a safe distance from the celebrity, which determines
the size of the cluster of people in his immediate vicinity.

The Millennium Simulation produces a simulated
distribution of galaxies, which is then compared to their actual
distribution so that astronomers can refine the model by altering
its proportions of dark matter and dark energy. However, as
with the problem of dark matter, the problem of dark energy
may indicate that there is something wrong with our theories
of gravity. Current evidence suggests that the Universe has
something like 26% of its density in matter (22 of the 26%
being dark matter) and 74% of its density in dark energy. This
indicates the scale of the problem – 96% of the contents of the
Universe are still a secret for us yet to discover.

64

Gravitational Waves

Whispers of neutron stars and the Big Bang

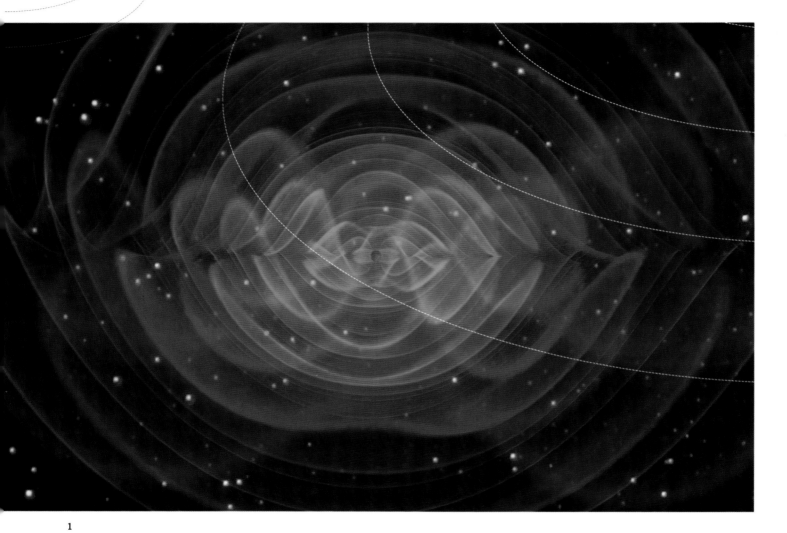

1

When the mass distribution of an object changes, the gravity around it changes, sending ripples through space at the speed of light. The ripples are called 'gravitational radiation' or 'gravitational waves'. Astronomers are trying to develop technology to study the waves, which could reveal what the Universe was like in the split seconds after the Big Bang.

2

Gravitational waves are emitted by revolving binary stars, by rotating stars that are not completely spherical, and by collapsing stars (if they do not collapse straight down, but splatter). Gravitational waves, which permeate the Universe, are waves *in* the gravity generated by an object (not to be confused with 'gravity waves', which are waves generated *by* gravity on an object: waves on the surface of the ocean or in the atmosphere of the Sun, for instance). Almost any mass that moves, such as a passing car, emits gravitational waves, though in the case of a car they are very weak because its mass is so small.

Gravitational waves are always relatively weak, even when generated by an object as massive as a star. About once a day on average, gravitational waves, generated by the collapse of a binary star system in a far distant galaxy, appear in the sky on Earth. These waves carry as much energy as the brightest stars or planets, but we do not notice them, because they pass right through the Earth, leaving behind almost no energy. Gravitational waves will only be fully understood in the future, when we have developed more sensitive equipment.

Although gravitational waves 'whisper' too quietly for us to hear them, their effects were discovered in the 1970s and 1990s as astronomers studied binary pulsars. The first of these, B1913+16, known as 'the' Binary Pulsar, was discovered by Russell Hulse and his PhD supervisor, Joseph Taylor, in 1973, using the Arecibo radio telescope in Puerto Rico; they were awarded the Nobel Prize in 1993 for the discovery. The Binary Pulsar is a pulsar in a highly eccentric, short-period (and therefore small) orbit around a second neutron star. Its pulses can be tracked very precisely by radio telescopes, arriving earlier or later when the pulsar is nearer to, or further from, the Earth in its orbit. From the start it was clear that the pulsar would show changes in its position so clearly that radio astronomers would be able to view the predicted effects of Einstein's theory of General Relativity on its orbit. By 1975, when Hulse was writing up his thesis, the first effects of General Relativity had already been detected in the pulsar timings, and by 1980 it was possible for Taylor to see the effects of its gravitational waves. As the gravitational waves radiated away, the loss of energy from the binary

1 **Gravitational waves from a binary black hole** As the binary star rotates in its orbit its gravity alters the structure of space-time around it, and the disturbances propagate outwards into space as gravitational waves that cause the stars to bob around like corks on the sea.

2 **Arecibo radio telescope** Its reflective dish is fixed in a hollow in the hills of Puerto Rico, and focuses radio waves from the zenith onto radio receivers that are suspended above. Building on the technology of Lord Rosse's telescope (Chapter 48, Fig. 3), the receivers can be made, by means of ropes and pulleys, to shift and track in order to find and follow an individual radio source.

3 **Joseph Taylor** American radio astronomer, Nobel Prize winner, known for his work on pulsars, particularly the Binary Pulsar (the first to be discovered).

4 **Russell Hulse** American physicist, Taylor's student, left astronomy after his Nobel Prize-winning discovery of the Binary Pulsar in Puerto Rico for the more stable life of a laboratory physicist.

3

4

system caused the pulsar's orbit to shrink by 1.5 cm per
revolution; it has shrunk by 500 m in total since its discovery.
A second binary pulsar, pulsar B1534+12, was discovered
in 1990 by Aleksander Wolszczan; although it has not been
observed for as long as the Hulse–Taylor pulsar, its pulses are
both stronger and narrower so it may turn out to be a better
laboratory for testing General Relativity. Measurements of the
effects of gravitational radiation in both pulsars confirm that
the calculations of General Relativity work amazingly well.

There is so much happening on Earth that can make
pendulums vibrate that gravitational-wave astronomers dream

Although gravitational radiation from space is weak,
ongoing attempts are being made to detect it on Earth.
The first detectors were made by Joseph Weber of the
University of Maryland in the 1970s; a ground-based
detector called LIGO is being brought into operation, with
twin installations in Hanford, Washington, and Livingston,
Louisiana, to prevent false readings from local seismic
disturbances; similar detectors are being constructed in
Cascina, Italy (VIRGO), Hanover, Germany (GEO 600),
and Tokyo (TAMA 300). All measure the distance between
two freely suspended pendulum mirrors; the passage of a
gravitational wave will cause the mirrors to bob like corks on
the sea, temporarily changing the distance between them.

There is so much happening on Earth that can make
pendulums vibrate that gravitational-wave astronomers dream
about the relatively quiet conditions in space. A space-borne
detector called LISA is being developed by the European Space
Agency and NASA. It will consist of three spacecraft travelling
in formation 5 million km apart, forming an equilateral triangle

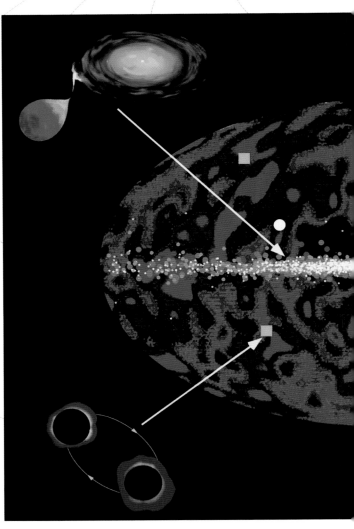

5

5 The sky in gravitational waves (artist's impression) On a background
of gravitational waves with fluctuations that originated in the Big Bang
(lower right) there are individual sources like black hole binary stars (lower
left), white-dwarf and neutron-star binary stars that will turn into Type Ia
supernovae (upper left) and stars that are falling into supermassive black
holes in the centres of galaxies (upper right).

6 LISA Three spacecraft orbiting 5 million km apart connect via laser beams
to measure the distance between them. As gravitational waves pass over
the spacecraft configuration they cause the spacecraft to oscillate in their
orbits. Concept sketch of what, if it comes to fruition, will be by far the
largest human construct ever built.

7 LIGO At present, the laser technology used in gravitational wave detectors
reaches only to kilometres, as in the LIGO project. The largest and most
ambitious project funded by the US National Science Foundation, LIGO is
based on two interferometers in Livingston, Louisiana (shown here), and
Hanover, Washington.

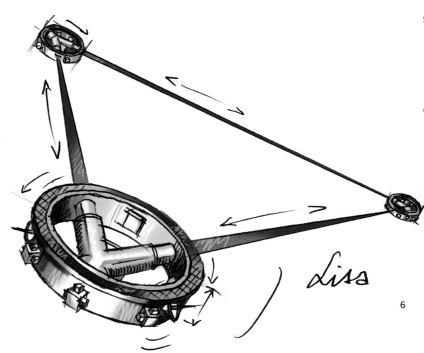

Lisa

6

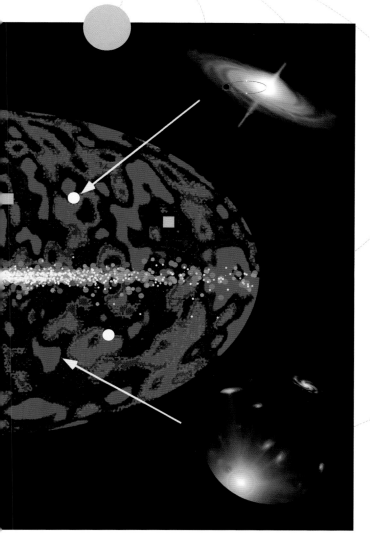

7

as they follow the Earth in its orbit. Each spacecraft will carry devices to counter the irregular effects of the solar wind, so that any variation in the distances between them (to be measured by lasers) will indicate the presence of gravitational waves. LISA will be especially sensitive not only because it avoids the Earth-tremors that shake ground-based detectors, but also because of its size. The key measurement is the fractional change in the distance between the mirrors, so if you can measure the distance to, say, 1 micron, an instrument in which the mirrors are a million km apart is 1 million times more sensitive than one in which they are 1 km apart. The earth-based detectors are between 300 m and 4 km long, but as a scientific instrument with dimensions of 5 million km on each, LISA will be the largest-ever man-made construction.

Only gravitational waves from binary stars have been detected so far, and even then only indirectly. Especially large quantities of gravitational waves are produced by a binary star if its two components merge into one; astronomers expect this eventually will happen to the binary pulsars. In theory, gravitational waves can also come from any spinning white dwarf or neutron star that has a bump on one side. Usually white dwarfs and neutron stars have strong gravity and are spherical, but some stars in binary systems sit under an infalling stream of gas that heats their surface and may raise a bump, perhaps metres high. Gravitational waves can also come in bursts from supernovae (**41**) that collapse to a neutron star (**35**) or to a black hole (**36**). Other gravitational waves were made in the early history of the Universe and may be able to show astronomers a picture of the conditions immediately after the Big Bang (when the Universe was 1 million million million millionth of a second old), just as the cosmic microwave background carries a picture of the Universe when it was 300,000 years old (**60**). LISA should be able to detect this gravitational wave cosmic background, but in the meantime, gravitational waves remain the last completely unexplored window on the Universe, through which new cosmic discoveries undoubtedly will be made.

GRAVITATIONAL WAVE COSMIC BACKGROUND

Perhaps my whisper was already born before my lips.
Osip Mandelstam, 1934

65.

Life in the Universe

Are we alone?

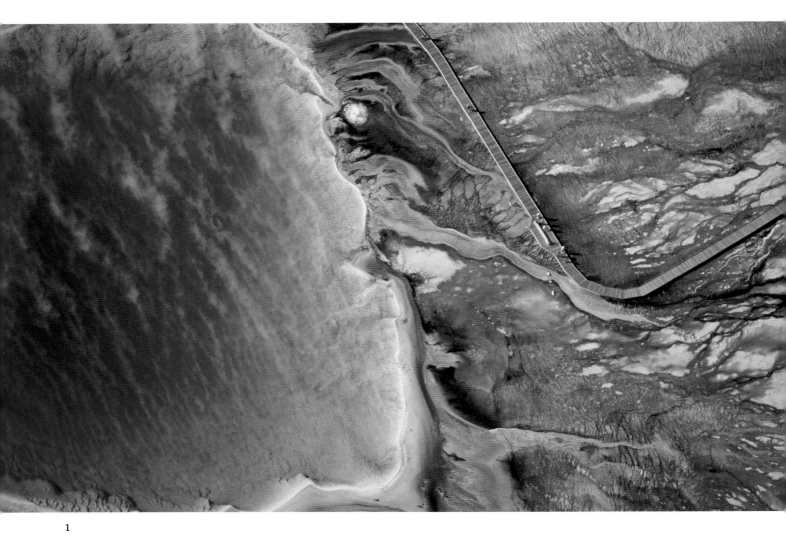

1

No one knows whether astronomers will ever discover life on other planets. If they do, it will prove that we are not alone in the Universe. Most people seem to like this idea, but the Universe has yet to reveal its secrets on the matter.

1 **Extremophiles** Life flourishes even in extreme environments. Here colonies of hot-water-loving cyanobacteria of different colours live at 46 °C in a hot spring in Yellowstone National Park, Wyoming, USA.

2 **Murchison meteorite** The larger piece of a meteorite that fell in Victoria, Australia, shows primitive chondrules (the light-coloured embedded bits) left over from the formation of the Sun and planets; laboratory analysis showed that it also contained organic chemicals like amino acids. Pre-biotic chemicals were present right from the start of the Solar System.

3 **Jean-Baptiste Biot** French astronomer and physicist who established that meteorites had fallen from space, and carried out work into the polarization produced by different chemicals.

4 **Louis Pasteur** French chemist, who demonstrated how disease was passed by microorganisms, and carried out work on the polarization of light.

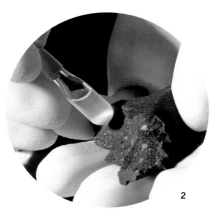

2

3

4

Life is based on carbon. Carbon is made in stars and is everywhere in the Universe. It is the only element capable of making complex 'organic' molecules, like DNA, which comprise the rich 'vocabulary' used in writing the 'script' of life. Radio astronomers have discovered about 150 varieties of organic molecules in interstellar space (**38**); comets also have organic 'tars' and 'crusts' on their surface (**18**).

In 1969 a meteorite broke up over Murchison, near Melbourne in Australia and pieces fell across the town. There were many eyewitnesses to the fall (churchgoers on a Sunday morning), who promptly collected about 80 kg of fragments. Deep inside the fragments were more than 90 amino acids, a type of organic compound essential for life, and it was clear that these substances were not terrestrial contaminants: they had been made in space. The implication is that basic biochemicals are made by natural processes throughout the Universe, and, as the meteorite showed, they can move from place to place.

The French chemists Jean-Baptiste Biot (1815) and Louis Pasteur (1848) discovered that certain biochemical molecules have a property called 'chirality': they occur in both 'left-' and 'right-handed' forms, only one of which is used by life. Thalidomide is a notorious example, in which one chirality,

used as a drug, is benign and the other actually very harmful, causing birth defects. Inorganic chemical processes usually make equal amounts of both chiralities. But if you shine strongly polarized light into a 'soup' in which there are equal numbers of left- and right-handed molecules, some can switch from one sort to the other, and the soup develops an excess of one form of chirality compared to the other, and may favour biochemical rather than the inorganic chemical reactions. Some astronomers think that chirality can be generated in the organic molecules in space when they pass a source of strongly polarized light, such as a pulsar, transforming them into 'seeds' for life (**35**).

Astronomers have discovered the processes that put the chiral chemicals on to the Earth, but geologists have identified the 'fire' that simmered them into life. Energy is required to fuel the chemical reactions that change the chiral chemicals into more complex life-forming molecules. On Earth, sunlight is the usual energy source – plants photosynthesize chemicals to create cells – but it is not the only source. 'Black smokers' are geothermal vents in the deepest, darkest parts of the ocean, which are colonized by oceanic life that feeds on geothermal energy rather than sunlight. The forms of life that flourish in such extreme environments are called archaeons, and they

6

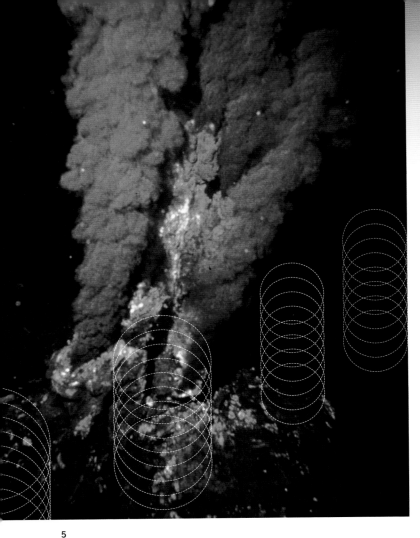

5

7

8

are the oldest living things on Earth. In fact life seems to have started under the oceans and only later evolved to harvest energy from sunlight on the surface.

Biochemistry needs a medium in which to operate: water. Although water is made of simple elements (hydrogen and oxygen) that are abundant everywhere in the Universe, liquid water can only survive on the surface of a planet in the 'goldilocks zone' of a planetary system: that is, where the temperature is neither so cold so that water freezes, nor so hot so that it evaporates, but just right. This is mainly a matter of the planet being at the right distance from its sun, but can be helped or hindered by any greenhouse effect in the planet's atmosphere (**23**). It is also possible for a planet outside the goldilocks zone to have liquid water because of geothermal heating, as on Mars or Europa (**25**).

The basic idea that organic chemicals developed in space and were brought to Earth by cosmic processes, to evolve in the oceans, was first proposed by Isaac Newton and given modern form in 1908–11 by the father-and-son team Thomas

C. Chamberlin and Rollin T. Chamberlin. The Chamberlins suggested that 'planetesimals,' the small bodies that merged into the planets (**50**), were the source of organic material from which life evolved, creating organic compounds as they collided and accreted with each other. The basic ideas of the theory were revived by Soviet astrobiologist Alexander Oparin in 1938, and reformulated with the aid of modern scientific knowledge by J. Oró and Chris Chyba, director of the SETI Institute (Search for Extraterrestrial Intelligence), and a student of Carl Sagan.

Another possible scenario for the early formation of life on Earth and elsewhere was discovered in the laboratory. In 1953, Stanley Miller, a PhD student at the University of Chicago, put simple organic chemicals into a flask of water from which oxygen gas had been removed, and then passed an electric spark in through the vapour. He did not have deep-sea geothermal vents in mind; he was guided by the idea that electric discharges (lightning) created biological molecules in the atmosphere. He found amino acids in the black sludge produced by his experiment. No-one knows how life first

5 **Black smoker** On the sea floor, a hydrothermal vent ('black smoker') belches clouds of iron sulphide. Even here, where atmospheric oxygen and light do not penetrate, life thrives, feeding on nutrients produced by the volcanic processes that energize the vents.

6 **Thomas C. Chamberlin** American geologist, creator of the planetesimal theory of the formation of the Solar System.

7 **David McKay** Astrobiologist and geologist of NASA's Johnson Space Center.

8 **Stanley Miller** American chemist who worked on the chemical origins of life on Earth.

9 **Carbonate minerals in a Martian meteorite** A false-colour microscopic image of the ALH84001 meteorite, showing the distribution of the carbonate globules (orange). These minerals form in liquid water and in this meteorite are very old. They add evidence that the environment on Mars was at one time able to sustain life.

10 **Christopher Chyba** American astrophysicist and global security expert at Princeton University.

11 **Enrico Fermi** Italian physicist, Nobel Prize winner known for his work on quantum mechanics and nuclear science, but also for posing the question about extraterrestrial life known as the Fermi Paradox.

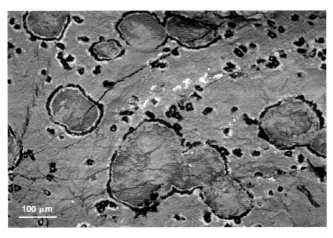

9

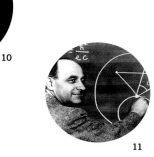

10

11

formed on Earth, but hundreds of similar experiments have since been carried out with a variety of chemicals and energy sources, showing similar results.

Scientists believe that the right chemical ingredients, a source of energy, and the presence of water as a solvent, at some point combined to transform inanimate organic chemicals into life on Earth, perhaps also on Mars and Europa (**25**). In 1996 David McKay and his coworkers discovered what appeared to be traces of bacteria in the Martian meteorite ALH 84001, which was found in Antarctica (**15**). The meteorite had been knocked off the surface of Mars by an asteroid impact 16 million years ago, and, after a period in orbit around the Solar System, fell to Earth 12,000 years ago. It contains molecules that were clearly made in liquid water, curious surface shapes that look like fossil bacteria, and mineral grains and molecules that are similar to some made organically here on Earth. McKay cautioned that 'none of these observations is in itself conclusive for the existence of past life', but asserted that 'when they are considered collectively, particularly in view of their spatial

association, we conclude that they are evidence for primitive life on Mars.' But this is a strong claim that needs stronger evidence.

What of life in the Universe beyond the Solar System? Although technology does not permit astronomers to identify any exoplanets much like the Earth (**44**), stars are numerous, planets orbit around many of them, and there has been a lot of time for complex life to develop. So why are we not in contact with intelligent extraterrestrials from all these suitable planets? This contradiction is called the Fermi Paradox after the physicist Enrico Fermi, who, speaking of extraterrestrial beings in 1950, asked 'Where are they all?'

We can indeed assume that many places in the Universe have the right conditions and ingredients for life. In these circumstances, single-celled life seems to evolve fairly quickly: the oldest fossils of primitive life on Earth are stromatolites, mats of blue-green algae (cyanobacteria) fossilized in rocks as much as 3.5 billion years old. Their age means that the algae represent what happened after the first 20% of the timeline of the history of the Earth. Life, simple life, may well be abundant in the Universe.

However, multicellular life forms and advanced intelligence need – crucially – time to evolve, and time of the right quality may be a much rarer commodity in the Universe than the right conditions and ingredients for single-celled life. Primates have evolved only in the last 85 million years, hominids less than 20 million years ago, and the earliest tool-making hominids 2 million years ago (less than 0.1% of the age of the Earth). Judging by the one example that we know, evolution needs to try out lots of experiments before it can come up with complex life, and requires a platform that is relatively stable for billions of years – a planet where temperature and climate change little, and where any sudden change is not too severe. The Earth has had several lucky escapes from catastrophic events (**20**, **21**, **22**). A freak accident (the creation of the Moon) gave the Earth an extra-large iron core and therefore a strong magnetosphere that defends it against the scouring and irradiating effects of the solar wind (**17**). The inner Solar System has also somehow survived the migration of Jupiter-sized planets that rampage through other planetary systems (**44**).

Your planet has survived all this and produced you. Other planets may not be so lucky, stable or fertile, so while simple life forms may be common in the Universe, complex, and therefore intelligent, life may be rather rare. The distances between the planets that have intelligent life, if there are indeed others, may consequently be so great that it is almost impossible for them to communicate with each other. This may be the explanation for the Fermi Paradox.

We may like to take comfort in thinking that there is life elsewhere in the Universe, but we also have to face the possibility that, at least in practical terms, we are indeed alone.

12 Stromatolites At Shark Bay, Australia, colonies of cyanobacteria and sediment represent the first examples of organized bacteria on the evolutionary trail towards multicellular life. They have been present on Earth for perhaps 3 billion years and may have been responsible for changing its atmosphere to its present oxygen-rich composition. They are very important in the evolution of life on Earth and it would be very significant to find similar life forms on another planet. However, if they represent the only kind of life that is present there, conversation with an extraterrestrial would probably be difficult and unrewarding.

Glossary,
Further Reading,
Acknowledgments,
Picture Credits
& Index

Glossary

accretion The process by which an astronomical body increases in size, by gathering particles of matter to itself, either by means of its gravitational pull or by the adherence of particles with which it collides. The additional matter orbits the main body in a flat 'accretion disc'. (50, 56)

active galaxy A **galaxy** with a powerful and energetic nucleus (that is, an 'active galactic nucleus' or AGN). (36, 55, 56)

Algol paradox In a **binary star** (such as the prototype, Algol), the anomaly that the less massive star is the more advanced in its evolution. (33)

ansae (Latin: 'handles') Ear-like protrusions from a celestial body. (27)

aperture synthesis interferometry A radio astronomy technique in which a line of stationary radio telescopes depends on the rotation of the Earth to gather radio waves in a way that simulates the activity of a much larger radio telescope. (Preface, 59)

asteroid A minor **planet**, of a size in the range of 1 m to 1,000 km, typically orbiting in the Solar System between Mars and Jupiter. An asteroid is not a **dwarf planet**, a **comet**, a **meteoroid**, or a **moon** or **satellite**. See also **asteroid belt** and **Kuiper Belt**. (12, 22)

asteroid belt The main band of **asteroids** of the inner Solar System, lying between the orbits of Mars and Jupiter; compare **Kuiper Belt**. (12, 22)

aurora Light produced by atoms and molecules in the atmosphere and caused by the impact of ionized particles accelerated in a magnetic field. The term is applied especially to the Earth's Northern and Southern Lights. (17, 26)

Becklin–Neugebauer Object (BN Object) A source of intense infrared radiation found in the Orion nebula and thought to be a newly formed star; it is named after the two astronomers who first observed it. (50)

Big Bang The dense, high-energy, explosive event that occurred at the beginning of the Universe. (31, 46, 59, 60)

Big Crunch The hypothetical high-energy event that will occur if the Universe eventually implodes. (60)

binary In astronomy, a system composed of two bodies in orbit around each other: for example, a binary planet, binary pulsar or **binary star**. (14, 33, 36, 64)

binary star A **double star**, particularly a close pair. (33, 64)

bipolar nebula A symmetrical **nebula** that has two lobes, pointing in opposite directions; often, but not exclusively, a **planetary nebula** of that shape. (49)

black hole An astronomical body that is both small and massive, thus exerting such a strong force of gravity that no light or other radiation can leave the surface. (32, 36, 40, 56, 57)

blink comparator A viewing device used in astronomy to compare two photographs: the viewing optics switch ('blink') quickly between the two images to reveal changes, such as the successive positions of a moving star or planet, or the different appearance of a **variable star**. (13)

Bode's Law (Titius–Bode Law) The law, popularized by Johannes Bode (first formulated by Johann Titius), by means of which the distances of the planets from the Sun can be calculated. (12)

caldera (Spanish: 'cauldron') A volcanic **crater**, created when magma sinks from the subsurface, which weakens the surface so that it can no longer support the weight of the material above and the volcano collapses inwards. (26)

canals of Mars (from Italian *canali*) A network of long straight markings apparently covering the surface of Mars; thought originally to be irrigation canals, they are, in fact, an optical illusion. (25)

catastrophism The theory that geological features are caused by unpredictable, large-scale events (catastrophes) such as floods or meteor impacts. Compare **gradualism**. (20)

celestial sphere In Ptolemaic astronomy, the imaginary sphere centred on the Earth (or the observer); the fixed stars appear to lie on its surface. (1)

Cepheid variable star A pulsating **variable star** that varies in brightness in a regular cycle, as in the case of the star Delta (δ) Cephei. (40, 43)

Chandrasekhar mass The maximum possible mass of a **white dwarf** before it collapses under its own weight (to become a **black hole**); by extension, the maximum mass of a **neutron star**. (34)

chaos A mathematical effect, the property of certain equations that predict future behaviour (such as the weather or planetary positions), which are so sensitive to the initial conditions that even the slightest change in the starting point eventually produces a completely different outcome. (29)

chirality Asymmetry of the type in which the mirror image of an object cannot be superimposed on the object itself, no matter how the image is rotated. It is a property of the human hand, and of some of the molecules that are key to biochemistry. (65)

chondrite A type of **meteorite** made of 'chondrules' (near-spherical globules) and other material, which have not been subject to melting, as has happened, for example, in the interior of a planet. (51)

chromosphere (from Greek: 'colour' and 'sphere') The pink-coloured lower atmosphere of the Sun, visible at a solar eclipse; it extends to a height of about 2,000 km and has a temperature of up to 20,000 °K. (28)

circum-nuclear disc (The outer part of) an **accretion disc** in orbit around the **nucleus** of an **active galaxy**. (57)

CNO cycle The nuclear fusion cycle in massive stars, in which carbon, nitrogen and oxygen nuclei progressively combine with protons and eject alpha particles, thus turning hydrogen in helium. (45)

coma (Latin: 'hair') The 'atmosphere' surrounding the solid nucleus of a **comet**; viewed through a telescope, it makes the comet look fuzzy, as if it has hair. (7, 18)

comet A small body in orbit in the Solar System; a comet resembles an **asteroid** but is made of icy material, the surface of which melts, surrounding the nucleus with a **coma** and producing a **tail**. (7, 18)

continuous creation A (no longer widely accepted) theory that material is continuously produced in space at such a rate as to ensure that the density of matter in the Universe is constant, even though space is expanding. (59)

Copernican theory The **heliocentric theory**, attributed to the astronomer Nicolaus Copernicus. (6)

core The innermost region of a **planet**, usually a dense sphere, surrounded by the **mantle**, or outer solid layers. (17, 21)

Coriolis force The apparent force (named after Gaspard-Gustave de Coriolis) that deflects objects moving on a rotating frame. Gives rise to the Coriolis effect, for example, in the northern hemisphere winds are deflected to the right because the Earth is rotating. (19)

corona (Latin: 'crown') The atmosphere of the Sun above the **chromosphere**, revealed in a solar eclipse. (32)

cosmic fireball The extremely intense radiation that occurred at the **Big Bang**, the relic of which is the **cosmic microwave background**. (60)

cosmic microwave background (CMB; cosmic background) An almost uniform source of infrared and microwave radiation, of cosmic origin (see **cosmic fireball**). (Preface, 60)

cosmic rays High-energy particles from the Sun and the Galaxy that permeate outer space. (47)

cosmological constant A term added by Einstein to his equations describing the gravitational force between galaxies in the Universe to prevent the Universe from collapsing. (59, 63)

crater A hole in the ground caused by an explosion or impact (volcano, meteorite impact, etc.). (8, 15, 21, 22, 24)

cubewano An object in the **Kuiper belt** that orbits undisturbed beyond Neptune, in the same plane as the major planets. (14)

dark ages In the early history of the Universe, the period after the **Big Bang** but before galaxies became visible. (53)

dark energy (vacuum energy) A hypothetical constituent of the Universe, which, released gradually into space, causes the expansion of the Universe to accelerate. (Preface, 63)

dark matter A hypothetical, unseen constituent of the Universe, which produces a gravitational attraction on galaxies and stars. (Preface, 62)

dark nebula A nebula of **interstellar dust** that makes its presence known by absorbing light from the stars that happen to lie behind it. (51)

degenerate matter Matter made of leptons (usually electrons) so dense that it shows quantum mechanical effects of pressure and temperature, as found in a **white dwarf**. (34)

double star A body made up of two stars in orbit around each other; a **binary star** is formally synonymous but in usage refers to a closer pair. (33, 38, 44)

dwarf planet (1) A **planet** in the Solar System – not one of the eight major planets, and not a **satellite** – large enough to have become spherical under its own gravity, like Vesta or Ceres, but not so large that it has cleared its neighbourhood of (other) minor planets (see **asteroid**) and **planetesimals**. (2) An **exoplanet** of the same type. (13)

eccentric orbit An orbit that is appreciably non-circular; hence 'eccentricity', the amount by which the orbit of a body around its host is non-circular. (13, 19)

eclipse A celestial event that occurs either when a body obscures another body behind it (as in a solar eclipse, or an eclipsing **variable star**), or when a body obscures the source of illumination of another body (as in a lunar eclipse). (1, 30, 33)

ecliptic The plane of the orbit of the Earth, and by extension the line that this plane makes when it intersects the **celestial sphere**. (19)

Edgeworth–Kuiper Belt See **Kuiper Belt**.

ellipse A geometrical figure having the shape of an elongated circle; the closed figure described by the orbit of a lone planet around a star. (6, 29)

elliptical galaxy (1) A **galaxy** of stars that has an elliptical shape in the sky. (2) A galaxy that has the form of a three-dimensional triaxial ellipsoid, which looks elliptical when projected onto the plane of the **celestial sphere**. (53)

epicycle In orbital theory, an imaginary circular orbit, which is itself performing a circular orbit. (6)

escape velocity The speed required for an object to escape from the gravitational pull of a planet, star, or other body if it is projected straight upwards. (36)

exoplanet A **planet** in a planetary system outside the Solar System. (44)

extra-galactic nebula A term, no longer in current use, for a galaxy outside our own **Galaxy**. (53)

Fermi paradox The contradiction, formulated by Enrico Fermi, that intelligent life may be ubiquitous in our Galaxy, but yet has not been identified by the inhabitants of the Earth. (65)

fireball A **meteor** that shows brightly in the sky. (20)

Fraunhofer lines Spectroscopic indications at various wavelengths in the spectrum of a star (especially the Sun) of the presence of various elements in the star's atmosphere. (28)

galaxy A system of stars like our own Milky Way **Galaxy**. (3, 43)

Galaxy The system of stars to which the Sun belongs. (38)

gamma rays The most energetic form of radiation. (32, 42)

gamma-ray burst A burst of celestial **gamma rays** from an astronomical object (called a 'gamma-ray burster'). (32, 58)

gas giant planet A planet such as Jupiter or Saturn, made mainly of gas; formed, and (in the Solar System but not in many of the known exoplanetary systems) still orbiting at a great distance from the Sun, they have no solid surface. Compare **terrestrial planet**. (27)

General Relativity Einstein's theory of gravitation. (Preface, 36, 59)

geocentric theory The theory that the Earth is at the centre of the Solar System of planets; see also Ptolemaic theory. (1)

geodesic The natural path taken by a light ray as it traverses a gravitational field. (36)

geothermal spring A source of water heated by energy within the Earth. (25)

giant molecular cloud (GMC) A large cloud of gas and dust, so dense that molecules survive within it; from these molecules, stars and planetary systems can form. (39)

glacial moraine Earth and rocks carried by a glacier and deposited off its end; the material accumulated by this method that is left behind if the glacier retreats. (25)

globular cluster A cluster of stars, in the form of a sphere or globule, reckoned to be old. (40)

gradualism The theory that geological formations are created incrementally. Contrast **catastrophism**; see also **uniformitarianism**. (20)

gravitational lens(ing) The effect of **gravity** on light rays, which produces distorted images. (30, 62)

gravitational radiation (gravitational waves) The disturbances in space produced by changes in the gravitational field of an object, such as a **binary** star. (Preface, 64)

gravitational redshift The loss of energy of radiation as it climbs out of a gravitational field; as visible light loses energy it becomes redder. (34)

gravity (gravitation) The attractive force produced by all matter, each particle attracting every other across space and tending to change its motion; the force giving 'weight' to 'mass'; the dominant long-range force and so the force which is most important in astronomy. (29)

greenhouse effect The warming effect produced by an atmosphere that is transparent to incoming light and opaque to outgoing **infrared radiation**, which thus becomes trapped. (Preface, 19, 23)

Hadley cell, wave A large zone or wave in the atmosphere, of global proportions. (19)

heliocentric theory The theory that the Sun is at the centre of the Solar System of planets; see also **Copernican theory**. (6)

helioseismology The study of the oscillations of the Sun. (40, 47)

Hubble Deep Fields Observations made by the Hubble Space Telescope of typical regions of cosmic space. (53)

hydrogen The simplest element and primary constituent of the Universe. (52)

infrared radiation A form of radiation, slightly less energetic than light, which is emitted by warm objects. (Preface, 23, 39)

interferometry A technique in which two separated detectors, each picking up radiation (radio waves, light, gravitational waves etc.), are used to simulate the capacity of a telescope equal in size to the distance between them. (Preface, 59, 64)

intergalactic cloud A cloud of gas that lies between galaxies. (62)

interstellar dust Dust particles that lie between the stars. (51)

interstellar hydrogen Hydrogen gas that lies between the stars. (52)

interstellar space Space that separates the stars. (51)

inverse square law The law that intensity or force (as in radiation or gravitational

pull) diminishes according to the square of the distance. (29, 61)

ionosphere The spherical layer (or layers) of ionized air lying at an altitude of approximately 100 km from the Earth. (31, 32)

irregular galaxy A **galaxy** that has no particular shape. (43, 53)

isothermal Of a body, having a homogeneous temperature. (60)

isotope (Greek: 'same' 'place' – in the Periodic Table of elements) A nucleus with a certain number of protons and a further number of neutrons. The number of protons in the nucleus of an atom defines the chemical element that it is (and therefore its place in the Periodic Table (148)); within a collection of atoms of a given element, the number of neutrons in the nucleus of each atom varies from atom to atom. Nuclear reactions may alter isotopes from one kind to another. (46)

isotropic Of an object, looking the same from all directions. (60)

jet A flow of material outwards in a straight line. (56)

jupiter A **planet** in another planetary system that resembles Jupiter in the Solar System; a **gas giant planet**. (44)

Kelvin degree A degree on the centigrade (Celsius) temperature scale, counted from a temperature of absolute zero. (22, 23)

KreeP A group of elements (potassium (K), rare earth elements (ree) and phosphorus (P) found together in some rocks in high abundance. (21)

Kuiper Belt The belt of **asteroids** and/or **comets**, or other **Trans-Neptunian Objects** which lie beyond Neptune; hence, 'Kuiper Belt Object', one of the bodies in the Kuiper Belt. (14)

Late Heavy Bombardment (lunar cataclysm) In the early history of the Solar System, an event in which a rain of meteorites crashed down on the surface of the planets. (22)

lava flow A flow of liquid magma from a volcano or other fissure in the surface of a planet. (26)

lenticular galaxy A **galaxy** that is lens-shaped. (53)

light year The distance that light travels in a year; hence also 'light minute', 'light second', etc. (Preface, 37, 47)

log N-log S In astronomy, a mathematical expression that counts and interprets the numbers of astronomical sources of a given brightness. (59)

lunar cataclysm See **late heavy bombardment**. (22)

lunar eclipse The event that occurs when the Moon passes into the shadow cast by the Earth in the light from the Sun. (1, 4)

lunar seas The magma plains that show as flat, dark areas on the Moon, once thought to be areas of liquid. (21)

magnetosphere The region on and around a **planet** influenced by its magnetic field. (17)

magnitude In astronomy, the brightness of a star. (33)

mantle In relation to the Earth or a similar

planet, the outer solid layers wrapped around the **core**. (21)

Medicean stars The name given by Galileo to the four satellites of Jupiter. (8)

meteor The phenomenon of a streak of light, radar echoes, etc. caused by a solid body (**meteoroid**) dropping from space into the Earth's or another planet's atmosphere. (Preface, 15, 20, 22)

meteor shower The appearance of many meteors over a period of hours or days, all coming from the same part of the sky. (16, 20)

meteorite The surviving part of the solid body, or **meteoroid**, that causes a **meteor**. (Preface, 15, 20, 22)

meteoroid A solid body orbiting in space, which, when it plunges into the Earth's or another planet's atmosphere, becomes a **meteor**. (15)

microwave radiation Radiation with a wavelength intermediate between **infrared radiation** and **radio waves**. (Preface, 51, 53, 60)

Milanković (Milankovitch) cycles Oscillations in the Earth's temperature caused by variations in the planet's orbital parameters (its eccentric orbit, tilt, etc). (19)

millimetre waves Radiation with a wavelength in the millimetre region: short-wavelength **microwave radiation**. (Preface, 51, 53)

minor planet Obsolescent term for **asteroid**. (12)

moon A **satellite** in orbit around a **planet** or **asteroid**. (8, 12, 21)

nebula (Latin: 'cloud') A body of gaseous material and/or dust grains in space, which emits or reflects light and other energy picked up and re-radiated from stars nearby. (38, 39)

Nebular Hypothesis The central tenet of a number of theories of the origin of the Solar System, according to which the planets formed from a nebula in orbit around the newly born Sun. (49, 50)

neutrino A particle with no electric charge, very little mass (formerly thought to have no mass at all), and spin ½, which is produced as a by-product of numerous nuclear reactions. There are three types or 'flavours' of neutrino. (Preface, 40, 42, 45, 47)

neutron star A star so small that its constituent material is made primarily of neutrons, as opposed to electrons and protons or other nuclei; see **pulsating radio star**. (35, 40)

nova (Latin: 'new [star]') A star that flares up because of an explosion on its surface and becomes temporarily bright where no star was noticeable before. See **supernova**. (31, 35, 41)

nuclear energy Energy released by nuclear processes; the source of star- and sunlight. (45)

nuclear fusion A nuclear process in which nuclei are fused together to form a heavier nucleus; the process that fuels the stars. (45)

oblate The term used to describe an ellipsoid that is flattened at the poles: the three-dimensional shape produced by rotating an ellipse around its minor axis. Compare **prolate**. (11)

obliquity of the ecliptic The angle (currently about 23.5°) between the Earth's equator and the plane of the Earth's orbit; thus, the tilt of the Earth's polar axis relative to its orbit. (19)

occultation An astronomical event in which a nearby body passes in front of and hides (occults) a more distant one, particularly but not exclusively when the Moon passes in front of a planet or a star. A total solar eclipse is an occultation of the Sun by the Moon. (9)

Olbers's Paradox The contradiction, popularized by Wilhelm Olbers, between the assumption that the Universe is infinite and eternal and the fact that it is dark at night. (61)

Oort Cloud The hypothetical region on the periphery of the Solar System from which long-period and sporadic, comets come (see **sporadic comet**). (18)

orbital plane The plane containing the orbit of a body that moves round another. (19)

parabola An open geometric figure described by the orbit of a body falling into the gravitational field of another, as in the case of a sporadic comet falling in towards the Sun from a very large distance (effectively 'from infinity'). Akin to an ellipse, which describes the orbit of a comet that is always contained within the Solar System. (18)

parallax The apparent shift in position of a distant object, caused by a change in the position from which it is observed. In astronomy, a shift of 1 arc second in the angular position of a star relative to its average position as the Earth moves around the Sun in one year is described as a parallax of 1 arc second, and means that the star is 1 parsec (3.26 light years) away. (37)

perihelion The point in a **planet**'s orbit at which it is closest to the Sun. (30)

permafrost Ground that is perpetually (or at least for an entire season) at a temperature below the freezing point of water. (25)

phase The appearance of a planetary body, such as the Moon or Venus, due to its illumination by the Sun and the angle at which it is viewed; in different phases such a body will appear as a crescent, a gibbous shape, or a complete circle. (9)

planet As defined by the International Astronomical Union, a celestial body orbiting its parent star, the body being of such a mass and size that it has become rounded and has also cleared the nearby orbital region of all the **planetesimals** among and from which it formed. Compare **dwarf planet**. (passim)

planetary nebula A **nebula** that resembles a **planet** in appearance; such nebulae were formed from material ejected by a star during its lifetime, and now illuminated by the star's core as it evolves to its end as a **white dwarf**. (34, 49)

planetary rings A system of **meteoroids** in orbit in a thin disc around a planet, as, for example, Saturn's rings. (27)

planetary system A system of planets in orbit around a star, as, for example, the Solar System. (44)

planetesimal In a newly formed planetary system, a small **planet**, formed from accreted dust and probably about to

merge with others to form a large planet. (12, 18, 50, 65)

plenum A material, hypothesized by René Descartes, that fills space and transmits force (**gravity**) from one body to another. (29)

plutino A **Trans-Neptunian Object**, whose orbit, like Pluto's, resonates with that of Neptune, and which makes two orbits to Neptune's three. (14)

pre-biotic chemistry In organic chemistry, a process that involves complex chemicals, which, though similar to those involved in biochemistry, are not necessarily or not perhaps as complex as biological ones. (65)

precession The gyrating motion of a planet (notably the Earth) as its rotational axis describes a cone in space; it resembles the motion of a spinning top. (5)

principle of equivalence The principle that the mass of a body participating in gravitational attraction and the mass that resists acceleration are identical. (30)

prolate The term used to describe an ellipsoid that is pointy at the poles: the shape produced by rotating an ellipse around its major axis. Compare **oblate**. (11)

proper motion The motion of a star through space across the line of sight; it is 'proper' in the sense of 'belonging to the star itself', as opposed to resulting from the way that it is viewed. (11)

proplyd A short form for **protoplanetary disc**. (50)

proton A fundamental particle, the nucleus of a typical hydrogen atom, of considerable mass, with positive charge and spin ½. (42)

protoplanet A newly forming **planet**. (44, 50)

protoplanetary disc (proplyd) A disc of dust and protoplanets in orbit around a newborn star. (44)

Ptolemaic theory The **geocentric theory**, attributed to Ptolemy. (1)

pulsar A short form for **pulsating radio star**. (31, 35, 44)

pulsating radio star (pulsar) A rotating **neutron star**, showing regular and rapid pulses of **radio waves**. (31, 36, 44)

quasar A short form for **Quasi-Stellar Radio Source**. (31, 35, 55)

Quasi-Stellar Radio Source (quasar; Quasi-Stellar Object, QSO) A point-like radio source that coincides with a star-like optical source in the sky, indicating the presence of a very distant **active galaxy**. (55)

radial velocity The motion of a star in the line of sight away from or towards the observer. (37)

radiant In a **meteor shower**, the vanishing point of the tracks of the individual meteors in their orbit around the Sun, as seen in perspective from the Earth. (16)

radiation pressure The pressure generated by light or other radiation on a solid surface. (18)

radio galaxy A **galaxy** that emits **radio waves**. (31)

radio waves Radiation with a wavelength upwards of about 1 m. (Preface, 31)

red giant A large star, red because it has a

low surface temperature; an evolved star. (34, 40, 46, 49)

reflection nebula A **nebula** of **interstellar dust**, which reflects the light of a very nearby or embedded star; essentially the same as a **dark nebula** except that there is a star to illuminate it. (51)

Roche limit The distance from a planet that constitutes the limit outside which a satellite will maintain its integrity, but within which it will break into pieces and form a **planetary ring**. (27)

satellite A body that is in orbit around, and subordinate to another; particularly a smaller body or **moon** in orbit around a larger **planet** or **asteroid** (8, 12, 21), but also a smaller galaxy in orbit around a larger (as with the Magellanic Clouds (54)).

scintillation Twinkling, as of a star, due to the atmosphere, or a radio-emitting **quasar**, due to interplanetary plasma. (35)

Seyfert galaxy An **active galaxy**, showing spectroscopic evidence of gas in an **accretion disc** and a **circum-nuclear disc**. (32, 56)

solar eclipse An **occultation** in which the Moon passes in front of the Sun, and may obscure it partially or totally. (28, 30)

solar flare An explosive event on the Sun's surface. (17)

solar nebula The **nebula** that surrounded the Sun at its formation. (50)

solar neutrino problem The question why there is a scarcity, below original expectations, of solar **neutrinos**, which is answered by the theory of neutrino oscillations. (47)

solar wind The outward flow of material from the solar surface into the rest of the Solar System. (17, 18, 21, 24)

space weather The influence of the **solar wind** on the terrestrial environment. (17)

Special Relativity Einstein's theory of phenomena that are particularly noticeable when something is moving near the speed of light. (Preface, 30)

spectroscopy The analysis of spectra (see **spectrum**); hence 'spectroscope' and 'spectrograph', instruments that perform such analyses. (28, 34, 46, 49)

spectrum The range of wavelengths across which an object emits radiation. (23, 28, 31, 49)

spiral arm An outward-curving component of a **spiral galaxy**. (40, 43)

spiral galaxy (spiral nebula) A **galaxy** of stars with a pronounced spiral pattern. (52, 62)

splosh crater A **crater** on Mars that is surrounded by a lobed pattern, apparently produced by the impact of a meteorite falling on an area of fluidized mud. (25)

sporadic meteor A meteor that has no correlation with others seen at the same time; compare **meteor shower**. (16)

standard candle A star or other luminous phenomenon that has the same brightness in all circumstances and everywhere; used in astronomy to determine the distance to distant star clusters or galaxies. (43)

star cluster A cluster of stars, formed together at the same time and continuing to exist together. (2, 40)

static (radio noise) An emission of radio energy across a broad band of frequencies. (31)

Steady-State theory The theory that the Universe has been and always will be the same; see also **continuous creation**. (59)

sungrazer A **comet** that approaches very close to the Sun – in some cases so close that it entirely melts. (7)

supermassive black hole A **black hole** formed from the amalgamation of many stars. (56)

supernova A major stellar explosion that causes a particularly large burst of light and a much brighter than usual **nova**. (Preface, 31, 41)

supernova of Type I A **supernova** that shows no sign of hydrogen; in particular, a supernova of Type Ia, which is the explosion of a **white dwarf**. (41, 43)

supernova of Type II A **supernova** involving hydrogen: the explosion of a massive star. (41)

tail In astronomy, the material ejected by a **comet** and pushed back behind the comet's nucleus by **radiation pressure** and other forces emanating from the Sun. (7, 18)

terrestrial planet A **planet**, such as the Earth, made of solids and having a solid surface; compare **gas giant planet**. (44, 65)

thermokarst A geological landform produced by the melting of a region of **permafrost**. (26)

time dilation An effect of relativity (see **Special Relativity**) that causes time to run slowly. (25)

Titius–Bode Law See Bode's Law. (12)

torus A ring shape, like that of a doughnut with a hole in the middle. (56)

Trans-Neptunian Object (TNO) An **asteroid** or **comet** that orbits beyond Neptune. (14)

Tychonic theory The theory, articulated by Tycho Brahe, that the Sun and the Moon orbit the Earth, which is stationary at the centre of the Solar `System, and the other planets orbit the Sun. (9)

unification model A theory that various sorts of active galactic nuclei (see **active galaxy**) are all fundamentally the same structure seen from different angles. (55)

uniformitarianism In geology, the theory that geological formations are created slowly and incrementally; the geological equivalent of **gradualism**, as opposed to **catastrophism**. (20)

Universe The entirety of creation; the largest object of scientific investigation, studied through the methods of cosmology. (Preface, 59, 61 and passim)

vacuum energy See **dark energy**.

Van Allen belts Zones of particle radiation (**cosmic rays**) that encircle the Earth, located in its **magnetosphere**. (17)

variable star One that changes brightness, such as a **Cepheid** (43), a **nova** (41) or a **binary star** that eclipses (33).

wavecut platform A shelf cut in a cliff wall by the action of waves. (25)

weakly interacting massive particle (WIMP) A hypothetical particle, not very interactive with others, with mass that could contribute to the amount of **dark matter** in the Universe. (62)

white dwarf A small hot star that supports itself by the pressure of **degenerate matter** in its core. (Preface, 41, 49, 64)

X-ray binary star A **double star**, at least one member of which is an **X-ray star** (36)

X-ray star A star that emits **X-rays**. (Preface, 32, 36)

X-rays Energetic radiation, a little less energetic than **gamma rays**. (Preface, 32)

zodiac The constellations that lie along the **ecliptic**. (2)

Further Reading

The following publications have been selected because they provide historical background for the discoveries recounted in this book; amplify and explain the science behind them; or are primary works mentioned in the text that are accessible to the general reader.

Books

HISTORY OF ASTRONOMY

Aratus of Soli, *Phaenomena*. The astronomical part of the book describes the classical (northern) constellations, many of which are still used today. It is available for free download at http://www.geocities.com/astrologysources/classicalgreece/phaenomena/index.htm

Galileo Galilei, *Sidereus nuncius*. Galileo describes the discoveries that he made with his first telescope. English translation (*Starry Messenger*) available for download at http://www.bard.edu/admission/forms/pdfs/galileo.pdf

Gleick, James, *Chaos: Making a New Science* (1988). A popular account of the history and science behind chaos theory.

Hockney, Thomas (ed.), *Biographical Encyclopaedia of Astronomers* (2007).

Hoskin, Michael (ed.) *The Cambridge Illustrated History of Astronomy* (1996).

Hoskin, Michael, *The Herschel Partnership* (2003). William Herschel, viewed by his sister Caroline.

Isaacson, Walter, *Einstein* (2007). Extensively researched biography detailing Einstein's life and work.

Jaki, Stephen, *The Milky Way: An Elusive Road for Science* (1973). A history of the theories explaining the Milky Way.

Jungk, Robert, *Brighter Than A Thousand Suns* (1970). A history of the atomic scientists, including the discoverers of nuclear fusion in stars.

King, Henry, *History of the Telescope* (1955).

Lubbock, Constance, *The Herschel Chronicle* (1933, repr. 2009). A history of the family by William Herschel's granddaughter.

Mitton, Simon, *Conflict in the Cosmos: Fred Hoyle's Life in Science* (2005). An illuminating scientific biography of the controversial 20th-century scientist.

Moore, Patrick, *The Planet Neptune: A Historical Survey before Voyager* (1966). Popular stories and popular science before the discoveries of the space age.

Needham, Joseph, *Science and Civilisation in China*, vol. 3, section 20: 'The Sciences of the Heavens' (1959). The first and most comprehensive history of astronomy in China, now showing its age, but still an essential first step for anyone interested in the topic.

Newton, Isaac, *Principia*. Not an easy read, but the preface, which describes Newton's outlook on science, is fundamental to the history of modern astronomy. Motte's English translation of 1846 is freely available from http://rack1.ul.cs.cmu.edu/is/newton/

Ronan, Colin, *Galileo* (1974). The story of the man and the astronomy behind his discoveries.

Struve, Otto and Velta Zebergs, *Astronomy of the Twentieth Century* (1962). A participant in some of the major discoveries gives a firsthand account of the half-century in which the astrophysics of the stars and galaxies emerged.

Toulmin, Stephen and June Goodfield, *The Fabric of the Heavens* (1961). Cosmology from Babylonian and Greek classical times to the age of Newton.

Westfall, Richard S., *Never at Rest: A Biography of Isaac Newton* (1980). The definitive biography of Newton; a big read.

GENERAL ASTRONOMY

Fischer, Daniel and Hilmar Duerbeck, *Hubble Revisited* (1998). Pictures from the Hubble Space Telescope, and their significance for science.

Maran, Stephen P., *Astronomy for Dummies* (1999). A clear and comprehensive basic guide to astronomy.

Mitton, Jacqueline, *Cambridge Illustrated Dictionary of Astronomy* (2007). Accessible, complete and comprehensive. Terms not defined in the glossary of the present book it can be found in this one.

Murdin, Paul, and Margaret Penston (eds.) *The Firefly (Canopus) Encyclopaedia of Astronomy Firefly* (2003). A condensed version of *The Encyclopaedia of Astronomy and Astrophysics*, suitable for the more general reader.

Murdin, Paul (ed.) *The Encyclopaedia of Astronomy and Astrophysics* (2001). Comprehensive encyclopaedia aimed at a professional readership, but each entry begins with an accessible instroduction and includes definitions, short biographies and historical articles.

COSMOLOGY

Coles, Peter, *Cosmology* (2001). A very short introduction to the subject, by a well-known teacher and cosmologist.

Rees, Martin, *Before the Beginning* (1988). Why the Universe is as it is, according to one of the world's leading astronomers.

Rees, Martin, *Just Six Numbers: The Deep Forces that Shape The Universe* (2000). Other reasons why the Universe is as it is.

Silk, Joseph, *A Short History of the Universe* (1997). Introduction for the general reader by a renowned cosmologist.

Singh, Simon, *The Big Bang* (2004). Accessible history and science of cosmology by a mathematician known for his popular-science writing.

Smoot, George, and Keay Davidson, *Wrinkles in Time* (1993). First-person account of the discovery of the fluctuations in the cosmic microwave background.

STARS, NEBULAE AND GALAXIES

Kaler, James, *Cambridge Encyclopedia of the Stars* (2006).

Kirshner, Robert P., *The Extravagant Universe: Exploding Stars, Dark Energy, and the Accelerating Cosmos* (2004). Supernovae and cosmology explained by someone intimately involved in the science.

Kwok, Sun, *Cosmic Butterflies* (2001). The colourful mysteries of planetary nebulae, described by an astrophysicist.

May, Brian; Patrick Moore; and Chris Lintott, *Bang!* (2006).

Modern astronomy explained by two astronomers and an astronomer-turned-rock-legend-turned-astronomer.

Melia, Fulvio, *The Black Hole at the Center of Our Galaxy* (2003). Our not-so-supermassive black hole.

Murdin, Paul, and Lesley Murdin, *Supernovae* (1985). The science and history of supernovae, and their place in literature.

Murdin, Paul, *End in Fire* (1990). An account of Supernova 1987A.

THE SUN AND THE TERRESTRIAL MAGNETOSPHERE

Eather, Robert H., *Majestic Lights* (1979). The aurora in science, history and the arts.

Lang, Kenneth (ed.), *The Cambridge Encyclopedia of the Sun* (2001). Comprehensive and well-illustrated account of the science surrounding the Sun.

Littmann, Mark; Fred Espenak; and Ken Willcox, *Totality*, 3rd ed. (2008). Everything about eclipses of the Sun.

PLANETS

Harland, David M., *Water and the Search for Life on Mars* (2004).

Harland, David M., *Exploring the Moon: The Apollo Expeditions* (1999).

Lang, Kenneth, *The Cambridge Guide to the Solar System* (2003). The Solar System in the space era.

Leutwyler, Kristin, *Moons of Jupiter* (2003). With stunning pictures and informative text.

Mackenzie, Dana, *The Big Splat* (2003). How our moon came to be.

Murdin, Paul, *Full Meridian of Glory* (2009). On the size and shape of the Earth.

COMETS AND METEORS

Mark, Kathleen, *Meteorite Craters* (1987). A comprehensive history of terrestrial craters.

Norton, O. Richard, *The Cambridge Encyclopedia of Meteorites* (2002). A comprehensive and well-illustrated account of the science surrounding meteorites.

Olsen, Roberta, *Fire and Ice* (1985). A history of comets in art.

Southgate, Nancy, *A Grand Obsession: Daniel Moreau Barringer and His Crater* (2002).

Yeomans, Donald, *Comets* (1991). A chronological history of comets: their observation, science, and significance in myth and folklore.

RADIO- AND X-RAY ASTRONOMY, PULSARS AND BLACK HOLES

Clark, David H., *The Quest for SS433* (1995). Firsthand account of a scientific discovery in optical, radio- and X-ray astronomy.

Ferguson, Kitty, *Prisons of Light* (1996). Black holes in stars and galaxies.

McNamara, Geoff, *Clocks in the Sky* (2008). The story of pulsars.

Sullivan, W. T., III, *The Early Years of Radio Astronomy* (1984). Reflections by the founders of radio astronomy, fifty years after Jansky's discovery.

Tucker, W. and Riccardo Giacconi, *The X-ray Universe* (1983). Inside story of the start of X-ray astronomy.

Verschuur, Gerrit L., *The Invisible Universe Revealed* (1987). The story of radio astronomy, by a radio astronomer.

EXOPLANETS AND EXTRATERRESTRIAL LIFE

Jones, Barrie, *Life in the Solar System and Beyond* (2004).

Mayor, Michel and Pierre-Yves Frei, *New Worlds in the Cosmos: The Discovery of Exoplanets* (2003). Direct account of the discovery of the first exoplanets, in a rather poor English translation.

Ward, Peter D. and Donald Brownlee, *Rare Earth* (2000). Why complex life is uncommon in the Universe.

Journals

A number of journals provide material of the same sort as the books that I have selected above, and their back issues and websites are worth browsing, even if only as a lucky dip.

Astronomy and Geophysics, the news and reviews journal of the Royal Astronomical Society (bimonthly). http://www.ras.org.uk (click on Publications)

Astronomy, a popular US-based astronomy journal (monthly). http://www.astronomy.com/asy/default.aspx

Astronomy Now, a popular UK-based astronomy journal (monthly). http://www.astronomynow.com/magazine.shtml

Journal for Astronomical History and Heritage, authoritative Australian-based journal (3 issues per year). http://www.jcu.edu.au/school/mathphys/astronomy/jah2/about.shtml

Journal for the History of Astronomy, authoritative UK-based journal (4 issues per year). http://www.shpltd.co.uk/jha.html

Mercury, the magazine of the Astronomical Society of the Pacific (4 issues per year, digital only). http://www.astrosociety.org/pubs/mercury/mercury.html

New Scientist, a magazine covering the whole of science at an accessible level (weekly).http://www.newscientist.com/

Sky and Telescope, US-based magazine for astronomy enthusiasts, covering news, sky-watching, astronomical science and history, with a high editorial standard (monthly). http://www.skyandtelescope.com/

Scientific American, US-based general science magazine in which professional scientists write about front-line research at an accessible level (monthly). http://www.sciam.com/

Websites

The SAO/NASA Astrophysics Data System (ADS)
http://adswww.harvard.edu/
One of the most valuable sources of astronomical literature, very little known outside the professional astronomy community, although freely available. It contains bibliographic entries covering virtually the entirety of astronomy, some with links to the full original articles (usually those more than a few years old). There are also extensive volumes of older material, and everything (even the older material) is text-searchable. Most of the entries are for professional research articles but there are also obituaries, review articles, and other more accessible material.

Nobel Prize
http://nobelprize.org/
The site contains extensive material including citations and autobiographies of Nobel Prize-winners.

Kavli Prize
http://www.kavliprize.no/
The Kavli Prize has been established to cover subjects outside the fields of the Nobel Prize, including astrophysics. The website contains citations and biographies, although the prize is newly established and the number of recipients is relatively small.

Astronomiae Historiae
http://www.astro.uni-bonn.de/~pbrosche/astoria.html
(German version at: http://www.astro.uni-bonn.de/~pbrosche/astoria-d.html).
Website for the history of astronomy, showing its age, but still useful. The site includes a comprehensive list of journals that publish on the history of astronomy:
http://www.astro.uni-bonn.de/~pbrosche/hist_astr/ha_pub_jour.html

Hubble Space Telescope
Highlights from the work of the Hubble Space Telescope are available through: http://hubblesite.org/.
There is an archive of press releases, and a gallery of pictures, as well as sections on astronomy and the telescope itself. The spectacular Hubble Heritage collection of pictures is available at http://heritage.stsci.edu/

Spitzer Telescope
Highlights from the Spitzer Telescope are available at: http://gallery.spitzer.caltech.edu/Imagegallery/chron.php?cat=Astronomical_Images

Chandra Telescope
Highlights from the Chandra Telescope are available at: http://chandra.harvard.edu/photo/index.html

European Southern Observatory
Information on scientific discoveries made with the ESO's telescopes are available through:
http://www.eso.org/public/outreach/pressmedia/index.html

National Radio Astronomy Observatory
http://www.nrao.edu/index.php/learn
Website covering the world of radio astronomy.

SOHO
http://sohowww.nascom.nasa.gov/
The Sun and space weather, in real time.

Wikipedia
Astronomy articles can be accessed through
http://en.wikipedia.org/wiki/Category:Astronomy
Astronomy is one of the higher-quality sections of Wikipedia, the amazing resource that can be freely edited by anyone, kept up-to-date by astronomy enthusiasts.

Astronomy Picture of the Day
The site has a searchable archive:
http://antwrp.gsfc.nasa.gov/apod/astropix.html
APOD is virtually an astronomy course seen through spectacular pictures, presented one a day since 1995.

Author's Acknowledgments

I would like to thank Robin Rees for his involvement with our recent book projects, including an extensive dialogue and picture research for this one; it is fun to work with him, as I have done for half a lifetime. Robin and I are especially grateful to Peter Hingley for his advice and assistance in researching images for the book. I would also like to thank the team at Thames and Hudson, a demanding, talented and hardworking group of people, including Ian Jacobs, Flora Spiegel, Gareth Walker, Avni Patel, Jo Walton and Philip Collyer; it has been a rewarding education to work with them, and the book is much better because of their participation. Finally I would like to thank the Institute of Astronomy, Cambridge University Library and Wolfson College at the University of Cambridge, and the Royal Astronomical Society for their support, especially from the libraries, while I worked on this book and other projects for the last several years.

Picture Credits

18 (1) Science Photo Library; 19 (2) Royal Belgian Institute of Natural Sciences; 19 (3) Pete Lawrence; 20 (4) British Museum, London; 20 (5) Dave Tyler; 22 (1) Royal Astronomical Society Library, London; 23 (2) British Library, London; 23 (3) Gabriel Seah; 24 (4), 25 (5) Royal Astronomical Society Library, London; 26 (1) James Symonds; 27 (2) National Gallery, London; 26–27 (3) Lund Observatory, Sweden; 28–29 (4) Alte Pinakothek, Munich; 29 (5) Anglo-Australian Observatory/David Malin Images; 30 (2) NASA; 31 (4) U.S. Navy photos by Mass Communication Specialist Seaman Joshua Valcarcel; 32 (5) James Symonds; 32 (6) NCAR; 33 (7) GFZ Potsdam PR; 32–33 (8) British Library, London; 34 (9) NASA/JPL-Caltech; 35 (10), 36 (1) James Symonds; 37 (2) Jef Maion; 37 (3) Josch Hambsch, www.astronomie.be; 38 (5) British Library, London; 39 (7) Observatoire de Paris; 40 (9) ESO; 42 (1), 42 (2) Royal Astronomical Society Library, London; 43 (3) James Symonds; 44 (4), 45 (5) Royal Astronomical Society Library, London; 46 (6) Science Photo Library; 46 (7) Courtesy of JAXA, NAOJ, PPARC and NASA; 47 (8) Manchester Art Gallery/Bridgeman Art Library; 47 (9) British Library, London; 50 (1) NASA/JPL-Caltech; 51 (2) Rice University, Houston; 51 (3) SOHO (ESA and NASA); 52 (4) Robin Rees; 52 (5), 53 (6), 54 (1), 54 (2), 55 (1) Royal Astronomical Society Library, London; 55 (5) Istituto e Museo di Storia della Scienza, Florence; 56 (6), 56 (7), 57 (8), 57 (9) NASA-JPL; 58 (1) Royal Astronomical Society Library, London; 59 (2) James Symonds; 59 (3) L. Esposito (University of Colorado, Boulder), and NASA; 60 Biblioteca Nazionale Marciana, Venice; 61 (4) Royal Astronomical Society Library, London; 61 (5) Cristiano Banti, *Galileo before the Inquisition*, 1857; 62 (1), 62 (2) Voyager 2 – NASA/JPL-Caltech; 63 (3) National Maritime Museum, London; 63 (4) Royal Astronomical Society Library, London; 64 (5) Erich Karkoschka (University of Arizona) and NASA; 65 (6) Royal Astronomical Society Library, London; 66 (1) James Symonds; 67 (2), 67 (3) Royal Astronomical Society Library, London; 67 (4), 68 (5) NASA-JPL; 70 (1) NASA, ESA, and Y. Momany (University of Padua) 70 (1b) NASA-JPL; 71 (2), 71 (3) INAF - Osservatorio di Palermjo Giuseppe S. Vaiana; 72 (4) Royal Astronomical Society Library, London; 72 (5) NASA/Johns Hopkins University Applied Physics Laboratory; 73 (6), 74 (2) NASA-JPL; 75 (3) Lowell Observatory; 76 (5) Jacqueline Mitton; 76 (6) Lowell Observatory; 76 (7) US Naval Observatory; 77 (8) NASA, ESA, H. Weaver (JHU/APL), A. Stern (SwRI), and the HST Pluto Companion Search Team; 77 (9) NASA, ESA and G. Bacon (STScI); 78 (1) NASA and G. Bacon (STScI); 79 (2) AIP Emilio Segrè Visual Archives, Physics Today Collection; 79 (3) NASA-JPL; 80 (4) NASA, ESA, and A. Schaller (for STScI); 81 (5) NASA, G. Bernstein and D. Trilling (University of Pennsylvania); 81 (6) Jing Li; 82 (1) James Symonds; 83 (2) Jo-H, http://www.flickr.com/photos/jo-h/; 84 (4) NASA; 84 (5) NASA-JPL; 85 (6) NASA; 85 (7) Courtesy Muhammad Mahdi Karim; 86 (1) James Symonds; 86 (2) Pete Lawrence; 88 (5) Katsuhiro Mauri and Shuji Kobayashi (Nagoya City Science Museum and Planetarium); 88 (6) Joseph Brimacombe; 90 (1) James Symonds; 91 (2) Pete Lawrence; 91 (3) AIP Emilio Segrè Visual Archives; 92 (4) SOHO (ESA and NASA); 92 (5) NASA; 93 (6) NASA/Don Pettit, ISS Expedition 6; 94 (7) NASA; 95 (8) SOHO (ESA and NASA); 95 (9) NASA-JSC; 96 (1) Sebastian Deiries/ESO; 97 (2) Royal Astronomical Society Library, London; 98 (3) ESA. Courtesy of MPAE, Lindau; 98 (4) James Symonds; 98 (5) NASA-JPL; 99 (6) R. Evans, J. Trauger, H. Hammel, the HST Comet Science Team and NASA; 100 (1) Courtesy the SeaWiFS Project, NASA/Goddard Space Flight Center, and ORBIMAGE; 101 (2), 101 (3) Royal Astronomical Society Library, London; 101 (4) James Symonds; 102 (2) National Ice Core Labs, USGS; 102 (5) Royal Astronomical Society Library, London; 103 (7) University Library, Belgrade University; 103 (8) NASA/Landsat; 104–5 (1) Jonathan S. Blair/National Geographic/Getty Images; 105 (2) Courtesy Sir Patrick Moore; 106 (3) ML Design after Lunar and Planetary Institute, University of Arizona; 106 (4) Courtesy Sir Patrick Moore; 107 (5) Dr David Kring/Science Photo Library; 107 (6) NASA-JPL; 108 (7) Bernhard Hampp; 109 (8) Caricature by John Kay, 1787; 110 (1) NASA-JSC; 110 (2) Lick Observatory; 111 (3) NASA-JSC; 112 (4), 112 (5) NASA Headquarters - Greatest Images of NASA; 112 (6) U.S.I.S.; 113 (7), 113 (8), 114 (9) NASA-JSC; 114 (10)NASA-LaRC; 114 (11) NASA-JSC; 115 (12), 116 (13) James Symonds; 117 (14) NASA-JSC; 118 (1) NASA-JPL; 118 (2) NASA; 119 (3) Goldstone/VLA/NRAO; 119 (4), 120 (5), 120 (6), 121 (7), 121 (8) NASA-JPL; 122 (1) NASA-GRC; 122 (2) NASA; 123 (3) Courtesy Sir Patrick Moore; 124 (4) James Symonds; 124 (5) NASA; 125 (6), 125 (7) NASA-JPL; 126 (1) NASA; 126 (2) Observatorio Astronomico Di Brera, Milan; 127 (3), 127 (4) Royal Astronomical Society Library, London; 128 (5) NASA/JPL/Texas A&M University/Cornell University; 128 (6) ESA/DLR/FU Berlin, G. Neukum et al.; 128 (7) NASA-JPL; 129 (8) NASA and The Hubble Heritage Team (STScI/AURA); 131 (1) ESA/DLR/FU Berlin, G. Neukum et al.; 131 (2) NASA-JPL; 132 (3) NASA/JPL/University of Arizona; 132 (4) NASA-JPL; 133 (5), 133 (6) NASA/JPL-Caltech/University of Arizona/Texas A&M University; 134 (7), 135 (8) NASA/JPL/Malin Space Science Systems; 135 (9), 135 (10), 135 (11), 136 (1), 137 (2), 137 (3) NASA-JPL; 137 (4) Courtesy Linda Morabito Kelly/NASA/JPL; 138 (5) NASA-JPL; 138 (6) NASA/ESA, John Clarke (University of Michigan); 139 (7) James Symonds; 139 (8) NASA-JPL; 140 (1) NASA-JPL; 141 (2), 141 (3), 141 (4) Royal Astronomical Society Library, London; 142 (5) NASA/JPL-Caltech; 142 (6) NASA-GSFC; 142 (7) Cassini Imaging Team, SSI, JPL, ESA, NASA; 143 (8) Dave Tyler; 146 (1) Canopus Publishing Limited; 147 (2) Courtesy the Oesper Collections in the History of Chemistry, University of Cincinnati; 148 (4) Robin Rees; 148 (5), 148 (6) Royal Astronomical Society Library, London; 148 (7) AIP Emilio Segrè Visual Archives, Brittle Books Collection; 148 (8) Robin Rees; 150 (1) James Symonds; 151 (2) Hans von Aachen, *c.* 1612; 151 (3) Royal Astronomical Society Library, London; 151 (4) Courtesy Sir Patrick Moore; 151 (5) British Library, London; 152 (6) James Symonds; 153 (7) AIP Emilio Segrè Visual Archives; 153 (8) American Geophysical Union, courtesy AIP Emilio Segrè Visual Archives; 155 (2) James Symonds; 155 (3) Royal Astronomical Society Library, London; 155 (4) NASA, ESA, A. Bolton (Harvard-Smithsonian CfA) and the SLACS Team; 155 (5) NASA, ESA, and STScI; 156 (6) NASA-JSC; 157 (7) James Symonds; 158 (1), 158 (2), 158 (3), 158 (4), 158 (5) NRAO/AUI; 160 (6) James Symonds; 161 (7), 161 (8) MERLIN: University of Manchester/STFC; 161 (9) Anthony Holloway/Jodrell Bank; 162 (1), 162 (3) James Symonds; 163 (3) G. L. Slater, G. A. Linford, S. L. Freeland and the Yohkoh Project; 164 (4) Courtesy Riccardo Giacconi; 164 (5) David Axe; 165 (6) NASA-JPL; 165 (7) NASA/CXC/UCLA/MIT/M. Muno et al.; 166 (1) British Library, London; 166 (2) James Symonds; 167 (4a) NASA/CXC/SAO/M. Karovska et al.; 167 (4b) CXC/M. Weiss; 168 (5), 168 (6) Royal Astronomical Society Library, London; 169 (7) James Symonds; 170 (1) NASA, H. E. Bond and E. Nelan (Space Telescope Science Institute, Baltimore), M. Barstow and M. Burleigh (University of Leicester), and J. B. Holberg (University of Arizona); 171 (2) Institute of Astronomy, Cambridge; 171 (3) Harvard College Observatory, courtesy AIP Emilio Segrè Visual Archives; 171 (4) AIP Emilio Segrè Visual Archives; 172 (5) NASA, ESA, C. R. O'Dell (Vanderbilt University), and M. Meixner, P. McCullough; 173 (6) Photograph by Dorothy Davis Locanthi, courtesy AIP Emilio Segrè Visual Archives; 173 (7), 173 (8) NASA, ESA, and H. Richer

(University of British Columbia); **174** (1) NASA/CXC/SAO; **175** (2) Jocelyn Bell Burnell; **175** (3) AIP Emilio Segrè Visual Archives, Physics Today Collection; **176** (4) Associated Press; **176** (5) AIP Emilio Segrè Visual Archives, Physics Today Collection; **177** (6), **178** (1) James Symonds; **179** (2) AIP Emilio Segrè Visual Archives; **179** (3), **180** (4) ESA; **181** (5) Courtesy Sir Patrick Moore; **184** (1) James Symonds; **185** (2) Tycho Brahe in his observatory, 1587, from *Tychonis Brahe astronomiae instauratae mechanica*, Bibliothèque Nationale, Paris; **187** (4) James Symonds; **188** (1) Forschungs- und Landesbibliothek Gotha; **188** (3) Orion from Galileo, *Sidereus nuncius*, 1610; **189** (3) Royal Astronomical Society Library, London; **190** (1) NASA/IRAS; **190** (5), **190** (6) From Thomas Wright, *An Original Theory or New Hypothesis of the Universe*, London, 1750; **191** (7) NASA; **193** (2), **193** (3) Royal Astronomical Society Library, London; **193** (4) Royal Astronomical Society/Science Photo Library; **194** (5) NASA-JPL; **195** (6) Royal Astronomical Society Library, London; **196** (1) Nik Szymanek; **197** (2) NASA, ESA, and The Hubble Heritage Team (STScI/AURA); **197** (3) Harlow Shapley, *Ad Astra Per Aspera, Through Rugged Ways to the Stars*, New York: Charles Scribner's Sons, 1969, courtesy AIP Emilio Segrè Visual Archives, Shapley Collection; **199** (4) Dr. Dorrit Hoffleit, Yale University, courtesy AIP Emilio Segrè Visual Archives, Tenn Collection; **199** (5) Davide De Martin (ESA/Hubble) and Edward W. Olszewski (University of Arizona, USA); **199** (6) Harvard College Observatory; **200** (1) NASA, ESA, and The Hubble Heritage Team (STScI/AURA); **201** (2) Royal Astronomical Society Library, London; **201** (3) From Tycho Brahe, *Astronomiae instauratae mechanica*, 1602, British Library, London; **201** (4) Royal Astronomical Society Library, London; **202** (5) James Symonds; **203** (6) Royal Astronomical Society Library, London; **203** (7) NASA/CXC/Rutgers/J. Warren and J. Hughes et al.; **204** (8) James Symonds; **204** (9) RGO; **205** (9) Maria Pilar Ruiz-Lapuente; **206** (1) ESO; **207** (2) Anglo-Australian Observatory/David Malin Images; **207** (3) Kamioka Observatory, Institute for Cosmic Ray Research, the University of Tokyo; **208** (4) NASA-MSFC; **208** (5) NASA, The Hubble Heritage Team (STScI/AURA); **208** (6) Joe Wampler; **209** (7) Dr Christopher Burrows, ESA/STScI and NASA; **209** (8) NASA, P. Challis, R. Kirshner (Harvard-Smithsonian Center for Astrophysics) and B. Sugerman (STScI); **210** (1) ESO; **211** (2) Harvard College Observatory; **211** (3) Royal Astronomical Society Library, London, photo courtesy Mt Wilson and Palomar Observatories; **212** (5) Photo Hans Petersen, Novdisk Pressefoto A/S, courtesy AIP Emilio Segrè Visual Archives, gift of Kaj Aage Strand; **212** (6) NASA, ESA, and The Hubble Heritage (STScI/AURA) - ESA/Hubble Collaboration; **213** (8) Dr Wendy Freedman, photo Skye Moorhead; **213** (9) Dr Wendy L. Freedman, Observatories of the Carnegie Institution of Washington, and NASA; **214** (1) David A. Hardy; **215** (2) C. R. O'Dell (Rice University, Houston) and NASA; **215** (3) Carlos Munoz-Yague/EURELIOS/Science Photo Library; **216** (4) OHP/CNRS; **216** (5) ESA and A. Vidal-Madjar (Institut d'Astrophysique de Paris, CNRS, France); **216** (6) NASA; **217** (7) Seth Shostak; **217** (8) James Symonds; **218** (9) ESA; **219** (10) NASA, ESA, P. Kalas, J. Graham, E. Chiang, E. Kite (University of California, Berkeley), M. Clampin (NASA-GSFC), M. Fitzgerald (Lawrence Livermore National Laboratory), and K. Stapelfeldt and J. Krist (NASA-JPL); **219** (11) Gemini Observatory; **220** (1) SOHO (ESA and NASA); **221** (2) Tate Britain, London; **222** (3) AIP Emilio Segrè Visual Archives, Brittle Books Collection; **222** (4) AIP Emilio Segrè Visual Archives, Physics Today Collection; **222** (5) Walker and Boutall, courtesy AIP Emilio Segrè Visual Archives, E. Scott Barr Collection; **222** (6) Radium Institute, courtesy AIP Emilio Segrè Visual Archives; **222** (7) Photo Bob Davis, courtesy AIP Emilio Segrè Visual Archives, Physics Today Collection; **222** (8) UK Atomic Energy Authority, courtesy AIP Emilio Segrè Visual Archives; **222** (9) Los Alamos National Laboratory; **224** (1) MS Oxford, St John's College 17, f. 7v; **225** (2) AIP Emilio Segrè Visual Archives, Margaret Russell Edmondson Collection; **226** (3), **226** (4) AIP Emilio Segrè Visual Archives, Physics Today Collection; **226** (5) AIP Emilio Segrè Visual Archives, E. E. Salpeter Collection; **226** (6) AIP Emilio Segrè Visual Archives, Clayton Collection; **227** (7) NASA and The Hubble Heritage Team (STScI/AURA); **228** (1) Pete Lawrence; **229** (2), **229** (3) SOHO (ESA and NASA); **230** (4) James Symonds; **231** (5) Courtesy of Brookhaven National Laboratory; **231** (6) SOHO (ESA and NASA); **231** (7) James Symonds; **232** (8) Nik Szymanek; **228** (9) James Symonds; **233** (10) A. Title (Stanford Lockheed Institute), TRACE, NASA; **234** (1) NASA, ESA, J. Hester and A. Loll (Arizona State University); **235** (2) Royal Astronomical Society Library, London; **235** (3) Science Museum, London; **236** (4) Hale Observatories, courtesy AIP Emilio Segrè Visual Archives; **236** (5) Doak Heyser; **236** (6) From *Ch'in-ting shu-ching t'u-shuo*; **237** (7) NASA/CXC/ASU/J. Hester et al.; **239** (1) NASA, ESA, HEIC, and The Hubble Heritage Team (STScI/AURA); **239** (2) James Symonds; **240** (3) Garrelt Mellema (Leiden University) et al., HST, ESA, NASA; **240** (4) NASA and The Hubble Heritage Team (STScI/AURA); **240** (5) The Hubble Heritage Team (AURA/STScI/NASA); **241** (6) NASA and The Hubble Heritage Team (STScI/AURA); **241** (7) Royal Astronomical Society/Science Photo Library; **241** (8) Hale Observatories, courtesy AIP Emilio Segrè Visual Archives; **241** (9) NASA and The Hubble Heritage Team (STScI/AURA); **241** (10) NASA, ESA and A. Zijlstra (UMIST, Manchester, UK); **242** (1) C. R. O'Dell (Rice University, Houston); NASA; **243** (3) AIP Emilio Segrè Visual Archives, Brittle Books Collection; **243** (4) AIP Emilio Segrè Visual Archives; **243** (5) NASA, ESA, D. Golimowski (Johns Hopkins University), D. Ardila (IPAC), J. Krist (JPL), M. Clampin (GSFC), H. Ford (JHU), and G. Illingworth (UCO/Lick) and the ACS Science Team; **244** (6) C. R. O'Dell (Rice University, Houston); NASA; **244** (7) Subaru Telescope, National Astronomical Observatory of Japan (NAOJ); **244** (8) NASA-GSFC; **245** (9) James Symonds; **246** (1) Adam Block/NOAO/AURA/NSF; **247** (2) Science Photo Library; **247** (3) AIP Emilio Segrè Visual Archives; **247** (4) NASA and The Hubble Heritage Team (STScI/AURA); **249** (5) ESO; **249** (6) James Symonds; **249** (7) ESA, image by D.Hardy; **249** (8) NASA-JPL; **252** (1) NASA, ESA, STScI, J. Hester and P. Scowen (Arizona State University); **252** (2) Carl Heiles; **253** (3) AIP Emilio Segrè Visual Archives; **253** (4) AIP Emilio Segrè Visual Archives, photo Ron Doel; **254** (5) NASA; **254** (6) Paul H. Donaldson, Harvard University, Cruft Laboratory Photographic Dept., courtesy AIP Emilio Segrè Visual Archives; **255** (7) NASA and The Hubble Heritage Team (STScI/AURA); **255** (8) Science Photo Library; **255** (9) Institute of Astronomy, Cambridge; **256** (1) NASA, ESA, and The Hubble Heritage Team (STScI/AURA); **257** (2) Royal Astronomical Society Library, London; **257** (3) AIP Emilio Segrè Visual Archives, E. Scott Barr Collection; **257** (4) California Institute of Technology; **258** (5) NASA, ESA, and The Hubble Heritage Team (STScI/AURA); **258** (6) Mt Wilson and Palomar Observatories; **258** (7) NASA, ESA, and M. Livio (STScI); **259** (8) NASA, ESA, The Hubble Heritage Team (STScI/AURA) - ESA/Hubble Collaboration, and K. Noll (STScI); **260** (9) NASA, ESA, S. Beckwith (STScI) and the Hubble Ultra Deep Fields Team; **261** (10) Royal Observatory Edinburgh (ROE); **262** (1) ESO; **263** (2) The Bridgeman Art Library/Istanbul University Library, Istanbul; **263** (3) National Library of Australia, Canberra, ms7860-2-s10-v; **264** (4) Mariner's Museum, Newport News, Virginia; **264** (5) White Images/Scala, Florence; **264** (6) Royal Astronomical Society Library, London; **264** (7)

Index

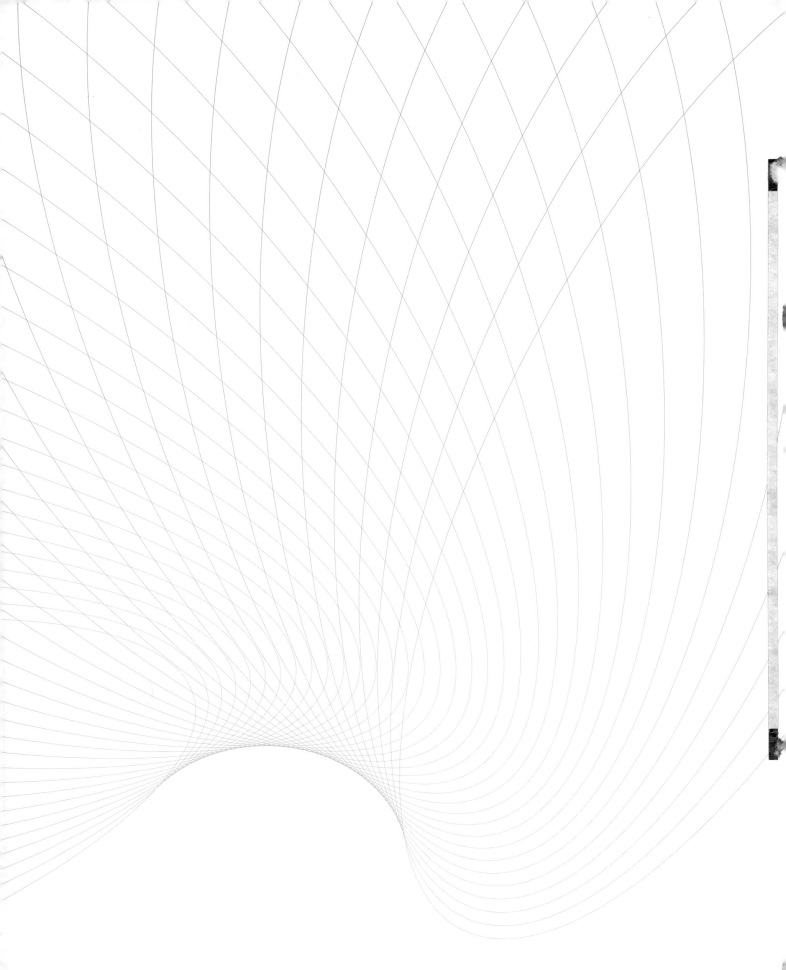

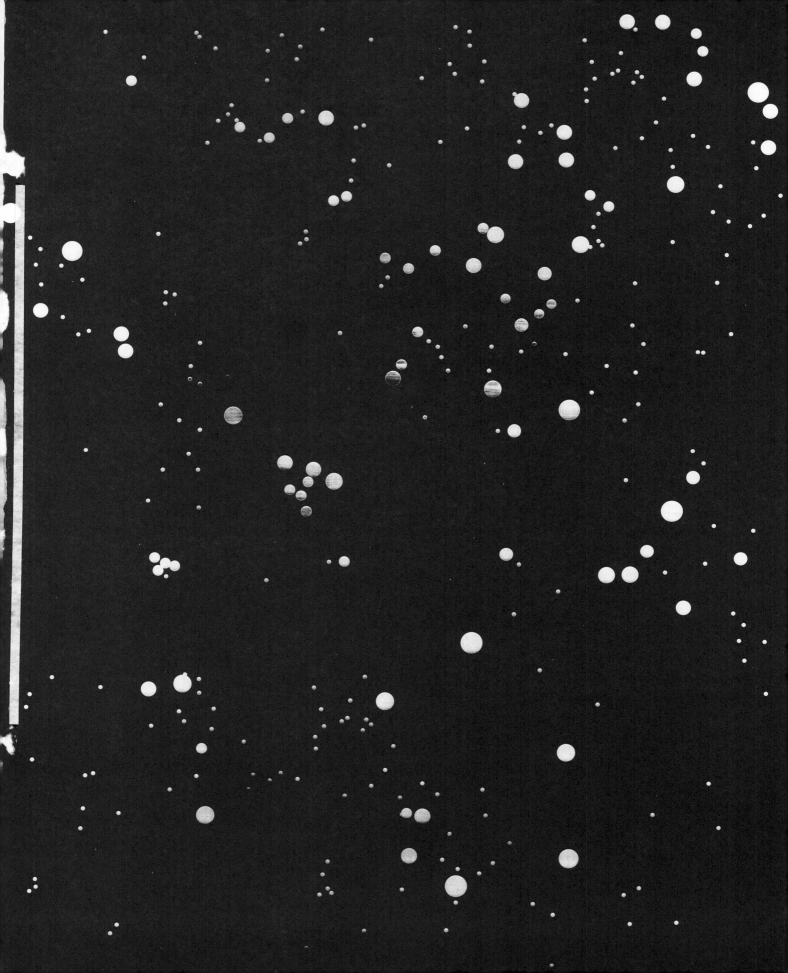